Taking Sides: Clashing Views on Global Issues, 9/e

James E. Harf and Mark Owen Lombardi

http://create.mheducation.com

ISBN-10: 1259826996   ISBN-13: 9781259826993

Contents

Detailed Table of Contents

Unit 1: Global Population and Resources

Neil Howe and Richard Jackson of the Center for Strategic and International Studies argue that global population aging is likely to have a profound and negative effect on global economic growth, living standards and conditions, and "the shape of the world order," particularly affecting China, Russia, Pakistan, Iran, and countries of the West. The Economist article argues that today's elderly are a much better educated class than past generations of elderly. Consequently, they will work longer and will represent a higher level of productivity than previous generations. Add to this, because of the cutting of pensions by governments around the world, there should emerge an unexpected degree of economic thrift, particularly in the developed world.

The UN Food and Agricultural Organization estimates that the number of undernourished around the globe has declined by 167 million over the last decade. The British government's report concludes that the world's existing food system is failing half of the world's population, with a billion people hungry and another billion suffering from "hidden hunger".

The report by UNICEF and WHO details how far the global community has come in the past 15 years in providing safe drinking water for its citizens, suggesting that over 90% of the world's now has access to good sources of drinking water. The UN-Water report predicts, however, that by 2025, the global demand for water is projected to increase by 55%, which will tax the planet's ability to meet that increased demand.

In this Internet post, Marc Lallanilla suggests that oil production will not reach a peak but, rather, a plateau, which will continue for some time before slowly declining. In this Internet post, Michael T. Klare argues that taking into account both political and physical constraints, peak oil "remains in our future,' although there will be a "gradual disappearance" of what the author calls "easy" oil.

The communication from the European Commission of the European Union spells out the major details of the climate change agreement, emphasizing the positive aspects of the accord and discusses the implementation of the agreement. The National Review's Rupert Darwall lays out several arguments against the treaty, such as the commitment of the West to paying billions of dollars to the developing world and to other binding obligations.

Unit 2: Expanding Global Forces and Movements

Issue: Will the International Community Be Able to Successfully Address the ZIKA Virus Pandemic?
Yes: World Health Organization, from "ZIKA: Strategic Response Framework & Joint Operations Plan," *World Health Organization (WHO)* (2016)
No: Matthew Weaver and Sally Desmond, from "Zika Virus Spreading Explosively, Says World Health Organization," *The Guardian* (2016)

The WHO report lays out a comprehensive global plan to address the pandemic threat posed by the Zika virus. The WHO-led effort is designed to provide help to affected countries, build the capacity to stop new outbreaks and to address them successfully when they do occur, and to engage in research to address all aspects of the pandemic. Matthew Weaver and Sally Desmond report that the Director General of WHO, in calling an emergency meeting, stated that the "level of alarm is extremely high."

Issue: Is the International Community Making Effective Progress in Securing Global Human Rights?
Yes: The Council on Foreign Relations, from "The Global Human Rights Regime," *Council on Foreign Relations* (2012)
No: Amnesty International, from "Amnesty International Report 2014/15: The State of the World's Human Rights," Amnesty International, London (2015)

The Council on Foreign Relations, an independent nonpartisan and essentially American think tank, in an Issue Brief summarizes the development of an elaborate global system of governmental and nongovernmental organizations developed primarily over the past few decades to promote human rights throughout the world, while recognizing that the task is still far from complete. Amnesty International's annual report on the state of human rights around the world suggests major failures in all regions, with suffering by many from conflict, displacement, discrimination, and/or repression.

Issue: Do Adequate Strategies Exist to Combat Human Trafficking?
Yes: Tierney Sneed, from "How Big Data Battles Human Trafficking," *U.S. News & World Report* (2015)
No: United Nations Office on Drugs and Crime, from "Global Report on Trafficking in Persons: 2014," *Global Report on Trafficking in Persons* (2014)

Tierney Sneed's article details how new technologies are being used to address the problem of human trafficking. The UNODC report spells out the magnitude of the problem with the compilation of major data collected about human trafficking.

Issue: Should the United States and the West Address the Syrian Refugee Crisis by Allowing Them to Migrate to the West?
Yes: Michael Ignatieff, from "The United States and the Syrian Refugee Crisis: A Plan of Action," Harvard Kennedy School Shorenstein Center on Media, Politics and Public Policy (2016)
No: Andrew C. McCarthy, from "The Controversy Over Syrian Refugees Misses the Question We Should Be Asking," *National Review* (2015)

Michael Ignatieff, Edward R. Murrow Professor of Press, Politics and Public Policy at Harvard University, argues that the United States should help its major European allies by offering generosity, vision and optimism to assist them in the resettlement of refugees. Andrew C. McCarthy suggests that even if the vetting process is perfect, there are two reasons why allowing a massive influx of refugees into the U.S. is "a calamity." The first is the vetting for terrorism ignores the real challenge, that of Islamic supremacism, of which violent terrorism is only a sub-set. The second is that the U.S. ignores the dynamics of jihadism, which suggests that there are individuals being admitted who are "apt to become violent jihadists."

Issue: Is Global Income Inequality on the Rise in the International Community?
Yes: Oxfam Report, from "An Economy for the 1%," Oxfam International (2016)
No: Max Roser, from "Income Inequality: Poverty Falling Faster than Ever but the 1% are Racing Ahead," *The Guardian* (2015)

The Oxfam report issued this year provides data to characterize global inequality as a crisis where the wealthiest 62% of people have amassed great wealth at the expense of billions of poor people. Max Roser argues that while a few have amassed great wealth, poverty as a whole is declining and tens of millions are moving into working and middle class positions.

Issue: Is Social Media Becoming the Most Powerful Force in International Politics?
Yes: Ritu Sharma, from "Social Media as a Formidable Force for Change," *Huffington Post* (2015)
No: Kathy Gilsinan, from "Is ISIS's Social-Media Power Exaggerated?" *The Atlantic* (2015)

Ritu Sharma sees social media as a powerful mobilizing force and uses many examples to illustrate that point. Kathy Gilsinan looks at ISIS's use of social media and questions its power and influence questioning the real reach of social media.

Issue: Are Cyber-Groups Terrorists or a potential Force for Good?
Yes: David Auerbach, from "The Sony Hackers are Terrorists," *Bitwise* (2014)
No: Evan Schuman, from "Anonymous Just Might Make All the Difference in Attacking ISIS," *ComputerWorld* (2015)

David Auerbach views hackers such as the recent Sony hack as terrorism because of the economic and social impact. Evan Schuman points out that hacker groups like Anonymous may indeed be forces for good in terms of combating real terrorists like ISIS.

Unit 3: The New Global Security Dilemma

Issue: Is Cyber Warfare the Future of War?
Yes: David Gewirtz, from "Why the Next World War will be a Cyberwar First, and a Shooting War Second," *ZDNet* (2015)
No: Thomas Rid, from "Cyberwar and Peace: Hacking Can Reduce Real World Violence," *Foreign Affairs* (2013)

David Gewirtz articulates the view that cyberwarfare will be a crucial and important element of the next and perhaps all future wars. Thomas Rid is skeptical of the notion that cyberwar is the wave of the future and in fact may act as an effective deterrent to real violence.

Issue: Are Russia and the United States in a New Cold War?
Yes: Andrej Krickovic and Yuval Weber, from "Why a New Cold War with Russia is Inevitable," The Brookings Institution (2015)
No: Matthew Rojansky and Rachel S. Salzman, from "Debunked: Why There Won't be Another Cold War," *The National Interest* (2015)

Andrej Krickovic and Yuval Weber argue that substantive differences between the US and Russia on NATO membership and the status of Ukraine among others will lead to further deterioration in the relationship. Matthew Rojansky and Rachel S. Salzman see the interconnections of social, economic and political as too deep for another true cold war to develop.

Issue: Can ISIS Be Defeated in the Near Future?
Yes: Max Boot, from "How ISIS can be Defeated," *Newsweek* (2015)
NO: Aaron David Miller, from "5 Reasons the U.S. Cannot Defeat ISIS," *RealClear World*, (2015)

Max Boot lays out a strong military strategy for the defeat of ISIS that relies on a coalition of ground forces. Aaron David Miller lays out the factors that will prevent a defeat of ISIS that currently exist.

<u>Issue: Is the Iran Nuclear Program Agreement Good for America and for the World?</u>
Yes: John Kerry, from "Remarks on Nuclear Agreement With Iran," U.S. Department of State (2015)
No: David E. Sanger and Michael R. Gordon, from "Future Risks of an Iran Nuclear Deal," *The New York Times* (2015)

Secretary of State John Kerry, whose team negotiated the Iran nuclear program agreement, lays out the rationale behind the support of the deal by many top nuclear scientists and other experts, and the reasons why he believes the world in general and the most relevant countries within that world will be safer as a consequence of the deal. David Sanger and Michael Gordon of The New York Times criticize the deal, focusing on the fact that in 15 years Iran will be free to produce massive quantities of uranium ad thus be in a position to produce nuclear bombs quickly.

<u>Issue: Is the European Union in Danger of Disintegrating?</u>
Yes: John Feffer, from "The European Union May be on the Verge of Collapse," *The Nation* (2015)
No: Kalin Anev Janse, from "How the Financial Crisis Made Europe Stronger," *World Economic Forum* (2016)

The latest crises facing Europe have caused some to question its viability. John Feffer sees an EU that may be one more challenge away from splintering. Kalin Anev Janse argues that the economic crisis has brought the EU together recognizing the shared benefits of unity.

<u>Issue: Is China the Next Superpower?</u>
Yes: Jonathan Watts, from "China: Witnessing the Birth of a Superpower," *The Guardian* (2012)
No: Jonathan Adelman, from "China's Long Road to Superpower Status," *U.S. News & World Report* (2014)

After living and reporting on China for over a decade, Jonathan Watts argues that China is becoming the next superpower. He cites its economic growth and expansion, government policies, and growing international influence as signs of this emerging status. Jonathan Adelman details the areas where China lags far beyond the US and other countries and shows no signs of solving.

<u>Issue: Should the International Community Pre-Empt against North Korea?</u>
Yes: Patrick M. Cronin, from "Time to Actively Deter North Korea," *The Diplomat* (2014)
No: Jeong Lee, from "North Korea: Don't Pick a Fight We Can't Win," *Small Wars Journal* (2015)

Patrick M. Cronin believes that the West must be more aggressive in its approach to North Korea and raise the threat of preemption against their nuclear forces. Jeong Lee argues that the nature of the North Korean regime means diplomacy and calm pressure will work best at changing behavior.

Preface

This volume reflects the dynamic revolutionary nature of the contemporary world in which we live. Not only are we now witnessing a dramatic leap in the scope of global change, but we are also experiencing a *rate* of change in the world unparalleled in recorded history. Change in the international system is not a new phenomenon. Since the creation in the early 1500s of a Euro-centric world system of sovereign nation-states that dominated political, economic, and social events throughout the known world, global change has been with us. But earlier manifestations of change were characterized by infrequent bursts of system-changing episodes followed by long periods of "normalcy," where the processes and structures of the international system demonstrated regularity or consistency.

First, the Catholic church sought to recapture its European dominance and glory during the Middle Ages in a last gasp effort to withstand challenges against its rule from a newly developed secularized and urbanized mercantile class, only to be pushed aside in a devastating continental struggle known as the Thirty Years' War and relegated to irrelevant status by the resultant Treaty of Westphalia in 1648. A century and a half later, the global system was again challenged, this time by a French general turned emperor, Napoleon Bonaparte, who sought to export the newly created utopian vision of the French Revolution beyond the boundaries of France to the rest of the world. Napoleon was eventually repelled by a coalition of major powers intent on preserving the world as these countries knew it.

Soon nineteenth-century Europe was being transformed by the intrusion of the Industrial Revolution on the daily lives of average citizens and national leaders alike. Technological advances enhanced the capacity of countries to dramatically increase their military capability, achieving the ability to project such power far beyond their national borders in a short time. Other threats to the existing world order also emerged. Nationalism began to capture the hearts of various country leaders who sought to impart such loyalties to their subjects, while new ideologies competed with one another as well as with nationalism to create a thirst for alternative world models to the existing nation-state system. The result was another failed attempt by a European power, this time Germany, to expand its influence via a major war, later to be called

World War I, throughout the continent. The postwar map of Europe reflected major consequences of the abortive German effort.

Almost immediately, the international system was threatened by a newly emergent virulent ideology of the left intent on taking over the world. Communism had gained a foothold in Russia, and soon its leaders were eager to transport it across the continent to the far corners of the globe, threatening to destroy the existing economic order and, by definition, its political counterpart. Shortly thereafter, a competing virulent ideology from the right, fascism emerged. Under its manipulation by the Nazis led by a new German leader, Adolf Hitler, and by the militarists in Japan, the international system was once again greatly threatened. Six long years of war and unthinkable levels of devastation and destruction followed, until the fascist threat was turned back. The communist threat persisted, however, until late in the millennium, when it also virtually disappeared, felled by its own weaknesses and excesses.

In the interim, new challenges to global order appeared in the form of a set of issues like no other during the 500-year history of the nation-state system. The nature of these global issues and the pace at which they both dominated the global agenda quickly reshaped the international system. This new agenda took root in the late 1960s, when astute observers began to identify disquieting trends: quickening population growth in the poorer sectors of the globe, growing disruptions in the world's ability to feed its population, increasing shortfalls in required resources, and expanding evidence of negative environmental impacts of human development. Some of these issues—like decreasing levels of adequate supplies of food, energy, and water—emerged as a result of both increased population growth and increased per capita levels of consumption. Dramatic population increases, in turn, resulted in changes in global population dynamics—increasing aged population or massive new urbanization patterns. The emergence of this set of new issues was soon followed by another phenomenon, globalization, which emphasized increasing flow of information, goods, and services through innovative technology and a resultant diffusion of regional cultures throughout the globe. Globalization not only affected the nature of the international system in these general ways but also influenced both the manner in which these global issues impacted the system and how both state and non-state actors addressed them. Most recently the emergence of the Internet has dramatically affected the ability of individuals to communicate with one another and to influence one another from any part of the globe to its far corners, and this influence may be for good or, conversely, for evil.

The major consequence of the confluence of these events was that the pace of change had greatly accelerated. No longer was the change measured in centuries or even decades. It was now measured in years or even months. No longer were likely solutions to such problems simple, known, confined to a relatively small part of the globe, and capable of being achieved by the efforts of one or a few national governments. Instead, these global issues were characterized by increased rapidity of change, increased complexity, increased geographical impact, increased resistance to solution, and increased fluidity.

One only has to compare the world of the 1960's to the world of today to grasp the difference. When student's first began to study these issues in the early 1960s, their written analysis was accomplished either by putting pen to pad or by engaging an unwieldy typewriter. Their experience with a computer was limited to watching a moon landing through the eyes of NASA Mission Control. The use of phones was relegated to a location where a cord could be plugged into a wall socket. Their written correspondence with someone beyond their immediate location had a stamp on it. Their reading of news, both serious and frivolous, occurred via a newspaper. Visual news invaded their space in 30-minute daily segments from three major TV networks. Being entertained required some effort, usually away from the confines of their homes or dorm rooms. Today, of course, the personal computer and its companions, the Internet, the Kindle, and the iPad, have transformed the way students learn, the way they communicate with one another, and the way they entertain themselves. Facebook, Twitter and YouTube have joined our vocabulary.

The age of globalization and the instant information age has accelerated, affecting and transforming trends that began several decades ago. No longer are nation-states the only actors on the global stage. Moreover, their position of dominance is increasingly challenged by an array of other actors—international governmental organizations, private international groups, multinational corporations, and important individuals—who might be better equipped to address newly emerging issues (or who might also serve as the source of yet other problems).

An even more recent phenomenon is the unleashing of ethnic pride, which manifests itself in both positive and negative ways. The history of post-cold war conflict is a history of intrastate, ethnically driven violence that has torn apart those countries that have been unable to deal effectively with the changes brought on by the end of the cold war. The most insidious manifestation of this emphasis on ethnicity is the emergence of terrorist groups who use religion and other aspects of ethnicity to justify bringing death and destruction on their perceived enemies. As national governments attempt to cope with this latest phenomenon, they too are changing the nature of war and violence. The global agenda's current transformation, brought about by globalization, demands that our attention turn toward the latter's consequences.

The recent economic collapse, or what some now call "the Great Recession," is evidence of this rapid globalization. Economic shifts were greatly accelerated throughout the global community by technology, interdependence, and connectivity such that governments, analysts, and the public at large were unable to comprehend the destabilizing events as they happened. Further, relations between states such as Russia, China, and the United States rapidly altered as a result. And the Middle East remains unsettled for a number of reasons. First, the vacuum created by the consequences of invasions by the United States into both Afghanistan and Iraq has led to utter chaos as ethnicity and religious zealotry have resulted in civil war among any number of factions, who are ever changing and whose relative importance is transitory. Second, throughout the Middle East autocratic governments are being threatened and some like Egypt have been toppled by their own citizens seeking either democracy or their own version of autocracy. Others like Syria find themselves locked in bitter confrontation between regime and revolutionary groups, each side helped by any number of global and regional powers who enter the fray in large part in pursuit of their own agendas. And third, add to this terrible mix the emergence of ISIS, the most extreme actor of all these groups, who is using the chaos of the conflict-torn region to advance its own cause of a caliphate based on principles alien to most of the rest of the world.

The format of *Taking Sides: Clashing Views on Global Issues*, Ninth Edition, follows the successful formula of other books in the Taking Sides series. The book begins with an introduction to the emergence of global issues, the new age of globalization, the effect of 9/11, the recent global economic crisis, a reemerging Russia, and the rise of revolutionary movements throughout North Africa and the Middle East—and the international community's response to these events—that characterize the second decade of the twentieth century. It then addresses current global issues such as population, which represents a global issue by itself, but which also affects the parameters of many other global issues; a range of problems associated with global resources and their environmental impact; widely disparate expanding forces and movements across national boundaries brought on by the dramatically increasing ease with which borders may be transgressed, and the emergence of cyber-related phenomena that manifest themselves in a variety of ways, from warfare to global crime to economic advantage.

Book Organization

Each issue has two selections, one pro and one con. The readings are preceded by *Learning Outcomes* and an issue *Introduction* that sets the stage for the debate by laying out both sides of the issue and briefly describes the two selections. Each issue concludes with *Critical Thinking and Reflection* questions, an *Is There Common Ground?* section discussing alternative viewpoints or approaches, as well as *Additional Resources* and *Internet References* for further exploration of the issue.

Changes to this edition   This ninth edition represents a significant revision. Twelve of the 19 issues are new issues and fully 35 of the 38 "pro-con" articles are new readings.

James E. Harf
Maryville University

Mark Owen Lombardi
Maryville University

Editors of This Volume

JAMES E. HARF currently serves as a professor of political science as well as associate vice president and director of the Center for Global Education at Maryville University in St. Louis. He spent most of his career at The Ohio State University, where he holds the title of professor emeritus. Among his over three-dozen authored and edited books are *The Wise World Traveler: Becoming a Part of All That You Meet* (Agapy LLC, 2015), *The Unfolding Legacy of 9/11* (University Press of America, 2004), *World Politics and You: A Student Companion to International Politics on the World Stage*,

5th ed. (Brown & Benchmark, 1995), and *The Politics of Global Resources* (Duke University Press, 1986). His first novel, *Memories of Ivy* (Ivy House Publishing Group, 2005), about life as a university professor, was published in 2005. He also coedited a four-book series on the global issues of population, food, energy, and the environment, as well as three other book series on national security education, international studies, and international business. As a staff member on President Jimmy Carter's Commission on Foreign Language and International Studies in the late 1970s, he was responsible for undergraduate education recommendations. He also served 15 years as executive director of the Consortium for International Studies Education. He has been a frequent TV and radio commentator on international issues.

MARK OWEN LOMBARDI is president and chief executive officer of Maryville University in St. Louis, Missouri. He is coeditor and author of *The Unfolding Legacy of 9/11* (University Press of America, 2004) and coeditor of *Perspectives of Third World Sovereignty: The Post-Modern Paradox* (Macmillian, 1996). Lombardi has authored numerous articles and book chapters on such topics as U.S. foreign policy, African political economy, the politics of the cold war, and higher education reform. Lombardi is a member of numerous civic organizations and boards locally and nationally and he has given over 200 speeches to local and national groups on topics ranging from higher education reform to U.S. politics, international affairs, and U.S. foreign policy. He has also appeared over 75 times as a political commentator for local and national news outlets.

Acknowledgments

We extend our heartfelt thanks to McGraw-Hill editor Mary Foust for her professionalism and flexibility in making the completion of this work possible.

To my daughter, Marie: May your world conquer those global issues left unresolved by my generation. (James E. Harf)

For Betty and Marty, who instilled a love of education and a need to explore the world. (Mark Owen Lombardi)

Academic Advisory Board Members

Members of the Academic Advisory Board are instrumental in the final selection of articles for each edition of TAKING SIDES. Their review of articles for content, level, and appropriateness provides critical direction to the editor and staff. We think that you will find their careful consideration well reflected in this volume.

Introduction

Threats of the New Millennium

As the new millennium dawned over fifteen years ago, the world witnessed two very different events whose impacts have been far reaching, profound, and in many ways have shaped the trajectory of global issues ever since. The first was the new era of terrorism, ushered in by the tragedy of 9/11. This act rocked the international community, occupying virtually every waking moment of national and global leaders throughout the world. The aftermath and the forces that it unleashed have shaped many if not all of the global security issues ever since. The focused interest of national policymakers was soon transformed into a war on terrorism, later transformed into a long continuous war on terrorists, while average citizens from the heartland of the United States, to the urban centers of Europe, the peoples of North Africa to the Sunni and Shi'ite communities of the Islamic world have been directly impacted. Unfortunately, as the millennium's first decade ended, other challenges to global welfare and security also emerged. At the global level, a severe financial crisis forced world leaders to question the major tenets of contemporary capitalism. At the national level, a reemerging Russian presence, flexing its new economic muscles based on energy and backed by a growing military might, brought back fears of a new cold war. Several countries long considered part of the developing world made significant strides toward the development of nuclear weapons. And throughout North Africa and the Middle East, citizens took to the streets to protest decades of autocratic rule by despotic rulers and to seek more democratic government.

The second event at the beginning of the millennium was less dramatic and certainly did not receive the same fanfare, but still has had both short- and long-term ramifications for the global community in the twenty-first century. This was the creation of a set of ambitious millennium development goals by the United Nations. In September 2000, 189 national governments committed to eight major goals in an initiative known as the UN Millennium Development Goals (MDG): eradicate extreme poverty and hunger; achieve universal primary education; promote gender equality and empower women; reduce child mortality; improve maternal health; combat HIV/ AIDS, malaria, and other diseases; ensure environmental sustainability; and develop a global partnership for development. This initiative was important not only because the United Nations was setting an actionable 15-year agenda against a relatively new set of global issues but also because it signified a major change in how the international community would henceforth address such problems confronting humankind. The new initiative represented recognition of (1) shared responsibility between rich and poor nations for solving such problems; (2) a link between goals; (3) the paramount role to be played by national governments in the process; and (4) the need for measurable outcome indicators of success. The UN MDG initiative went virtually unnoticed by much of the public, although governmental decision makers involved with the United Nations understood its significance. As we approach the 15-year timeline for implementation of these millennium goals, the success rate has been mixed at best as the commitments of time and money made earlier have not materialized as planned.

These two major events, although vastly different, symbolize the world in which we now find ourselves, a world far more complex and more violent than either the earlier one characterized by the cold war struggle between the United States and the Soviet Union, or the post–cold war era of the 1990s, where global and national leaders struggled to identify and then find their proper place in the post–cold war world order. Consider the first event, the 9/11 tragedy. This act reminds us all that the use of power in pursuit of political goals in earlier centuries is still an option for those throughout the world who believe themselves disadvantaged or under attack but what is different and profound is the global reach of terrorism and the expanding array of tools available to those groups. Formally declared wars fought by regular national military forces publicly committed (at least on paper) to the tenets of just war theory have now been replaced by a plethora of "quasi-military tactics" whose defining characteristics are asymmetrical, civilian targeted and designed to undermine the very foundation of open societies, freedom. At the same time, a few rogue states, Iran and North Korea, for example, have accelerated and in some respects attained the ultimate in weaponry (nuclear) to further their own national goals.

On the other hand, the second event of the new century, the UN MDG initiative, symbolizes the other side of the global coin, the recognition that the international community is also beset with a number of problems that transcend national security and call into question the very viability of the Earth's ecosystem. The past four decades have witnessed the emergence and thrust to prominence of a number of new problems relating to social, economic, and environmental characteristics of the citizens who inhabit this planet. These problems impact the basic quality of life of global inhabitants in ways very different from the scourges of military violence. But their impact is just as profound.

At the heart of this global change affecting the global system and its inhabitants for good or for ill is a phenomenon called globalization.

The Age of Globalization

The cold war era, marked by the domination of two superpowers in the decades following the end of World War II, has given way to a new era called globalization. This new epoch is characterized by a dramatic shrinking of the globe in terms of travel and communication, increased participation in global policymaking by an expanding array of national and nonstate actors, and an exploding set of integrated problems with ever growing consequences. While the tearing down of the Berlin Wall twenty-five years ago dramatically symbolized the end of the cold war era, the creation of the Internet, with its ability to connect around the world, and the fallen World Trade Center, with its dramatic illustration of vulnerability, symbolize the new paradigm of connectivity, impact, cooperation and violence.

Globalization is a fluid and complex phenomenon that manifests itself in thousands of wondrous and equally disturbing ways. In the past couple of decades, national borders have shrunk or disappeared, with a resultant increase in the movement of ideas, products, resources, finances, and people throughout the globe. This reality has brought with it great advances and challenges. For example, the ease with which people and objects move throughout the globe has greatly magnified fears like the spread of disease. The term "epidemic" has been replaced by the phrase "global pandemic," as virulent scourges unleashed in one part of the globe now have greater potential to find their way to the far corners of the planet. The world has also come to fear an expanded potential for terrorism, as new technologies combined with increasing cultural friction and socioeconomic disparities have conspired to make the world far less safe than it had been. The pistol that killed the Austrian Archduke in Sarajevo in 1914, ushering in World War I, has been replaced by the jumbo jet used as a missile to bring down the World Trade Center, snuffing out the lives of thousands of innocent victims. We now live in an era of global reach for both good and ill, where a small group or a single individual can touch the hearts of people around the world with acts of kindness or can shatter their dreams with acts of terror.

This increase in the movement of information and ideas has ushered in global concerns over cultural imperialism and religious/ethnic wars. The ability both to retrieve and to disseminate information in the contemporary era will have an impact in this century as great as, if not greater than, the telephone, radio, and television in the last century. The potential for global good or ill is mind-boggling. Finally, traditional notions of great power security embodied in the cold war rivalry have given way to concerns about terrorism, genocide, nuclear proliferation, cultural conflict, rogue states, and the diminishing role of international law.

Globalization heightens our awareness of a vast array of global issues that will challenge individuals as well as governmental and nongovernmental actors. Everyone has become a global actor and so each has policy impact. This text seeks to identify those issues that are central to the discourse on the impact of globalization. The issues in this volume provide a broad overview of the mosaic of global issues that will affect students' daily lives.

What Is a Global Issue?

We begin by addressing the basic characteristics of a *global issue*.[1] By definition, the word *issue* suggests disagreement among several related dimensions:

1. Whether a problem exists and how it comes about.
2. The characteristics of the problem.
3. The preferred future alternatives or solutions.
4. How these preferred futures are to be obtained.

These problems are real, vexing, and controversial because policymakers bring to their analyses different historical experiences, values, goals, and objectives. These differences impede and may even prevent successful problem solving. In short, the key ingredient of an issue is disagreement.

The word *global* in the phrase *global issue* is what makes the set of problems confronting the human race today far different from those that challenged earlier generations. Historically, problems were confined to a village, city, or region. The capacity of the human race to fulfill its daily needs was limited to a much smaller space:

the immediate environment. In 1900, 90 percent of all humanity was born, lived, and died within a 50-mile radius. Today, a third of the world's population travel to one or more countries. In the United States, 75 percent of people move at least 100 miles away from their homes and most travel to places their grandparents could only dream about.

What does this mobility mean? It suggests that a vast array of issues are now no longer only local or national but are global in scope, including but not limited to food resources, trade, energy, health care, the environment, disease, natural disasters, conflict, cultural rivalry, populism, rogue states, democratic revolutions, and nuclear Armageddon.

The character of these issues is thus different from those of earlier eras. First, they transcend national boundaries and impact virtually every corner of the globe. In effect, these issues help make national borders increasingly meaningless. Environmental pollution or poisonous gases do not recognize or respect national borders. Birds carrying the avian flu and nuclear radiation leaking from disabled power plants have no knowledge of political boundaries.

Second, these new issues cannot be resolved by the autonomous action of a single actor, be it a national government, international organization, or multinational corporation. A country cannot guarantee its own energy or food security without participating in a global energy or food system. Third, these issues are characterized by a wide array of value systems. To a family in the developing world, giving birth to a fifth or sixth child may contribute to the family's immediate economic well-being. But to a research scholar at the United Nations Population Fund, the consequence of such an action multiplied across the entire developing world leads to expanding poverty and resource depletion.

Fourth, these issues will not go away. They require specific policy action by a consortium of local, national, and international leaders. Simply ignoring the issue cannot eliminate the threat of chemical or biological terrorism, for example. If global warming does exist, it will not disappear unless specific policies are developed and implemented.

These issues are also characterized by their persistence over time. The human race has developed the capacity to manipulate its external environment and, in so doing, has created a host of opportunities and challenges. The accelerating pace of technological change suggests that global issues will proliferate and will continue to challenge human beings throughout the next millennium.

In the final analysis, however, a global issue is defined as such only through mutual agreement by a host of actors within the international community. Some may disagree about the nature, severity, or presence of a given issue. These concerns then become areas of focus after a significant number of actors (states, international organizations, the United Nations, and others) begin to focus systematic and organized attention on the issue itself.

The Nexus of Global Issues and Globalization

Since 1989, the world has been caught in the maelstrom of globalization. Throughout the 1990s and into the twenty-first century, scholars and policymakers have struggled to define this new era. As the early years of the new century ushered in a different and heightened level of violence, a sense of urgency emerged. At first, some analyzed the new era in terms of the victory of Western or American ideals, the dominance of global capitalism, and the spread of democracy versus the use of religious fanaticism by the have-nots of the world as a ploy to rearrange power within the international system. But recent events call into question assumptions about Western victory or the dominance of capitalism. Others have defined this new era simply in terms of the multiplicity of actors now performing on the world stage, noting how states and their sovereignty have declined in importance and impact vis-à-vis . . . others such as multinational corporations and nongovernmental groups like Greenpeace and Amnesty International. Still others have focused on the vital element of technology and its impact on communications; information storage and retrieval; global exchange; and attitudes, culture, and values.

Whether globalization reflects one, two, or all of these characteristics is not as important as the fundamental realization that globalization is the dominant element of a new era in international politics. The globalization revolution now shapes and dictates the agenda. To argue otherwise is frankly akin to insisting on using a rotary phone in an iPhone world. This new period is characterized by several basic traits that greatly impact the definition, analysis, and solution of global issues. They include the following:

- An emphasis on information technology.
- The increasing speed of information and idea flows.
- The ability of global citizens to access information at rapidly growing rates and thus empower themselves for good or for ill.

- A need for greater sophistication and expertise to manage such flows.
- The control and dissemination of technology.
- The cultural diffusion and interaction that comes with information expansion and dissemination.

Each of these areas has helped shape a new emerging global issue agenda. Current issues remain important and, indeed, these factors help us understand them on a much deeper level. Yet globalization has created a new array of problems that is reshaping the international landscape and the dialogue, tools, strategies, and approaches that all global actors will take.

For example, the spread of information technology has made ideas, attitudes, and information more available to people throughout the world. Americans in Columbus, Ohio, had the ability to log onto the Internet and speak with their counterparts in Kosovo to discover when NATO bombing had begun and to gauge the accuracy of later news reports on the bombing. Norwegian students can share values and customs directly with their counterparts in South Africa, thereby experiencing cultural attitudes firsthand without the filtering mechanisms of governments or even parents and teachers. Scientific information that is available through computer technology can now be used to build sophisticated biological and chemical weapons of immense destructive capability, or equally to promote the dissemination of drugs and medicines outside of "normal" national channels. Ethnic conflicts and genocide between groups of people are now global news, forcing millions to come to grips with issues of intervention, prevention, and punishment. And terrorists in different parts of the globe can communicate readily with one another, transferring plans and even money across national and continental boundaries with ease. And antagonists against autocratic regimes can also communicate with their counterparts within their own society as well as those in neighboring countries, as witnessed by communication flows throughout North Africa and the Middle East among groups and individuals seeking democracy in the early months of 2011.

Globalization is an international system but it is also a revolutionary force that is rapidly adapting and changing. Because of this fluid nature and the fact that it is both relatively new and largely fueled by the amazing speed of technology, continuing issues are constantly being transformed and new issues are emerging regularly. The nexus of globalization and global issues has now become, in many ways, the defining dynamic of understanding global issues. Whether dealing with new forms of terrorism and new concepts of security, expanding international law, solving ethnic conflicts, dealing with mass migration, coping with individual freedom and access to information, or addressing cultural clash and cultural imperialism, the transition from a cold war world to a globalized world helps us understand in part what these issues are and why they are important. But most importantly, this fundamental realization shapes how governments and people can and must respond.

Identifying the New Global Issue Agenda

The organization of this text reflects the centrality of globalization. The first unit focuses on the continuing global agenda of the post-cold war era. The emphasis is on global population and environmental issues, and the nexus between these two phenomena. The next unit addresses the consequences of the decline of national boundaries and the resultant increased international flow of information, ideas, money, and material things in this globalization age, some for good and others for evil. The last unit addresses the new global security dilemma that has developed as a consequence of both the end of the cold war and 9/11.

The revolutionary changes of the last few decades present us with serious challenges unlike any others in human history. However, as in all periods of historic change, we possess significant opportunities to overcome problems. The task ahead is to define these issues, explore their context, and develop solutions that are comprehensive in scope and effect. The role of all researchers in this field, or any other, is to analyze objectively such problems and search for workable solutions. As students of global issues, your task is to educate yourselves about these issues and become part of the solution.

Note

1. The characteristics are extracted from James E. Harf and B. Thomas Trout, *The Politics of Global Resources* (Duke University Press, 1986, pp. 12–28).

James E. Harf
Maryville University

Mark Owen Lombardi
Maryville University

Unit 1

UNIT

Global Population and Resources

*I*t is not a coincidence that many contemporary global issues—environmental degradation, resources depletion, global warming and climate change, hunger, health pandemics, and the like—arrived at about the same time as world population growth was exploding, about half a century ago. No matter what the issue was, the resultant presence of a large and fast-growing population occurring simultaneously with initial cracks in the ability of global consumers to easily find required resources not only contributed to the creation of this new set of problems but it also transformed each of their basic characteristics and exacerbated the ability of national and global policymakers to address them.

As the world entered the new millennium, however, population birth rates changed again, this time showing declining growth rates caused by lowering birth rates, particularly in the developing world. This, in turn, made the challenge of addressing the newly emerging global agenda initially easier. But these declining rates following decades of significant growth unfortunately led to other problems, such as an aging global population and a declining work force. The late 20th century problem of a too large young population vis-à-vis the older generations was reversed as the latter started to grow at a faster pace than the former. The emergence of a growing graying population throughout the globe immediately began to have a significant impact, which has only continued to grow as the demographics of birth and death rates over the past 40+ years play themselves out now and in the coming years and decades. In short, the ability of the global community to respond to any given global issue is diminished by certain population conditions, be it an extremely young consuming population in a poor country in need of producers or an ever-growing senior population for whom additional services are needed.

Nowhere is the problems associated with population more evident than when we consider the planet's resource base. We live on a planet with finite resources, conceptualized in the Spaceship Earth picture taken from outer space. The availability of these resources—water, food, oil, and the like—and the manner in which and the pace at which the planet's inhabitants use them characterize a major component of today's global agenda.

Typically, a wide range of opinion exists about the environment and its resource base—some driven by ideology, some by science, some by economic self-interest, and some by simple noninterest. Disagreements abound about the existence of the problem, the nature of the problem, the scope of the problem, and how to solve the problem. And except for the noninterest group, none are shy about using the media and other venues to advance their own perceptions of what constitutes reality. Nowhere is this more evident, for example, than the debate over climate change and its effects on the ability to take maximum advantage of the planet's resource base.

For some, those "environmentalists" who take a pessimistic view of future resource availability are simply alarmists who have allowed ideology rather than science to drive their convictions. For others, many policy-makers and other nonbelievers are to be called out because they chose to ignore overwhelming scientific evidence in order to advance their own political or economic agendas. And among the population are also many uncaring individuals who simply ignore warning signs because they choose to put their heads in the sand out of total noninterest.

Not only does the availability of these resources (production) pose a potential issue, but how we distribute the resources and how we consume them also are thought by many to leave their marks on the planet. A basic set of questions relates to whether these impacts are permanent, too degrading to the planet, too damaging to one's quality of life, or simply beyond a threshold of acceptability.

Selected, Edited, and with Issue Framing Material by:
James E. Harf, *Maryville University*
and
Mark Owen Lombardi, *Maryville University*

ISSUE

Is Global Aging a Major Problem?

YES: Neil Howe and Richard Jackson, from "Global Aging and the Crisis of the 2020s," *Current History* (2011)

NO: "Age Invaders," from *The Economist* (2014)

Learning Outcomes
After reading this issue, you will be able to:
• Gain an understanding of why the 2020s are said to be an upcoming decade of global population aging and population decline.
• Appreciate how a global aging and declining population will affect economic growth, living standards, and the shape of the world order.
• Understand how future demographic conditions will negatively influence developing societies differently.
• Discuss the need for national planners to consider future population trends in planning national strategy.
• Appreciate why the United States may be better able to cope with the effects of population aging and decline better than its Western counterparts.
• Understand the three main channels through which demography influences the economy and their implications for an aging population.

ISSUE SUMMARY

YES: Neil Howe and Richard Jackson of the Center for Strategic and International Studies argue that global population aging is likely to have a profound and negative effect on global economic growth, living standards and conditions, and "the shape of the world order," particularly affecting China, Russia, Pakistan, Iran, and countries of the West.

NO: *The Economist* article argues that today's elderly are a much better-educated class than past generations of elderly. Consequently, they will work longer and will represent a higher level of productivity than previous generations. Add to this, because of the cutting of pensions by governments around the world, an unexpected degree of economic thrift might emerge, particularly in the developed world.

Today one in eight inhabitants of this planet is 60 years or older, and by 2030 it is projected to be one in six persons, as the elderly population is projected to grow by 56 percent. At that time there will be more older people than persons under the age of nine. Aging is happening in virtually all regions and all countries of the globe, and is especially evident most recently in the developing world.

The developed world has been faced with an aging population for some time now as one in five in Europe and North America is aged 60 or older and expected to increase to one in four by 2030. This phenomenon has been brought on by declining birth rates and an increasing life expectancy, and it is now being followed by the developing world. Between now and 2030, the number of elderly will grow the fastest in Latin America and the Caribbean, with an estimated

71 percent increase in those over 60 years of age, and the percentages for other regions of the developing world are not far behind. The phenomenon first appeared during the last quarter of the previous century with the demographic transition from high birth and death rates to lower rates in more affluent countries. The drop in death rates in these countries was a function of two basic factors: (1) the dramatic decline in both infant mortality (within the first year of birth) and child mortality (within the first five years) due to women being healthier during pregnancy and nursing periods, and due to the virtually universal inoculation of children against principal childhood diseases and (2) longer life spans, once people reach adulthood, in large part because of medical advances against key adult illnesses, such as cancer and heart disease.

Declining mortality rates yield an aging population in need of a variety of services—heath care, housing, and guards against inflation, for example—provided, in large part, by the tax dollars of the younger, producing sector of society. As the "grey" numbers of society grow, the labor force is increasingly called upon to provide more help for this class.

Declining birth and death rates mean that significantly more services will be needed to provide for the aging populations of the industrialized world, while at the same time, fewer individuals will be joining the workforce to provide the resources to pay for these services. However, some experts say that the new work force will be able to take advantage of the skills of the more aged, unlike previous eras. In order for national economies to grow in the information age, an expanding workforce may not be as important a prerequisite as it once was. Expanding minds, not bodies, may be the key to expanding economies and increased abilities to provide public services.

However, the elderly and the young are not randomly distributed throughout society, which is likely to create a growing set of regional problems. In the United States, for example, the educated young are likely to leave the "grey belt" of the north for the Sun Belt of the south, southwest, and west. Who will be left in the older, established sectors of the country that were originally at the forefront of the industrial age to care for the disproportionately elderly population? Peter G. Peterson introduces the phrase "the Floridization of the developed world," where retirees continue to flock in unprecedented numbers to more desirable locations, in order to capture the essence of the problems associated with the changing age composition in industrial societies. What will happen 30–40 years from now, when the respective sizes of the young and the elderly populations throughout the developed world will yield a much larger population at the twilight of their existence? Although the trend has been most evident in the richer

part of the globe, people are now also living longer in the developing world, primarily because of the diffusion of modern medical practices. But unless society can accommodate their skills of later years, they may become an even bigger burden in the future for their national governments.

A 2001 report, *Preparing for an Aging World: The Case for Cross-National Research* (National Academy of Sciences), identified a number of areas in which policymakers need a better understanding of the consequences of aging and resultant appropriate policy responses. Unless national governments of the developed world can effectively respond to these issues, the economic and social consequences can have a significant negative impact in the aging population cohort as well as throughout the entire society. This theme was reiterated in a major report of the Population Reference Bureau in March 2005 (*Global Aging: The Challenge of Success*), suggesting three major challenges of an aging population: (1) economic development issues, (2) health and well-being issues, and (3) the challenge of enabling and supportive environments.

The issue of the changing age composition in the developed world was foreseen a few decades ago but its heightened visibility is more recent. This visibility culminated in a UN-sponsored conference on aging in Madrid in April 2002. Its plan of action commits governments to address the problem of aging and provides them with a set of 117 specific recommendations covering three basic areas: older individuals and development, advancing health and well-being into old age, and ensuring enabling and supportive environments.

With the successful demographic transition in the industrial world, the percentage of those older than 60 years is on the rise, whereas the labor force percentage is decreasing. In 1998, 19 percent of the first world population fell into the post-60 category (10 percent worldwide). Children younger than 15 years also make up 19 percent of the developed world's population, whereas the labor force is at 62 percent. With birth rates hovering around 1 percent or less, and life expectancy increasing, the percentages will likely continue to grow toward the upper end of the scale.

Paul Peterson has argued that the costs of global aging will outweigh the benefits, and the capacity of the developed world to pay for these costs is questionable at best. He suggests that the economic burden on the labor force will be "unprecedented" and offers a number of solutions ("Grey Dawn: The Global Aging Crisis" in *Foreign Affairs*, January/February 1999).

A U.S. Department of Commerce study, *An Aging World: 2008,* suggests nine trends that will likely present aging challenges throughout the globe. (1) The population is aging, as people 65 years and older will outnumber those younger than 5 years. (2) Life expectancy is increasing, raising questions about human lifespan. (3) The number of the oldest, those older than 80 years, is rising, and will more than double within 30 years. (4) Some countries are experiencing aging populations while their total populations decline. (5) Noncommunicable diseases are becoming an increasing burden as they now are the major cause of death among old people. (6) Because of longer lifespans and fewer children, family structures are changing and care options for the elderly may change. (7) There are shrinking ratios of workers to pensioners, as people live longer in retirement, taxing health and pension systems. (8) Social insurance systems are becoming less sustainable. (9) Population aging is having huge effects on social entitlement programs.

A World Bank study (Ronald Lee et al., *Some Economic Consequences of Global Aging,* December 2010) outlines concerns about the effects of aging on societies: slower economic growth, poverty among the elderly, generational equity, inadequate investment in physical and human capital, inefficiency in labor markets, suboptimal consumption profiles, and unsustainable public transfer systems.

The authors of one of this issue's readings, Neil Howe and Richard Jackson, describe the geopolitical implications of global aging for the highly developed societies in *The Graying of the Great Powers* (Center for Strategic and International Studies, 2008). The first is that both the population and the GDP of the developed world will decline as a percentage of global totals, thus leading to a loss of influence. Within the developed world, though, the U.S. share will rise, leading to increased influence. Most nations in sub-Saharan Africa and those in the Muslim world will experience large youth bulges, leading to a chronically unstable situation until at least the 2030s. Many nations in North Africa, the Middle East, South and East Asia, and the former Soviet bloc are experiencing rapid demographic change that could lead to either civil collapse or a reactive neo-authoritarianism. The threat of ethnic and religious conflict will continue as a security challenge throughout the world. The 2020s will be a decade of maximum political danger. The aging developed world will have shortages of young-adult manpower. And finally, this world may struggle to remain culturally attractive and politically relevant to younger peoples.

Alternatively, some analysts are looking at an aging population from a lens that is not half empty but that has some positive aspects to it. One such positive possibility is a decrease in military conflict throughout the globe as societies face manpower shortages while dealing with social services for the elderly. Another approach is to observe how the experience of age is brought to bear in a productive way in modern societies that can reap the benefits of productive labor achieved through brains rather than brawn.

Finally, the 2015 UN report, *World Population Ageing 2015* (UN Department of Economic and Social Affairs), offered a balanced view of how the world must react to this ever increasing global phenomenon. The report makes the obvious argument that preparing for this change in global age cohorts is critical to the international community's ability to achieve its sustainable development goals. Pension systems are particularly at risk as old-age support ratios decline. Health care systems represent another important area that must adjust to the needs of the growing numbers of the elderly, including lifelong health and preventive care. The very elderly will be in greater need of home-based and facility-based around-the-clock care. And the increasing need to eliminate age-related discrimination will become more pressing. Easier access to infrastructure and services will remain a taunting challenge to governments. The UN report concludes with the belief that governments have the capacity to determine current and future needs, and thus will be able to "ensure the well-being and full socio-economic integration of older persons while maintaining the fiscal solvency of pension and health care systems and promoting economic growth."

The YES selection by Neil Howe and Richard Jackson of the Center for Strategic and International Studies argues that global population aging is likely to have a profound and negative effect on global economic growth, living standards and conditions, and "the shape of the world order," particularly affecting China, Russia, Pakistan, Iran, and countries of the West. For them, the critical decade is the 2020s, when the ability of the developed world to maintain global security will be brought into question. In the NO selection, the basic thesis is that because today's senior citizens are better educated than past generations of the elderly, their potential for greater productivity is also greater.

YES ←

**Neil Howe and
Richard Jackson**

Global Aging and the Crisis of the 2020s

From the fall of the Roman and the Mayan empires to the Black Death to the colonization of the New World and the youth-driven revolutions of the twentieth century, demographic trends have played a decisive role in many of the great invasions, political upheavals, migrations, and environmental catastrophes of history. By the 2020s, an ominous new conjuncture of demographic trends may once again threaten widespread disruption. We are talking about global aging, which is likely to have a profound effect on economic growth, living standards, and the shape of the world order.

For the world's wealthy nations, the 2020s are set to be a decade of rapid population aging and population decline. The developed world has been aging for decades, due to falling birthrates and rising life expectancy. But in the 2020s, this aging will get an extra kick as large postwar baby boom generations move fully into retirement. According to the United Nations Population Division (whose projections are cited throughout this article), the median ages of Western Europe and Japan, which were 34 and 33 respectively as recently as 1980, will soar to 47 and 52 by 2030, assuming no increase in fertility. In Italy, Spain, and Japan, more than half of all adults will be older than the official retirement age—and there will be more people in their 70s than in their 20s.

Falling birthrates are not only transforming traditional population pyramids, leaving them top-heavy with elders, but are also ushering in a new era of workforce and population decline. The working-age population has already begun to contract in several large developed countries, including Germany and Japan. By 2030, it will be stagnant or contracting in nearly all developed countries, the only major exception being the United States. In a growing number of nations, total population will begin a gathering decline as well. Unless immigration or birthrates surge, Japan and some European nations are on track to lose nearly one-half of their total current populations by the end of the century.

These trends threaten to undermine the ability of today's developed countries to maintain global security. To begin with, they directly affect population size and GDP size, and hence the manpower and economic resources that nations can deploy. This is what RAND scholar Brian Nichiporuk calls "the bucket of capabilities" perspective. But population aging and decline can also indirectly affect capabilities—or even alter national goals themselves.

Rising pension and health care costs will place intense pressure on government budgets, potentially crowding out spending on other priorities, including national defense and foreign assistance. Economic performance may suffer as workforces grey and rates of savings and investment decline. As societies and electorates age, growing risk aversion and shorter time horizons may weaken not just the ability of the developed countries to play a major geopolitical role, but also their will.

The weakening of the developed countries might not be a cause for concern if we knew that the world as a whole were likely to become more pacific. But unfortunately, just the opposite may be the case. During the 2020s, the developing world will be buffeted by its own potentially destabilizing demographic storms. China will face a massive age wave that could slow economic growth and precipitate political crisis just as that country is overtaking America as the world's leading economic power. Russia will be in the midst of the steepest and most protracted population implosion of any major power since the plague-ridden Middle Ages. Meanwhile, many other developing countries, especially in the Muslim world, will experience a sudden new resurgence of youth whose aspirations they are unlikely to be able to meet.

The risk of social and political upheaval could grow throughout the developing world—even as the developed world's capacity to deal with such threats declines. Yet, if the developed world seems destined to see its geopolitical stature diminish, there is one partial but important exception to the trend: the United States. While it is fashionable to argue that US power has peaked, demography suggests

America will play as important a role in shaping the world order in this century as it did in the last.

Graying Economies

Although population size alone does not confer geopolitical stature, no one disputes that population size and economic size together constitute a potent double engine of national power. A larger population allows greater numbers of young adults to serve in war and to occupy and pacify territory. A larger economy allows more spending on the hard power of national defense and the semi-hard power of foreign assistance. It can also enhance what political scientist Joseph Nye calls "soft power" by promoting business dominance, leverage with nongovernmental organizations and philanthropies, social envy and emulation, and cultural clout in the global media and popular culture.

The expectation that global aging will diminish the geopolitical stature of the developed world is thus based in part on simple arithmetic. By the 2020s and 2030s, the working-age population of Japan and many European countries will be contracting by between 0.5 and 1.5 percent per year. Even at full employment, growth in real GDP could stagnate or decline, since the number of workers may be falling faster than productivity is rising. Unless economic performance improves, some countries could face a future of secular economic stagnation—in other words, of zero real GDP growth from peak to peak of the business cycle.

Economic performance, in fact, is more likely to deteriorate than improve. Workforces in most developed countries will not only be stagnating or contracting, but also graying. A vast literature in the social and behavioral sciences establishes that worker productivity typically declines at older ages, especially in eras of rapid technological and market change.

Economies with graying workforces are also likely to be less entrepreneurial. According to the Global Entrepreneurship Monitor's 2007 survey of 53 countries, new business start-ups in high-income countries are heavily tilted toward the young. Of all "new entrepreneurs" in the survey (defined as owners of a business founded within the past three and one-half years), 40 percent were under age 35 and 69 percent under age 45. Only 9 percent were 55 or older.

At the same time, savings rates in the developed world will decline as a larger share of the population moves into the retirement years. If savings fall more than investment demand, as much macroeconomic modeling suggests is likely, either businesses will starve for investment funds or the developed economies' dependence on capital from higher-saving emerging markets will grow. In the first case, the penalty will be lower output. In the second, it will be higher debt service costs and the loss of political leverage, which history teaches is always ceded to creditor nations.

Even as economic growth slows, the developed countries will have to transfer a rising share of society's economic resources from working-age adults to nonworking elders. Graying means paying—more for pensions, more for health care, more for nursing homes for the frail elderly. According to projections by the Center for Strategic and International Studies, the cost of maintaining the current generosity of today's public old-age benefit systems would, on average across the developed countries, add an extra 7 percent of GDP to government budgets by 2030.

Yet the old-age benefit systems of most developed countries are already pushing the limits of fiscal and economic affordability. By the 2020s, political conflict over deep benefit cuts seems unavoidable. On one side will be young adults who face stagnant or declining after-tax earnings. On the other side will be retirees, who are often wholly dependent on pay-as-you-go public plans. In the 2020s, young people in developed countries will have the future on their side. Elders will have the votes on theirs.

Faced with the choice between economically ruinous tax hikes and politically impossible benefit cuts, many governments will choose a third option: cannibalizing other spending on everything from education and the environment to foreign assistance and national defense. As time goes by, the fiscal squeeze will make it progressively more difficult to pursue the obvious response to military manpower shortages—investing massively in military technology, and thereby substituting capital for labor.

Diminished Stature

The impact of global aging on the collective temperament of the developed countries is more difficult to quantify than its impact on their economies, but the consequences could be just as important—or even more so. With the size of domestic markets fixed or shrinking in many countries, businesses and unions may lobby for anticompetitive changes in the economy. We may see growing cartel behavior to protect market share and more restrictive rules on hiring and firing to protect jobs.

We may also see increasing pressure on governments to block foreign competition. Historically, eras of stagnant population and market growth—think of the 1930s—have been characterized by rising tariff barriers, autarky, corporatism, and other anticompetitive policies that tend to shut the door on free trade and free markets.

This shift in business psychology could be mirrored by a broader shift in social mood. Psychologically, older societies are likely to become more conservative in outlook and possibly more risk-averse in electoral and leadership behavior. Elder-dominated electorates may tend to lock in current public spending commitments at the expense of new priorities and shun decisive confrontations in favor of ad hoc settlements. Smaller families may be less willing to risk scarce youth in war.

We know that extremely youthful societies are in some ways dysfunctional—prone to violence, instability, and state failure. But extremely aged societies may also prove dysfunctional in some ways, favoring consumption over investment, the past over the future, and the old over the young.

Meanwhile, the rapid growth in ethnic and religious minority populations, due to ongoing immigration and higher-than-average minority fertility, could strain civic cohesion and foster a new diaspora politics. With the demand for low-wage labor rising, immigration (at its current rate) is on track by 2030 to double the percentage of Muslims in France and triple it in Germany. Some large European cities, including Amsterdam, Marseille, Birmingham, and Cologne, may be majority Muslim.

In Europe, the demographic ebb tide may deepen the crisis of confidence that is reflected in such best-selling books as *France Is Falling* by Nicolas Baverez, *Can Germany Be Saved?* by Hans-Werner Sinn, and *The Last Days of Europe* by Walter Laqueur. The media in Europe are already rife with dolorous stories about the closing of schools and maternity wards, the abandonment of rural towns, and the lawlessness of immigrant youths in large cities. In Japan, the government has half-seriously projected the date at which only one Japanese citizen will be left alive.

Over the next few decades, the outlook in the United States will increasingly diverge from that in the rest of the developed world. Yes, America is also graying, but to a lesser extent. Aside from Israel and Iceland, the United States is the only developed nation where fertility is at or above the replacement rate of 2.1 average lifetime births per woman. By 2030, its median age, now 37, will rise to only 39. Its working-age population, according to both US Census Bureau and UN projections, will also continue to grow through the 2020s and beyond, both because of its higher fertility rate and because of substantial net immigration, which America assimilates better than most other developed countries.

The United States faces serious structural challenges, including a bloated health care sector, a chronically low savings rate, and a political system that has difficulty making meaningful trade-offs among competing priorities. All of these problems threaten to become growing handicaps as the country's population ages. Yet, unlike Europe and Japan, the United States will still have the youth and the economic resources to play a major geopolitical role. The real challenge facing America by the 2020s may not be so much its inability to lead the developed world as the inability of the other developed nations to lend much assistance.

Perilous Transitions

Although the world's wealthy nations are leading the way into humanity's graying future, aging is a global phenomenon. Most of the developing world is also progressing through the so-called demographic transition—the shift from high mortality and high fertility to low mortality and low fertility that inevitably accompanies development and modernization. Since 1975, the average fertility rate in the developing world has dropped from 5.1 to 2.7 children per woman, the rate of population growth has decelerated from 2.2 to 1.3 percent per year, and the median age has risen from 21 to 28.

The demographic outlook in the developing world, however, is shaping up to be one of extraordinary diversity. In many of the poorest and least stable countries (especially in sub-Saharan Africa), the demographic transition has failed to gain traction, leaving countries burdened with large youth bulges. By contrast, in many of the most rapidly modernizing countries (especially in East Asia), the population shift from young and growing to old and stagnant or declining is occurring at a breathtaking pace—far more rapidly than it did in any of today's developed countries.

Notwithstanding this diversity, some demographers and political scientists believe that the unfolding of the transition is ushering in a new era in which demographic trends will promote global stability. This "demographic peace" thesis, as we dub it, begins with the observation that societies with rapidly growing populations and young age structures are often mired in poverty and prone to civil violence and state failure, while those with no or slow population growth and older age structures tend to be more affluent and stable. As the demographic transition progresses—and population growth slows, median ages rise, and child dependency burdens fall—the demographic peace thesis predicts that economic growth and social and political stability will follow.

We believe this thesis is deeply flawed. It fails to take into account the huge variation in the timing and pace of the demographic transition in the developing world. It tends to focus exclusively on the threat of state failure, which indeed is closely and negatively correlated with the degree of demographic transition, while ignoring the threat of "neo-authoritarian" state success, which is more likely to occur in societies in which the transition is well under way. We are, in other words, not talking just about a hostile version of the Somalia model, but also about a potentially hostile version of the China or Russia model, which appears to enjoy growing appeal among political leaders in many developing countries.

More fundamentally, the demographic peace thesis lacks any realistic sense of historical process. It is possible (though by no means assured) that the global security environment that emerges after the demographic transition has run its course will be safer than today's. It is very unlikely, however, that the transition will make the security environment progressively safer along the way. Journeys can be more dangerous than destinations.

Economists, sociologists, and historians who have studied the development process agree that societies, as they move from the traditional to the modern, are buffeted by powerful and disorienting social, cultural, and economic crosswinds. As countries are integrated into the global marketplace and global culture, traditional economic and social structures are overturned and traditional value systems are challenged.

Along with the economic benefits of rising living standards, development also brings the social costs of rapid urbanization, growing income inequality, and environmental degradation. When plotted against development, these stresses exhibit a hump-shaped or inverted-U pattern, meaning that they become most acute midway through the demographic transition.

The demographic transition can trigger a rise in extremism. Religious and cultural revitalization movements may seek to reaffirm traditional identities that are threatened by modernization and try to fill the void left when development uproots communities and fragments extended families. It is well documented that international terrorism, among the developing countries, is positively correlated with income, education, and urbanization. States that sponsor terrorism are rarely among the youngest and poorest countries; nor do the terrorists themselves usually originate in the youngest and poorest countries. Indeed, they are often disaffected members of the middle class in middle-income countries that are midway through the demographic transition.

Ethnic tensions can also grow. In many societies, some ethnic groups are more successful in the marketplace than others—which means that, as development accelerates and the market economy grows, rising inequality often falls along ethnic lines. The sociologist Amy Chua documents how the concentration of wealth among "market-dominant minorities" has triggered violent backlashes by majority populations in many developing countries, from Indonesia, Malaysia, and the Philippines (against the Chinese) to Sierra Leone (against the Lebanese) to the former Yugoslavia (against the Croats and Slovenes).

We have in fact only one historical example of a large group of countries that has completed the entire demographic transition—today's (mostly Western) developed nations. And their experience during that transition, from the late 1700s to the late 1900s, was filled with the most destructive revolutions, civil wars, and total wars in the history of civilization. The nations that engaged in World War II had a higher median age and a lower fertility rate—and thus were situated at a later stage of the transition—than most of today's developing world is projected to have over the next 20 years. Even if global aging breeds peace, in other words, we are not out of the woods yet.

Storms Ahead

A number of demographic storms are now brewing in different parts of the developing world. The moment of maximum risk still lies ahead—just a decade away, in the 2020s. Ominously, this is the same decade when the developed world will itself be experiencing its moment of greatest demographic stress.

Consider China, which may be the first country to grow old before it grows rich. For the past quarter-century, China has been "peacefully rising," thanks in part to a one-child-per-couple policy that has lowered dependency burdens and allowed both parents to work and contribute to China's boom. By the 2020s, however, the huge Red Guard generation, which was born before the country's fertility decline, will move into retirement, heavily taxing the resources of their children and the state.

China's coming age wave—by 2030 it will be an older country than the United States—may weaken the two pillars of the current regime's legitimacy: rapidly rising GDP and social stability. Imagine workforce growth slowing to zero while tens of millions of elders sink into indigence without pensions, without health care, and without large extended families to support them. China could careen toward social collapse—or, in reaction, toward an authoritarian clampdown. The arrival of China's age wave, and the turmoil it may bring, will coincide with its expected displacement of the United States as the world's largest economy in the 2020s. According to "power transition" theories of global conflict, this moment could be quite perilous.

By the 2020s, Russia, along with the rest of Eastern Europe, will be in the midst of an extended population decline as steep or steeper than any in the developed world. The Russian fertility rate has plunged far beneath the replacement level even as life expectancy has collapsed amid a widening health crisis. Russian men today can expect to live to 60—16 years less than American men and marginally less than their Red Army grandfathers at the end of World War II. By 2050, Russia is due to fall to 16th place in world population rankings, down from 4th place in 1950 (or third place, if we include all the territories of the former Soviet Union).

Prime Minister Vladimir Putin flatly calls Russia's demographic implosion "the most acute problem facing our country today." If the problem is not solved, Russia will weaken progressively, raising the nightmarish specter of a failing or failed state with nuclear weapons. Or this cornered bear may lash out in revanchist fury rather than meekly accept its demographic fate.

Of course, some regions of the developing world will remain extremely young in the 2020s. Sub-Saharan Africa, which is burdened by the world's highest fertility rates and is also ravaged by AIDS, will still be racked by large youth bulges. So will a scattering of impoverished and chronically unstable Muslim-majority countries, including Afghanistan, the Palestinian territories, Somalia, Sudan, and Yemen. If the correlation between extreme youth and violence endures, chronic unrest and state failure could persist in much of sub-Saharan Africa and parts of the Muslim world through the 2020s, or even longer if fertility rates fail to drop.

Meanwhile, many fast-modernizing countries where fertility has fallen very recently and very steeply will experience a sudden resurgence of youth in the 2020s. It is a law of demography that, when a population boom is followed by a bust, it causes a ripple effect, with a gradually fading cycle of echo booms and busts. In the 2010s, a bust generation will be coming of age in much of Latin America, South Asia, and the Muslim world. But by the 2020s, an echo boom will follow—dashing economic expectations and perhaps fueling political violence, religious extremism, and ethnic strife.

These echo booms will be especially large in Pakistan and Iran. In Pakistan, the decade-over-decade percentage growth in the number of people in the volatile 15- to 24-year-old age bracket is projected to drop from 32 percent in the 2000s to just 10 percent in the 2010s, but then leap upward again to 19 percent in the 2020s. In Iran, the swing in the size of the youth bulge population is projected to be even larger: minus 33 percent in the 2010s and plus 23 percent in the 2020s. These echo booms will be occurring in countries whose social fabric is already strained by rapid development. One country teeters on the brink of chaos, while the other aspires to regional hegemony. One already has nuclear weapons, while the other seems likely to obtain them.

Pax Americana Redux?

The demographer Nicholas Eberstadt has warned that demographic change may be "even more menacing to the security prospects of the Western alliance than was the cold war for the past generation." Although it would be fair to point out that such change usually presents opportunities as well as dangers, his basic point is incontestable: Planning national strategy for the next several decades with no regard for population projections is like setting sail without a map or a compass. It is likely to be an ill-fated voyage. In this sense, demography is the geopolitical cartography of the twenty-first century.

Although tomorrow's geopolitical map will surely be shaped in important ways by political choices yet to be made, the basic contours are already emerging. During the era of the Industrial Revolution, the population of what we now call the developed world grew faster than the rest of the world's population, peaking at 25 percent of the world total in 1930. Since then, its share has declined. By 2010, it stood at just 13 percent, and it is projected to decline still further, to 10 percent by 2050.

The collective GDP of the developed countries will also decline as a share of the world total—and much more steeply. According to new projections by the Carnegie Endowment for International Peace, the Group of 7 industrialized nations' share of the Group of 20 leading economies' total GDP will fall from 72 percent in 2009 to 40 percent in 2050. Driving this decline will be not just the slower growth of the developed world, as workforces age and stagnate or contract, but also the expansion of large, newly market-oriented economies, especially in East and South Asia.

Again, there is only one large country in the developed world that does not face a future of stunning relative demographic and economic decline: the United States. Thanks to its relatively high fertility rate and substantial net immigration, its current global population share will remain virtually unchanged in the coming decades. According to the Carnegie projections, the US share of total G-20 GDP will drop significantly, from 34 percent in 2009 to 24 percent in 2050. The combined share of Canada, France, Germany, Italy, Japan, and the United Kingdom, however, will plunge from 38 percent to 16 percent.

By the middle of the twenty-first century, the dominant strength of the US economy within the developed world will have only one historical parallel: the immediate aftermath of World War II, exactly 100 years earlier, at the birth of the "Pax Americana."

The UN regularly publishes a table ranking the world's most populous countries over time. In 1950, six of the top twelve were developed countries. In 2000, only three were. By 2050, only one developed country will remain—the United States, still in third place. By then, it will be the only country among the top twelve committed since its founding to democracy, free markets, and civil liberties.

All told, population trends point inexorably toward a more dominant US role in a world that will need America more, not less.

Neil Howe, an historian, economist, and demographer, is a senior associate at the Center for Strategic and International Studies, and coauthor of *The Graying of the Great Powers: Demography and Geopolitics in the 21st Century* (2008). He is a recognized authority on global aging.

Richard Jackson is a senior fellow at the Center for Strategic and International Studies, and coauthor of *The Graying of the Great Powers: Demography and Geopolitics in the 21st Century* (2008), and author of *Global Aging and the Future of Emerging Markets* (2011).

The Economist (2014)    **NO**

Age Invaders

A generation of old people is about to change the global economy. They will not all do so in the same way.

In the 20th century the planet's population doubled twice. It will not double even once in the current century, because birth rates in much of the world have declined steeply. But the number of people over 65 is set to double within just 25 years. This shift in the structure of the population is not as momentous as the expansion that came before. But it is more than enough to reshape the world economy.

According to the UN's population projections, the standard source for demographic estimates, there are around 600m people aged 65 or older alive today. That is in itself remarkable; the author Fred Pearce claims it is possible that half of all the humans who have ever been over 65 are alive today. But as a share of the total population, at 8%, it is not that different to what it was a few decades ago.

By 2035, however, more than 1.1 billion people—13% of the population—will be above the age of 65. This is a natural corollary of the dropping birth rates that are slowing overall population growth; they mean there are proportionally fewer young people around. The "old-age dependency ratio"—the ratio of old people to those of working age—will grow even faster. In 2010 the world had 16 people aged 65 and over for every 100 adults between the ages of 25 and 64, almost the same ratio it had in 1980. By 2035 the UN expects that number to have risen to 26.

In rich countries it will be much higher. Japan will have 69 old people for every 100 of working age by 2035 (up from 43 in 2010), Germany 66 (from 38). Even America, which has a relatively high fertility rate, will see its old-age dependency rate rise by more than 70% to 44. Developing countries, where today's ratio is much lower, will not see absolute levels rise that high; but the proportional growth will be higher. Over the same time period the old-age dependency rate in China will more than double from 15 to 36. Latin America will see a shift from 14 to 27.

Three Ways Forward

The big exceptions to this general greying are south Asia and Africa, where fertility is still high. Since these places are home to almost 3 billion people, rising to 5 billion by mid-century, their youth could be a powerful counter to the greying elsewhere. But they will slow the change, not reverse it. The emerging world as a whole will see its collective old-age dependency rate almost double, to 22 per 100, by 2035.

The received wisdom is that a larger proportion of old people means slower growth and, because the old need to draw down their wealth to live, less saving; that leads to higher interest rates and falling asset prices. Some economists are more sanguine, arguing that people will adapt and work longer, rendering moot measures of dependency which assume no one works after the age of 65. A third group harks back to the work of Alvin Hansen, known as the "American Keynes," who argued in 1938 that a shrinking population in America would bring with it diminished incentives for companies to invest—a smaller workforce needs less investment—and hence persistent stagnation.

The unexpected baby boom of 1946–64 messed up Hansen's predictions, and unforeseen events could undermine today's demographic projections, too—though bearing in mind that the baby boom required a world war to set the stage, that should not be seen as a source of hope. But if older people work longer and thus save longer, while slowing population growth means firms have less incentive to invest, something close to what Hansen envisaged could come about even without the sort of overall population decline he foresaw. A few months ago Larry Summers, a Harvard professor and former treasury secretary, argued that America's economy appeared already to be suffering this sort of "secular stagnation"—a phrase taken directly from Hansen.

Who is right? The answer depends on examining the three main channels through which demography influences the economy: changes in the size of the workforce, changes in the rate of productivity growth, and changes in the pattern of savings. The result of such examination is not conclusive. But, for the next few years at least, Hansen's worries seem most relevant, not least because of a previously unexpected effect: the tendency of those with higher skills to work for longer, and more productively, than they have done to date.

The first obvious implication of a population that is getting a lot older without growing much is that, unless the retirement age changes, there will be fewer workers. That means less output, unless productivity rises to compensate. Under the UN's standard assumption that a working life ends at 65, and with no increases in productivity, ageing populations could cut growth rates in parts of the rich world by between one-third and one-half over the coming years.

Have Skills, Will Work

Amlan Roy, an economist at Credit Suisse, has calculated that the shrinking working-age population dragged down Japan's GDP growth by an average of just over 0.6 percentage points a year between 2000 and 2013, and that over the next four years that will increase to 1 percentage point a year. Germany's shrinking workforce could reduce GDP growth by almost half a point. In America, under the same assumptions, the retirement of the baby-boomers would be expected to reduce the economy's potential growth rate by 0.7 percentage points.

The real size of the workforce, though, depends on more than the age structure of the population; it depends on who else works (women who currently do not, perhaps, or immigrants) and how long people work. In the late-20th century that last factor changed little. An analysis of 43 mostly rich countries by David Bloom, David Canning, and Günther Fink, all of Harvard University, found that between 1965 and 2005 the average legal retirement age rose by less than six months. During that time male life expectancy rose by nine years.

Since the turn of the century that trend has reversed. Almost 20% of Americans aged over 65 are now in the labor force, compared with 13% in 2000. Nearly half of all Germans in their early 60s are employed today, compared with a quarter a decade ago.

This is in part due to policy. Debt-laden governments in Europe have cut back their pension promises and raised the retirement age. Half a dozen European countries, including Italy, Spain, and the Netherlands, have linked the statutory retirement age to life expectancy. Personal financial circumstances have played a part, too. In most countries the shift was strongest in the wake of the 2008 financial crisis, which hit the savings of many near-retirees.

The move away from corporate pension plans that provided a fraction of the recipient's final salary in perpetuity will also have kept some people working longer.

But an even more important factor is education. Better-educated older people are far more likely to work for longer. Gary Burtless of the Brookings Institution has calculated that, in America, only 32% of male high-school graduates with no further formal education are in the workforce between the ages of 62 and 74. For men with a professional degree the figure is 65% (though the overall number of such men is obviously smaller). For women the ratios are one-quarter versus one-half, with the share of highly educated women working into their 60s soaring. In Europe, where workers of all sorts are soldiering on into their 60s more than they used to, the effect is not quite as marked, but still striking. Only a quarter of the least-educated Europeans aged 60–64 still work; half of those with a degree do.

It is not a hard pattern to explain. Less-skilled workers often have manual jobs that get harder as you get older. The relative pay of the less skilled has fallen, making retirement on a public pension more attractive; for the unemployed, who are also likely to be less skilled, retirement is a terrific option. Research by Clemens Hetschko, Andreas Knabe, and Ronnie Schöb shows that people who go straight from unemployment to retirement experience a startling increase in their sense of well-being.

Higher-skilled workers, on the other hand, tend to be paid more, which gives them an incentive to keep working. They are also on average healthier and longer-lived, so they can work and earn past 65 and still expect to enjoy the fruits of that extra labor later on.

This does not mean the workforce will grow. Overall work rates among the over-60s will still be lower than they were for the same cohort when it was younger. And even as more educated old folk are working, fewer less-skilled young people are. In Europe, jobless rates are highest among the least-educated young. In America, where the labor participation rate (at 63%) is close to a three-decade low, employment has dropped most sharply for less-skilled men. With no surge in employment among women, and little appetite for mass immigration, in most of the rich world the workforce looks likely to shrink even if skilled oldies stay employed.

Legacy of the Void

A smaller workforce need not dampen growth, though, if productivity surges. This is not something most would expect to come about as a result of an aging population. Plenty of studies and bitter experience show that most physical and many cognitive capacities decline with age. A new analysis by a trio of Canadian academics based on the video game "StarCraft II," for instance, suggests that raw brainpower peaks at 24. And aging societies may ossify. Alfred Sauvy, the French thinker who coined the term "third world," was prone to worry that the first world would become "a society of old people, living in old houses, ruminating about old ideas." Japan's productivity growth slowed sharply in the 1990s when its working-age population began to shrink; Germany's productivity performance has become lacklustre as its population ages.

But Japan's slowed productivity growth can also be ascribed to its burst asset bubble, and Germany's to reforms meant to reduce unemployment; both countries, aging as they are, score better in the World Economic Forum's ranking for innovation than America. A dearth of workers might prompt the invention of labor-saving capital-intensive technology, just as Japanese firms are pioneering the use of robots to look after old people. And a wealth of job experience can counter slower cognitive speed. In an age of ever-smarter machines, the attributes that enhance productivity may have less to do with pure cognitive oomph than motivation, people skills, and managerial experience.

Perhaps most important, better education leads to higher productivity at any age. For all these reasons, a growing group of highly educated older folk could increase productivity, offsetting much of the effect of a smaller workforce.

Evidence on both sides of the Atlantic bears this out. A clutch of recent studies suggest that older workers are disproportionately more productive—as you would expect if they are disproportionately better educated. Laura Romeu Gordo of the German Centre of Gerontology and Vegard Skirbekk of the International Institute for Applied Systems Analysis in Austria have shown that in Germany older workers who stayed in the labor force have tended to move into jobs which demanded more cognitive skill. Perhaps because of such effects, the earnings of those over 50 have risen relative to younger workers.

Saving Graces

This could be good news for countries with well-educated people currently entering old age—but less so in places that are less developed. Nearly half of China's workers aged between 50 and 64 have not completed primary school. As these unskilled people age, their productivity is likely to fall. Working with his IIASA colleagues Elke Loichinger and Daniela Weber, Mr. Skirbekk tried to gauge this effect by creating a "cognition-adjusted dependency ratio." They compared the cognitive ability of people aged 50 and over across rich and emerging economies by means of an experiment which tested their ability to recall words, and used the results to weigh dependency ratios. This cognition-adjusted ratio is lower in northern Europe than it is in China, even though the age-based ratio is far higher in Europe, because the elderly in Europe score much more strongly on the cognitive-skill test. Similarly adjusted, America's dependency ratio is better than India's.

If skill and education determine how long and how well older people work, they also have big implications for saving, the third channel through which ageing affects growth. A larger group of well-educated older people will earn a larger share of overall income. In America the share of male earnings going to those aged 60–74 has risen from 7.3% to 12.7% since 2000 as well-educated baby-boomers have moved into their 60s. Some of these earnings will finance retirement, when those concerned finally decide to take it; more savings by people in their 60s will be matched by more spending when they reach their 80s. But many of the educated elderly are likely to accumulate far more than they will draw down towards the end of life. Circumstantial evidence supports this argument. Thomas Piketty, a French economist, calculates that the average wealth of French 80-year-olds is 134% that of 50- to 59-year-olds, the highest gap since the 1930s. For the next few years at least, skill-skewed ageing is likely to mean both more inequality and more private saving.

At the same time governments across the rich world (and particularly in Europe) are trimming their pension promises and cutting their budget deficits, both of which add to national saving. Reforms designed to trim future pensions mean that, regardless of their skill level, those close to retirement are likely to save more and that governments will spend less per old person. The European Commission's latest forecasts suggest overall pension spending in the EU will fall by 0.1% of GDP between 2010 and 2020, before rising by 0.6% in the subsequent decade. That is not insignificant, but it is far less than some of the breathless commentary about the "burden" of ageing implies.

Taken together, the net effect of high saving by educated older workers and less-generous pensions is likely to be an unexpected degree of thrift in the rich world, at least for the next few years. If the money saved finds productive investment opportunities, it has the potential to boost long-run growth. But where will these opportunities be? In principle, two possibilities stand out. One is rapid innovation in advanced economies. The second is fast growth in emerging economies—especially younger, poorer ones.

Unfortunately, more capital currently flows out of emerging economies into the rich world than the other way. The most successful emerging economies have built up huge stashes of foreign currency; many are leery of depending too much on foreign borrowing. Even if that were to change, the youthful economies of south Asia and Africa are too small to absorb huge flows of capital from those countries that are aging fast.

And in the rich world, despite lots of obvious innovation, particularly in computer technology, both productivity growth and investment have been tepid of late. That may be a hangover from the financial crisis. But it could also be a structural change. The price of capital goods, notably anything to do with computers, has fallen sharply; it may be that today's innovation is simply less investment-intensive than it was in the manufacturing age. And the aging population itself may deter investment. Fewer workers, other things being equal, means the economy needs a smaller capital stock, even if some of those workers are clever old sticks. And an aging population spends differently. Old people buy fewer things that require heavy investment—notably houses—and more services, whether in health care or tourism.

Not Destiny, but not Nothing

Demographic trends will shape the future, but they do not render particular outcomes inevitable. The evolution of the economy will depend on the way policymakers respond to the new situation. But those policy reactions will themselves be shaped by the priorities of older people to a greater extent than has previously been the case; they will be a bigger share of the population and in democracies they tend to vote more than younger people do.

On both sides of the Atlantic, recent budget decisions appear to reflect the priorities of the aging and affluent. Annuities reform in Britain increased people's freedom to spend their pension pots; the disappearance of property-tax reform spared homeowning older Italians a new burden; America's budget slashed spending on the young and poor while failing to make government health and pension spending any less generous to the well-off. Few rich-country governments have shown any appetite for large-scale investment, despite low interest rates.

A set of forces pushing investment down and pushing saving up, with no countervailing policy response, makes the impact of aging over the next few years look like the world that Hansen described: one of slower growth (albeit not as slow as it would have been if older folk were not working more), a surfeit of saving and very low interest rates. It will be a world in which aging reinforces the changes in income distribution that new technology has brought with it: the skilled old earn more, the less-skilled of all ages are squeezed. The less-educated and jobless young will be particularly poorly served, never building up the skills to enable them to become productive older workers.

Compared with the dire warnings about the bankrupting consequences of a "grey tsunami," this is good news. But not as good as all that.

EXPLORING THE ISSUE

Is Global Aging a Major Problem?

Critical Thinking and Reflection

1. What do analysts really mean when they talk about a global aging population?
2. Why is the 2020s considered to be a decade of reckoning because of an aging global population?
3. What are differences and similarities regarding global aging between developed and developing societies?
4. Should national governments be concerned that their populations are about to go through a demographic transition that will lead to aging populations?
5. Will an aging population lead to economic stagnation or growth over the coming decades?

Is There Common Ground?

There is now a consensus that demographic trends matter with respect to a wide range of public policy issues. There is also agreement on the nature of future trends, particularly as they relate to global aging. And societies are coming to agree on the economic implications of such aging.

Additional Resources

Jackson, Richard and Howe, Neil, *The Graying of the Great Powers* (Center for Strategic and International Studies, 2008)

This report describes global aging trends and geopolitical consequences.

Kunkel, Suzanne R., *Global Aging: Comparative Perspectives on Aging and the Life Course* (Springer Publishing Company, 2013)

This book examines the demands that an aging population places on a society.

Lee, Ronald, *Population Aging and the Generational Economy: A Global Perspective* (Edward Elgar Publisher, 2011)

This book describes the implications for population aging on the macroeconomy. It was the result of a seven-year research project involving over 50 economists and demographers.

Lee, Ronald D., *Global Population Aging and Its Economic Consequences* (AEI Press, 2007)

This book, first appearing as a lecture, describes how the risks of aging can be contained through foresight and public policy.

Lee, Ronald et al., *Some Economic Consequences of Global Aging* (The World Bank, December 2010)

As the title implies, this study spells out economic implications of global aging.

Magnus, George, *The Age of Aging: How Demographics Are Changing the Global Economy and Our World* (Wiley, 2008)

This book is an easily read analysis of basic demographic trends, particularly aging, and their consequences.

McMorrow, Kleran, *The Economic and Financial Market Consequences of Global Ageing* (Springer, 2010)

This book examines the effects of population changes on the size of the labor force and the broader financial implications.

National Institute on Aging, *Why Population Aging Matters: A Global Perspecive* (Amazon Digital Services, 2011)

This book describes population trends, including known trends in global aging. Nine emerging trends are identified.

Powell, Jason L., *The Global Dynamics of Aging* (Nova Science Pub Inc, 2012)

The book explores the impact of aging on a number of countries and continents throughout the world.

Powell, Jason L., "The Power of Global Aging," *Ageing International* (Vol. 35, March 2010)

This article analyzes the dramatic growth in the proportion of the elderly across the global and the factors behind it.

Robinson, Mary et al. (eds.), *Global Health and Global Aging* (Jossey-Bass, 2007)

This book describes the basic aspects of global aging, using real-world models from countries and regions that demonstrate best practices.

Rowland, Donald T., *Population Aging: The Transformation of Societies* (Springer, 2012)

The book presents a synthesis of interdisciplinary research on population aging in countries where the phenomenon is occurring.

Uhlenberg, Peter, *International Handbook of Population Aging* (Springer, 2009)

This book examines research on various aspects of global aging, including how the world is changing as a consequence.

UNFPA, *Ageing in the Twenty-First century: A Celebration and A Challenge* (UNFPA, 2012)

United Nations Department of Economic and Social Affairs, *World Population Ageing 2015* (United Nations, 2015)

This report describes key trends in population aging and policy implications of these trends.

United Nations Population Fund and HelpAge International, *Ageing in the Twenty-First Century: A Challenge and a Celebration* (United Nations, 2012)

This publication draws on the work of over 20 United Nations entities and major international organizations working in the field of aging to review policies and action taken by governments and other stakeholders since the Second World Assembly on Ageing in 2002.

U.S. Department of Commerce, *An Aging World: 2008* (U.S. Census Bureau, 2008)

This comprehensive report describes global demographic trends and their implications, particularly as they relate to global aging.

U.S. Department of State, *Why Population Aging Matters: A Global Perspective* (2007)

This report examines the impact of population on nations as well as describing population trends.

Internet References . . .

Eurolink Age

www.eurolinkage.org

Global Action on Aging

www.globalaging.org

Helpage International

www.helpage.org/Home

International Association of Gerontology

www.iagg.com.br/webforms/index.aspx

International Federation on Ageing (IFA)

www.ifa-fiv.org

Population Reference Bureau

www.prb.org

The CSIS Global Aging Initiative

www.csis.org/gai/

The Population Council

www.popcouncil.org

United Nations Population Fund (UNFPA)

www.unfpa.org

UN Program on Ageing

www.un.org/esa/socdev/ageing

Selected, Edited, and with Issue Framing Material by:
James E. Harf, *Maryville University*
and
Mark Owen Lombardi, *Maryville University*

ISSUE

Will the World Be Able to Feed Itself in the Foreseeable Future?

YES: Food and Agricultural Organization of the United Nations, from "The State of Food Insecurity in the World," *United Nations Publications* (2015)

NO: The British Government Office for Science, from "The Future of Food and Farming: Challenges and Choices for Global Sustainability," *Foresight* (2011)

Learning Outcomes

After reading this issue, you will be able to:

- Gain an understanding of the extent of undernourishment and hunger globally.
- Describe progress made in addressing global undernourishment in the past 25 years.
- Understand the role played by both economic growth and social protection programs in fostering progress toward hunger and poverty targets.
- Gain an understanding of the factors (increased competition for land, water, and energy) that will adversely affect the world's ability to produce food over the next 40 years.
- Understand how many current systems of food production are unsustainable.
- Discuss why policy-makers must look at the entire food system and its uniqueness when considering action.

ISSUE SUMMARY

YES: The UN Food and Agricultural Organization estimates that the number of undernourished around the globe has declined by 167 million over the past decade.

NO: The British government's report concludes that the world's existing food system is failing half of the world's population, with a billion people hungry and another billion suffering from "hidden hunger."

The lead editorial in *The New York Times* on March 3, 2008 began with the sentence: "The world's food situation is bleak. . . ." The primary culprit, according to the editorial, is the rising cost of wheat. The blame, in turn, was placed on the growing impact of biofuels. Others echoed the same message, adding climate change and the rising cost of shipping to the list of culprits. The UN Food and Agricultural Organization (FAO) also issued a series of warnings in late 2007 and early 2008 about the growing food crisis. Nine days later, UN Secretary-General Ban Ki-moon in *The Washington Post* also sounded the

global food alarm, alluding to high food costs and food insecurity.

Nineteen months later in October 2009, the UN FAO Director-General Jacques Diouf in a major speech suggested that agriculture would have to become more productive if the globe's growing population was to be fed. He suggested that future production growth would be found in increased yields and better crop intensity. Further, he argued that a major problem was that food was not being produced by 70 percent of the world's poor who worked in agriculture.

In November 2010, FAO issued another edition of *Food Outlook*. In it, the UN organization suggested that another food crisis was upon us, caused primarily by less than anticipated food production. At the same time, FAO also targeted higher grain prices. The International Food Policy Research Institute (IFPRI) echoed the need for more production, blaming growing world population and negative production results because of climate change and suggesting that rising food prices were simply a result of the increased demand.

These studies followed pre-2008 reports about food supply and demand. For example, the IFPRI 2007 report, The World Food Situation: New Driving Forces and Required Actions, found these "new driving forces" on both the supply and demand side. Rising incomes and urbanization led to the call for better food (meat and milk). On the supply side, increased food production-related energy prices and the large-scale diversion of corn from food to energy (ethanol) production as well as less than desirable weather all contribute to a lowered supply.

And the warnings continued. In 2013, the UN Food and Agricultural organization (FAO) reported that one in eight individuals suffered from chronic undernourishment. As 2015 ended, concerns of international governmental organizations continued. The UN Food Programme revealed that approximately 800 million people throughout the globe, 98 percent of whom were in the developing world, were suffering from hunger as a result of global conditions. FAO also issued dire warnings about the effect of weather on the global food situation.

Private individuals share the same view, such as Robert G. Lewis, former U.S. government administrator, who once oversaw farm price support programs. Lewis, writing in World Policy Institute (2008), suggested that despite short-term problems relating to pricing, the real enduring problem is that demand for more and better food is growing at a fast pace. And the demand is occurring in places whose population can less afford to purchase food. Respected observers of the global food problem echo this message. Lester Brown argued recently in a new book that the world is undergoing a major transition from food abundance to food scarcity. Several bloggers have picked up a common theme, suggesting 20 signs that a terrible food crisis is just on the horizon.

Of course, not everyone believed that for whatever the reason, food shortage was a myth despite the fact that one in six people are going hungry. One such individual is John McCabe, who makes his case on the blog, *Give It to Me Raw*.

A 2012 report by the International Food Policy Research Institute echoed these dire statements, spelling out serious weaknesses that continue to plague the global food system—"lack of ability to respond to volatile food

prices, extreme weather, and inadequate response to food emergencies." But this study also reveals some encouraging signs. After years of neglect, both agriculture and food security are back on the world's political agenda. China and India have increased spending in these two areas. About 20 African countries have adopted national investment plans where 10 percent of their budgets are being devoted to achieving an annual agricultural growth rate of 6 percent.

And the U.S. Agency for International Development (USAID) has pushed forward with its Feed the Future Initiative, while the World Bank continues its commitment of about $6 billion to agriculture.

And the world is finally beginning to see positive results. World hunger has declined somewhat since 1990, although the situation remains serious. Fifteen countries have made substantial progress, with the largest improvements in Angola, Bangladesh, Ethiopia, Malawi, Nicaragua, Niger, and Vietnam, according to the 2012 Global Hunger Index (GHI).

Visualize two pictures. One is a group of people in Africa, including a significant number of small children, who show dramatic signs of advanced malnutrition and even starvation. The second picture shows an apparently wealthy couple finishing a meal at a rather expensive restaurant. The waiter removes their plates still half-full of food, and deposits them in the kitchen garbage can. These scenarios once highlighted a popular film about world hunger. The implication was quite clear. If only the wealthy would share their food with the poor, no one would go hungry. Today the simplicity of this image is obvious. And yet the recent food crisis of 2008 said nothing about an inadequate or maldistributed supply of food.

This issue addresses the question of whether or not the world will be able to feed itself by the middle of the twenty-first century. A prior question, of course, is whether or not enough food is grown throughout the world today to handle current nutritional and caloric needs of all the planet's citizens. News accounts of chronic food shortages somewhere in the world seem to have been appearing with regular consistency for close to 40 years. This time has witnessed graphic accounts in news specials about the consequences of insufficient food, usually somewhere in sub-Saharan Africa. Also, several national and international studies have been commissioned to address world hunger. An American study organized by President Carter, for example, concluded that the root cause of hunger was poverty.

One might deduce from all of this activity that population growth had outpaced food production and that the planet's agricultural capabilities are no longer sufficient, or that the poor have been priced out of the marketplace. Yet the ability of most countries to grow enough food has not yet been challenged. During the 1970–2000 period, only one region of the globe, sub-Saharan Africa, was unable to have its own food production keep pace with population growth.

This is instructive because, beginning in the early 1970s, a number of factors conspired to lessen the likelihood that all humans would go to bed each night adequately nourished. Weather in major food-producing countries turned bad; a number of countries, most notably Japan and the Soviet Union, entered the world grain-importing business with a vengeance; the cost of energy used to enhance agricultural output rose dramatically; and less capital was available to poorer countries as loans or grants for purchasing agricultural inputs or the finished product (food) itself. Yet the world has had little difficulty growing sufficient food, enough to provide every person with two loaves of bread per day as well as other commodities.

Why then did famine and other food-related maladies appear with increasing frequency? The simple answer is that food has been treated as a commodity, not a nutrient. Those who can afford to buy food or grow their own do not go hungry. However, the world's poor became increasingly unable to afford either to create their own successful agricultural ventures or to buy enough food.

The problem for the next half-century, then, has several facets to it. First, can the planet physically sustain increases in food production equal to or greater than the ability of the human race to reproduce itself? This question can only be answered by examining both factors in the comparison—likely future food production levels and future fertility scenarios. A second question relates to the economic dimension—will those poorer countries of the globe that are unable to grow their own food have sufficient assets to purchase it, or will the international community create a global distribution network that ignores a country's ability to pay? And third, will countries that want to grow their own food be given the opportunity to do so?

Three alternative perspectives are relevant to a discussion over the world's future ability to feed its population. The most basic alternative is the question of the target of international action. At its most basic, the root cause of hunger is poverty, first suggested by President Jimmy Carter's hunger commission. If the truth is in the ability to afford to pay for food, why not simply focus on eliminating poverty, hence solving the food problem?

Two other alternative perspectives are found in the literature. One is a report by the Rabobank group, "Sustainability and Security of the Global Food Supply Chain" (undated). Although it concludes that there is sufficient global potential to produce the 70 percent more food needed for 2050, the key factor is the global food system, a long supply chain that "encompasses different countries and numerous participants and stakeholders." To this group, less important are balancing food shortages across regions/countries and changing dietary habits and needs.

A very different alternative approach has recently been advanced by the Worldwatch Institute (*The State of the World*, 2011). Instead of growing more food, the key to addressing the food crisis is to "encourage self-sufficiency and waste reduction." That is, the emphasis should not be on the food production side of the equation but, rather, on the food consumption side. This was earlier seen in the energy issue where attention away from production to consumption changed the entire global mindset in the mid-1970s. This view is shared by Julian Cribb, award-winning journalist, who suggests that rarely have we been advised "of the true ecological costs of eating" (*The Coming Famine*, 2010). He argues the need for a world diet that "is sparing of energy, water, land, and other inputs and has minimal impact on the eider environment." He adds other "big-picture solutions": curbing waste, sharing knowledge, recarbonizing, and movement toward a world farm.

The selections for this issue address the specific question of the planet's continuing ability to grow sufficient food for its growing population. In the YES selection, FAO focuses on the positive trends in the global food crisis. The number of undernourished has declined, particularly in developing areas, almost in half. More than half of the developing countries have reached their hunger targets for the Millennium Development Goal. In the NO selection, the British government's report suggests that "the global food system will experience an unprecedented confluence of pressures over the next 40 years." Any one of these pressures could trigger a major challenger to food security. The report concludes with a call for decisive and immediate action.

YES ↵

Food and Agriculture Organization of the United Nations

The State of Food Insecurity in the World

Meeting the 2015 International Hunger Targets: Taking Stock of Uneven Progress

Undernourishment Around the World in 2015

The Global Trends

Progress continues in the fight against hunger, yet an unacceptably large number of people still lack the food they need for an active and healthy life. The latest available estimates indicate that about 795 million people in the world—just over one in nine—were undernourished in 2014–16. The share of undernourished people in the population, or the prevalence of undernourishment (PoU),[1] has decreased from 18.6 percent in 1990–92 to 10.9 percent in 2014–16, reflecting fewer undernourished people in a growing global population. Since 1990–92, the number of undernourished people has declined by 216 million globally, a reduction of 21.4 percent, notwithstanding a 1.9 billion increase in total population over the same period. The vast majority of the hungry live in the developing regions,[2] where an estimated 780 million people were undernourished in 2014–16. The PoU, standing at 12.9 percent in 2014–16, has fallen by 44.5 percent since 1990–92.

Changes in large populous countries, notably China and India, play a large part in explaining the overall hunger reduction trends in the developing regions.[3] Rapid progress was achieved during the 1990s, when the developing regions as a whole experienced a steady decline in both the number of undernourished and the PoU (Figure 1). This was followed by a slowdown in the PoU in the early 2000s before a renewed acceleration in the latter part of the decade, with the PoU falling from 17.3 percent in 2005–07 to 14.1 percent in 2010–12. Estimates for the most recent period, partly based on projections, have again seen a phase of slower progress, with the PoU declining to 12.9 percent by 2014–16.

Measuring Global Progress Against Targets

The year 2015 marks the end of the monitoring period for the two internationally agreed targets for hunger

reduction. The first is the World Food Summit (WFS) goal. At the WFS, held in Rome in 1996, representatives of 182 governments pledged "... *to eradicate hunger in all countries, with an immediate view to reducing the number of undernourished people to half their present level no later than 2015.*"[4] The second is the Millennium Development Goal 1 (MDG 1) hunger target. In 2000, 189 nations pledged to free people from multiple deprivations, recognizing that every individual has the right to dignity, freedom, equality and a basic standard of living that includes freedom from hunger and violence. This pledge led to the formulation of eight Millennium Development Goals (MDGs) in 2001. The MDGs were then made operational by the establishment of targets and indicators to track progress, at national and global levels, over a reference period of 25 years, from 1990 to 2015. The first MDG, or MDG 1, includes three distinct targets: halving global poverty, achieving full and productive employment and decent work for all, and cutting by half the proportion of people who suffer from hunger[5] by 2015. FAO has monitored progress towards the WFS and the MDG 1c hunger targets, using the three-year period 1990–92 as the starting point.

The latest PoU estimates suggest that the developing regions as a whole have almost reached the MDG 1c hunger target. The estimated reduction in 2014–16 is less than one percentage point away from that required to reach the target by 2015 (Figure 1).[6] Given this small difference, and allowing for a margin of reliability of the background data used to estimate undernourishment, the target can be considered as having been achieved. However, as indicated in the 2013 and 2014 editions of this report, meeting the target exactly would have required accelerated progress in recent years. Despite significant progress in many countries, the needed acceleration does not seem to have materialized in the developing regions as a whole.

The other target, set by the WFS in 1996, has been missed by a large margin. Current estimates peg the number of undernourished people in 1990–92 at a little less

Figure 1

The trajectory of undernourishment in developing regions: actual and projected progress towards the MDG and WFS targets

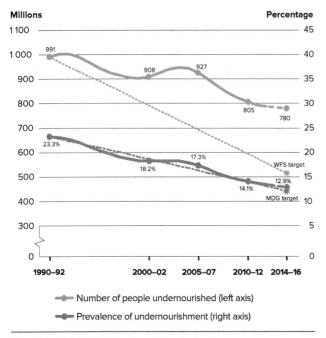

Note: Data for 2014–16 refer to provisional estimates.

Source: FAO.

highest level since the Second World War. These developments have taken their toll on food security in some of the most vulnerable countries, particularly in sub-Saharan Africa, while other regions such as Eastern and South-Eastern Asia have remained unaffected or have been able to minimize the adverse impacts.

The changing global economic environment has challenged traditional approaches to addressing hunger. Social safety nets and other measures that provide targeted assistance to the most vulnerable population groups have received growing attention. The importance of such targeted measures, when combined with long-term and structural interventions, lies in their ability to lead to a virtuous circle of better nutrition and higher labour productivity. Direct interventions are most effective when they target the most vulnerable populations and address their specific needs, improving the quality of their diet. Even where policies have been successful in addressing large food-energy deficits, dietary quality remains a concern. Southern Asia and sub-Saharan Africa remain particularly exposed to what has become known as "hidden hunger"—the lack of, or inadequate, intake of micronutrients, resulting in different types of malnutrition, such as iron-deficiency anaemia and vitamin A deficiency. . . .

Wide Differences Persist Among Regions

Progress towards improved food security continues to be uneven across regions. Some regions have made remarkably rapid progress in reducing hunger, notably the Caucasus and Central Asia, Eastern Asia, Latin America and Northern Africa. Others, including the Caribbean, Oceania and Western Asia, have also reduced their PoU, but at a slower pace. Progress has also been uneven within these regions, leaving significant pockets of food insecurity in a number of countries. In two regions, Southern Asia and sub-Saharan Africa, progress has been slow overall. While some countries report successes in reducing hunger, undernourishment and other forms of malnutrition remain at overall high levels in these regions.

The different rates of progress across regions have brought about changes in the regional distribution of hunger since the early 1990s. Southern Asia and sub-Saharan Africa now account for substantially larger shares of global undernourishment.[7] The shares for Oceania and Western Asia also rose, albeit by much smaller margins and from relatively low levels. In tandem, faster-than-average progress in Eastern Asia and Latin America and the Caribbean means that these regions now account for much smaller shares of global undernourishment.

than a billion in the developing regions. Meeting the WFS goal would have required bringing this number down to about 515 million, that is, some 265 million fewer than the current estimate for 2014–16. However, considering that the population has grown by 1.9 billion since 1990–92, about two billion people have been freed from a likely state of hunger over the past 25 years.

Significant progress in fighting hunger over the past decade should be viewed against the backdrop of a challenging global environment: volatile commodity prices, overall higher food and energy prices, rising unemployment and underemployment rates and, above all, the global economic recessions that occurred in the late 1990s and again after 2008. Increasingly frequent extreme weather events and natural disasters have taken a huge toll in terms of human lives and economic damage, hampering efforts to enhance food security. Political instability and civil strife have added to this picture, bringing the number of displaced persons globally to the

Progress Towards the International Hunger Targets
It differs how the various developing regions fare with respect to these targets. The estimates suggest that Africa as a whole, and sub-Saharan Africa in particular, will not achieve the MDG 1c target. Northern Africa, by contrast, has reached the target.[8] The more ambitious WFS goal, however, appears to be out of reach for Africa as a whole, as well as for all its subregions. Asia as a region has already achieved the MDG 1c hunger target, but would need a further reduction of about 140 million undernourished people to reach the WFS goal—an achievement that is unlikely to materialize in the near future. Latin America and the Caribbean, considered together, have achieved both the MDG 1c hunger target and the WFS goal in 2014–16. Finally, Oceania has reached neither the MDG 1c hunger target nor the WFS goal.

Some countries have met both international targets. Based on the latest estimates, a total of 72 developing countries have achieved the MDG 1c hunger target by 2014–16.[9] Of these, 29 countries have also reached the WFS goal. Another 31 developing countries have reached only the MDG 1c hunger target, either by reducing the PoU by 50 percent or more, or by bringing it below 5 percent. Finally, a third group of 12 countries is also categorized alongside those that have reached the MDG 1c hunger target, as they have maintained their PoU close to or below 5 percent since 1990–92.

Sub-Saharan Africa: Some Success Stories, but the International Hunger Targets Are Far from Being Met
In sub-Saharan Africa, just under one in every four people, or 23.2 percent of the population, is estimated to be undernourished in 2014–16. This is the highest prevalence of undernourishment for any region and, with about 220 million hungry people in 2014–16, the second highest burden in absolute terms. In fact, the number of undernourished people even increased by 44 million between 1990–92 and 2014–16. Taking into account the region's declining PoU, this reflects the region's remarkably high population growth rate of 2.7 percent per year. The slow pace of progress in fighting hunger over the years is particularly worrisome. While the PoU fell relatively rapidly between 2000–02 and 2005–07, this pace slowed in subsequent years, reflecting factors such as rising food prices, droughts and political instability in several countries.

In the Central African subregion,[10] the number of undernourished people more than doubled between 1990–92 and 2014–16, while the PoU declined by 23.4 percent. The divergence between the increase in absolute numbers and the decline in the PoU is explained by the Central Africa's rapid population growth. The lack of

progress in absolute terms reflects prevailing problems in the subregion, notably political instability, civil strife and outright war, as is the case in the Central African Republic.

Eastern Africa remains the subregion with the biggest hunger problem in absolute terms, being home to 124 million undernourished people. As in Central Africa, the region continues to experience rapid population growth. While the share of undernourished has fallen by 33.2 percent, the number of hungry people has risen by nearly 20 percent over the MDG monitoring period. A more favourable picture emerges in Southern Africa, where the PoU has fallen by 28 percent since 1990–92 and a little more than 3 million people remain undernourished. The most successful subregion in reducing hunger is Western Africa, where the number of undernourished people has decreased by 24.5 percent since 1990–92, while the PoU is projected to be less than 10 percent in 2014–16. This success has been achieved despite a combination of limiting factors, such as rapid population growth—Nigeria is the most populated country in the region—drought in the Sahel and the high food prices experienced in recent years.

A total of 18 countries in sub-Saharan Africa have achieved the MDG 1c hunger target, and four more are close to reaching it (i.e. they are expected to do so before 2020 if current trends persist). Of these, seven countries have also achieved the more ambitious WFS goal (Angola, Cameroon, Djibouti, Gabon, Ghana, Mali and Sao Tome and Principe), and two more (South Africa and Togo) are close to doing so. While these are welcome developments, progress mostly started from high levels of undernourishment, and many of these countries are still burdened with high hunger levels. The more populous countries that have reached the MDG 1c hunger target include Angola, Cameroon, Ethiopia, Ghana, Malawi, Mozambique, Nigeria and Togo. In addition, many smaller countries, including Benin, the Gambia, Mauritius and the Niger have reached MDG 1c. Others, including Chad, Rwanda and Sierra Leone, are close to reaching the MDG 1c hunger target, even if the hunger burden in these countries remains high, both in relative and absolute terms. However, most countries in sub-Saharan Africa show lack of progress towards the international targets, and many countries, including the Central African Republic and Zambia, still face high PoU levels.

Many of the countries that have made good progress in fighting hunger have enjoyed stable political conditions, overall economic growth and expanding primary sectors, mainly agriculture, fisheries and forestry. Many had policies in place aimed at promoting and protecting access to food. Moreover, many of these countries have experienced high population growth rates, yet have still achieved the MDG 1c target and even the WFS goal.[11] This shows that hunger reduction can be achieved even where populations are increasing rapidly, if adequate policy and institutional conditions are put in place. By contrast, countries where progress has been insufficient or where hunger rates have deteriorated are often characterized by weak agricultural growth and inadequate social protection measures. Many are in a state of protracted crisis. The number of such countries extends beyond those for which data are provided. The lack of reliable information on food availability and access prevents a sound analysis of the PoU for countries such as Burundi, the Democratic Republic of the Congo, Eritrea and Somalia, hence their exclusion, but food security indicators for which data are available suggest that their levels of undernourishment remain very high.

Northern Africa: International Hunger Targets Are Met, Despite Potential Instability

Trends and levels of undernourishment in Northern Africa are very different from those in the rest of the continent. The region has attained PoU levels below 5 percent according to the projections for 2014–16.[12] The positions of individual countries vis-à-vis the international hunger targets are more or less consistent. While 5 percent of the population can still amount to a considerable number of people in Algeria, Egypt, Morocco and Tunisia, the generally low PoU indicates that, based on current trends, the region is close to eradicating severe food insecurity.

Subsidized access to food is a central policy element in the region, with prices for basic foods remaining low in many countries, even when world prices spiked. While the sustainability of these measures can be questioned, they have helped keep levels of undernourishment low, by supplying large amount of calories affordably. The focus on calories, however, has left dietary quality concerns largely unaddressed, giving rise to other forms of malnutrition, including a rising prevalence of overweight and obesity. Moreover, the region remains exposed to potential and actual economic and political instability. Some countries are heavily dependent on food imports, and their limited resource base, coupled with rapid population growth, suggests that import dependence will remain a feature of the region in the future, notwithstanding efforts to increase agricultural productivity.

Southern Asia: Some Progress, but too Slow to Meet the International Hunger Targets

The highest burden of hunger in absolute terms is to be found in Southern Asia. Estimates for 2014–16 suggest that about 281 million people are undernourished in the region, marking only a slight reduction from the number in 1990–92. But there has been noticeable progress in relative terms: the PoU has declined from 23.9 percent in 1990–92 to 15.7 percent in 2014–16. The region is on a trajectory towards a more manageable hunger burden. Most importantly, progress has accelerated over the last decade, notwithstanding higher prices on international commodity markets. The evolution of hunger trends in India, in particular, has a significant influence on results for the region. Higher world food prices, observed since the late 2000s, have not been entirely transmitted into domestic prices, especially in large countries such as India. In this country, the extended food distribution programme also contributed to this positive outcome. Higher economic growth has not been fully translated into higher food consumption, let alone better diets overall, suggesting that the poor and hungry may have failed to benefit much from overall growth.

Most countries in Southern Asia have made progress towards the international hunger targets, even if the pace has been too slow for them to reach either the WFS or the MDG targets, including, for example, Afghanistan, India, Pakistan and Sri Lanka. As these countries constitute a large share of the region's population, they account for the low overall performance—India still has the second-highest estimated number of undernourished people in the world. A notable exception in terms of performance is Bangladesh, which has made faster progress and has already reached the MDG 1c hunger target, thanks also to the comprehensive National Food Policy framework adopted in the mid-2000s. Nepal, also, has not only reached the MDG 1c hunger target, but has almost reached the 5 percent threshold. One more country in the region, the Islamic Republic of Iran, has already brought the PoU below 5 percent, and has thus reached the MDG 1c target.

Eastern and South-Eastern Asia: Rapid and Generalized Progress Towards the International Hunger Targets

The most successful subregions in fighting hunger have been Eastern and South-Eastern Asia. The number of undernourished people in Eastern Asia has fallen from 295 million in 1990–92 to 145 million in 2014–16, a 50.9 percent reduction. Over the same period, the PoU dropped from 23.2 percent at the beginning of the monitoring period, to 9.6 percent in 2014–16, a reduction of more than 60 percent.

In South-Eastern Asia, the number of undernourished people has continued its steady decline, from 137.5 million in 1990–92 to 60.5 million by 2014–16, a 56 percent reduction overall. The PoU has shrunk by a remarkable 68.5 percent, falling from 30.6 percent in

1990–92 to less than 10 percent in 2014–16. Most countries in South-Eastern Asia are making rapid progress towards international targets. Cambodia, Indonesia, the Lao People's Democratic Republic, Malaysia, Myanmar, the Philippines, Thailand and Viet[nam] all account for this positive performance. No country in the region shows lack of progress with respect to the international targets. Brunei Darussalam and Malaysia have reduced their PoU to below the 5 percent threshold, which means they are close to having eradicated hunger.

As discussed in more detail in the section "Food security and nutrition: the drivers of change," much of the success of Eastern and South-Eastern Asia was possible due to high overall economic growth. Unlike Southern Asia, these subregions experienced more inclusive growth, with more of the poor and vulnerable sharing the benefits. Rapid productivity growth in agriculture, since the Green Revolution, has boosted food availability and significantly improved access to food for the rural poor.

China's achievements in reducing hunger dominate the overall performance of Eastern Asia. The country accounts for almost two-thirds of the reduction in the number of undernourished people in the developing regions between 1990–92 and 2014–16. China and the Republic of Korea have achieved both the MDG 1c hunger target and the WFS goal. Nevertheless, given the sheer size of its population, China is still home to an estimated 134 million people facing hunger, and the country with the highest number of undernourished people. The prospects of continued growth, the increasing orientation of the economy towards the domestic market, the expansion of economic opportunities in internal areas of the country and the growing ability of the poor to benefit from these developments, have been and will continue to be key factors in hunger reduction. Again, given its size, this also holds at the regional level and has a marked influence on global results. The only major exception to overall favourable progress in the region is the Democratic People's Republic of Korea, which is burdened by continuously high levels of undernourishment and shows little prospect of addressing its problems any time soon.

Caucasus and Central Asia: Rapid Recovery from the Transition to the Market Economy Enabled the International Hunger Targets to be Met

A combination of factors accounts for progress in the Caucasus and Central Asia, including rapid economic growth, a resource-rich environment and remittances. After a difficult transition in the early 1990s, often characterized by political instability and economic austerity, economic conditions have improved significantly and the political situation has stabilized. This progress has translated into lower hunger burdens throughout the region. Latest estimates point to a steady decline in the PoU, which has contracted from 14.1 percent in 1990–92 to 7.0 percent for 2014–16. The number of undernourished people is much lower than in other Asian subregions—5.8 million in 2014–16, down from 9.6 million in 1990–92.

Progress has been sufficiently rapid to enable both the region as a whole and most countries to achieve the MDG 1c hunger target. Indeed, most countries have attained PoU levels close to, or below, the 5 percent threshold. Armenia, Azerbaijan, Georgia, Kyrgyzstan and Turkmenistan have achieved the WFS goal, while Kazakhstan and Uzbekistan have achieved the MDG 1c hunger target. The only country still lagging behind is Tajikistan,[13] which is making insufficient progress to reach the international targets, and is burdened by a relatively high PoU (33.2 percent in 2014–16).

Western Asia: No Progress Towards the International Hunger Targets, Despite Low Undernourishment Levels in Several Countries

A less encouraging picture emerges from Western Asia, where very different patterns can be observed. Some countries, including Iraq and Yemen, show high levels of food insecurity and have made slow progress towards improving this situation. Most other countries, on the contrary, have long since attained solid levels of food security, after having brought undernourishment levels below 5 percent. These include politically stable, resource-rich economies, such as Kuwait, Saudi Arabia and the United Arab Emirates, together with Jordan, Lebanon and Oman—all of which have achieved the MDG 1c hunger target; Kuwait and Oman have also achieved the WFS goal. The group also includes rapidly growing and politically stable countries, such as Turkey. In Iraq and Yemen, as well as other countries in the region for which no reliable information is available, political instability, war and civil strife, as well as fragile institutions, are the main factors underlying the lack of progress.[14]

Despite a relatively low number of undernourished people, Western Asia saw an increase in undernourishment throughout the monitoring period: the PoU rose by 32.2 percent between 1990–92 and 2014–16, from 6.4 to 8.4 percent. In parallel, rapid population growth has brought about a dramatic increase in the number of undernourished people, from 8 million to nearly 19 million. The region in its entirety, therefore, has not made progress towards reaching either of the international hunger targets, as a result of the polarized situation across countries.

Latin America and the Caribbean: International Hunger Targets have been Met, due to Rapid Progress in South America

In Latin America, the PoU has declined from 13.9 percent in 1990–92 to less than 5 percent in 2014–16. In parallel, the number of undernourished people fell from 58 million to fewer than 27 million. As in most regions, stark differences can be found across countries and subregions. The Central American subregion, for instance, saw much less progress compared with that of South America and even Latin America overall. While South America has been able to bring undernourishment down by more than 75 percent and eventually to below the 5 percent mark, the PoU for Central America has declined by only 38.2 percent over the MDG monitoring period.

Despite divergent developments within the region, Latin America has achieved both the MDG 1c and WFS targets by large margins. The overall achievements are to a large extent also a reflection of robust progress in its most populous countries. Good overall economic performance, steady output growth in agriculture and successful social protection policies are among the main correlates of progress in the region. The combination of safety nets with special programmes for family farmers and smallholders and targeted support to vulnerable groups, together with broad-based food security interventions such as school-feeding programmes, have contributed significantly to improving food security in the region. At the continental level, important commitments started in 2005 with the Hunger-Free Latin America and the Caribbean Initiative and, through various other initiatives, eventually led to the Plan for Food Security, Nutrition and Hunger Eradication 2025 of the Community of Latin American and Caribbean States (CELAC),[15] adopted by all countries of the region in January 2015 during its third Presidential Summit.

Hunger rates are currently below the 5 percent threshold in Argentina, Brazil, Chile, Costa Rica, Mexico, Uruguay and the Bolivarian Republic of Venezuela, and the WFS hunger goal has been achieved in Argentina, Brazil, Chile, Guyana, Nicaragua, Peru, Uruguay, and the Bolivarian Republic of Venezuela. In all, 13 countries in Latin America have achieved the MDG 1c hunger target. Beyond those listed above, these include the Plurinational State of Bolivia, Guyana, Panama, Peru and Suriname. Another four countries, including Colombia, Ecuador, Honduras and Paraguay, are on track to reach the MDG 1c target over the next few years, if current trends persist. Even if some countries, such as Guatemala or El Salvador, appear to be off-track for reaching the international targets, no country in the region has a PoU higher than 20 percent.

The Caribbean as a whole, like Central America, has failed to meet the MDG 1c target. Unlike Central America, however, the remaining hunger burden in almost all Caribbean countries is lower and thus more manageable. The PoU has dropped from 27.0 percent in 1990–92 to 19.8 percent in 2014–16, a 26.6 percent decrease in relative terms. Many individual Caribbean countries, however, have achieved the international targets or are at least close to reaching them. Barbados, Cuba, the Dominican Republic and Saint Vincent and the Grenadines have all attained the MDG 1c hunger target. The latter three have also reached the more demanding WFS goal. Jamaica and Trinidad and Tobago are also very close to reaching the MDG 1c target. The explanation for the region as a whole lagging behind lies in the severe and still largely unabated problems experienced by Haiti—a country hit by recurrent natural disasters, still facing slow growth in food availability *vis-à-vis* population growth and burdened by an increasingly degraded resource base as well as a fragile national economy.[16]

Oceania

The developing countries of Oceania have experienced slow progress towards improved food security. The overall PoU in the region fell by less than 10 percent between 1990–92 and 2014–16. This corresponds to an increase in the number of undernourished people of about 0.5 million, or 50 percent. Being largely small island developing states characterized by high dependency on food imports, food security in most countries can be severely affected by external shocks, including international price volatility, adverse weather events and sudden changes in the availability of a few important staples, such as rice. The Pacific Islands face multiple burdens of malnutrition; while hunger has fallen slowly, overweight, obesity and, as a consequence, non-communicable diseases, such as type 2 diabetes and coronary heart disease, are taking a growing toll on the region's health and economic status.

Several countries in the Oceania region covered by this report have achieved the MDG 1c hunger target, including Fiji, Kiribati, Samoa and the Solomon Islands, while Vanuatu has not. Samoa has also reached the more ambitious WFS goal. The situation in Vanuatu has deteriorated dramatically since Cyclone Pam hit the islands in March 2015.[17] Before this catastrophic event, the country had been showing consistent progress in reducing hunger. In the case of Papua New Guinea, by far the most populous country in the region, a detailed assessment has not been possible due to the lack of reliable background data. Overall progress notwithstanding, there is considerable uncertainty about the situation in the country, where the information needed to reliably estimate undernourishment is largely absent. Anecdotal evidence indicates that the country's food security situation is far from resolved.

Inside the Hunger Target: Comparing Trends in Undernourishment and Underweight in Children

Progress towards the MDG "hunger target," or MDG target 1c, which requires halving, between 1990 and 2015, the proportion of people who suffer from hunger, is measured by two different indicators: the prevalence of undernourishment (PoU), monitored by FAO, and the prevalence of underweight children under five years of age (CU5), monitored by the United Nations Children's Fund (UNICEF) and the World Health Organization (WHO). The end of the MDG monitoring period offers a good opportunity to look back at the evolution of these indicators and to identify common trends, but also to understand the reasons for possible deviations.

Common trends should be discernible as both indicators were approved by the international community to measure the hunger target. Deviations, however, could arise from the different methods used to compile them[18] and the different dimensions of food insecurity that they are expected to capture.

Understanding the different trends of the two indicators across regions and over time is important, as it may offer insights into the complexity of food security, and possibly lead to more targeted policy interventions. Underweight can be caused by a range of different factors—not only calorie or protein deficiency, but also poor hygiene, disease or limited access to clean water. All these factors impede the body's ability to absorb nutrients from food and eventually result in manifestations of nutrient deficits such as stunting, wasting or underweight. For this reason, the two indicators do not always reflect the same underlying problem. Where lack of sufficient food is the main cause of underweight, the PoU and the CU5 should move synchronously. Where poor food utilization prevails instead, the two indicators are likely to diverge.

Considering the developing regions as a whole for the entire MDG monitoring period, the two indicators show consistent trends. From 1990 to 2013, the CU5 moved from 27.4 percent to 16.6 percent, a 39.3 percent reduction, while the PoU declined by 44.5 percent between 1990–92 and 2014–16.[19] The annual rate of decline is similar.

Regional Patterns

The parallel progress of the two indicators for the developing regions as a whole is not always evident when the analysis focuses on individual regions. In some, the PoU and CU5 indicators show different rates of reduction. Within sub-Saharan Africa, for instance, the PoU and the CU5 only move together for Eastern Africa, while they diverge over time for almost all other subregions. By contrast, trends in the subregions in Asia and in Latin America and the Caribbean largely move in parallel. The rest of this section will analyse these divergences and similarities in trends.

Northern Africa

The region's problems are well captured by MDG hunger indicators. Both the PoU and CU5 show low absolute levels of food insecurity, even more so than for other developing regions. In particular, the CU5 declined rapidly over the monitoring period, with a reduction from 9.5 to 4.8 percent. Food utilization conditions appear favourable in the region, with more than 90 percent of the population having access to clean water and improved sanitation facilities in 2012. The PoU has remained below the 5 percent threshold since 1990–92. Many countries of the region have not only sufficient, but excessive levels of calorie availability. Just as in Western Africa, much of the problem lies in unbalanced diets with too many calories from carbohydrates, which are mostly derived from cereals and sugar. Food consumption subsidies, which are granted in several Northern African countries, have played a part in maintaining undernourishment at low levels, while at the same time favouring an excessive consumption of energy-intense foods, potentially leading to increased risks of non-communicable diseases and obesity.

Sub-Saharan Africa

For the region as a whole, undernourishment and child underweight were looming large at the beginning of the 1990s, with both indicators exceeding 25 percent. Since then, the PoU and CU5 have decreased at a similarly slow pace.

During the 1990s, per capita GDP decreased in a number of sub-Saharan countries, and the region's Human Development Index was the lowest in the world.[20] These factors explain the slow decline in undernourishment, as well as the sluggish investment in infrastructure and health.[21] On average, during the 1990s, only one in four people had access to electricity, compared with a world average of one in three. Likewise, there were only 0.15 physicians available for every thousand people, compared with a world average of 1.3.

Over the 2000s, the food security situation in sub-Saharan Africa gradually improved. Economic growth resumed in several countries, resulting in a decline of the PoU, but major challenges remained unaddressed, especially in terms of addressing the region's inadequate hygiene conditions and quality of diets. This divergence appears particularly evident for Western Africa. Here, the PoU has fallen by over 60 percent since 1990–92, owing to the progress of large countries such as Ghana and Nigeria. These changes, however, were largely brought about by the higher availability of staple foods, which did not address the dietary imbalances in the region. While the PoU for Western Africa fell rapidly, the CU5 remained stubbornly high, at levels of more than 20 percent.

Sub-Saharan Africa's problems not only illustrate the multifaceted nature of food security, but also suggest that different dimensions require different approaches to successfully improve food security. For instance, making even more carbohydrates available is unlikely to further improve overall food security. Rather, new measures should focus on the ability of poor people to access balanced diets and on overall living conditions, to prevent negative health outcomes such as underweight, wasting and stunting in children.

Caucasus and Central Asia

The region has had overall low rates and has made good progress over time for both the PoU and CU5 indicators. The economic and political transitions of the early 1990s and, later, the economic crisis of the early 2000s only seem to have influenced the PoU, which exhibited marked swings during these periods. The two indicators were again moving in parallel by the early 2000s, with improvements in living conditions. In recent years, the CU5 has maintained levels below 5 percent in most countries, with the exception of Tajikistan, where it remains at about 15 percent. Since the early 1990s, only few countries have occasionally presented CU5 values above 10 percent. At the same time, the transition turmoil barely affected the region's overall health and hygiene conditions. The proportion of the population with access to clean water and improved sanitation facilities has always been higher than 85 percent and 90 percent, respectively, throughout the monitoring period. These conditions, together with the improvement in nutrition experienced during the past decade, explain the steady downward trend in the CU5. It is worth highlighting that the high poverty rates experienced by most countries were limited to relatively short periods of time and did not significantly worsen food utilization.

Eastern Asia

Steady and rapid progress for both indicators is observed in Eastern Asia. At the beginning of the monitoring period, the PoU declined slightly faster than the CU5. The region's average PoU saw some minor ups and downs in the 1990s and the early 2000s, while the reduction in undernourishment accelerated again after 2006.

The more consistent decline in the CU5 can be traced to the steady improvement of hygiene conditions in several countries. Access to safe water, for example, increased by 37 percent over the monitoring period, while access to improved sanitation facilities has increased by 153 percent since the early 1990s. These factors have had a strong positive impact on food utilization, and support both the low CU5 levels and its rapid improvement over time.

Southern Asia

Southern Asia is the region with the highest historical CU5 levels, but is also the region where rapid progress has been made in reducing underweight among young children. The prevalence of underweight children declined from 49.2 percent in 1990 to 30.0 percent in 2013, with a 39.0 percent reduction over the MDG monitoring period. By contrast, the PoU in Southern Asia made less progress overall, resulting in a convergence between the two indicators over time.

There is growing evidence that helps explain the relatively rapid decline of the CU5. Many countries in the region have experienced robust economic growth over the past 25 years, bringing down poverty rates. While the steady decline in child underweight is consistent with the decrease in poverty, undernourishment only went down from 23.9 percent to 15.7 percent between 1990–92 and 2014–16. This different pattern is largely due to India, the country that more directly affects the regional picture as a result of its large population. Explanations offered for the inconsistency between food consumption and income levels in India range from increasing inequalities, to poor data, to the challenges of capturing the changing energy requirements of the population.[22] But the puzzle still seems to be unresolved; and, as noted in the previous section, calorie consumption is lower than what per capita incomes and poverty rates would suggest.

The reasons for CU5 progress include enhanced access to safe water and sanitation and, as a consequence, better hygiene and health conditions. For instance, household access to improved sanitation nearly doubled from 23 percent to 42 percent between 1990 and 2012. Over the same period, access to safe water rose from 73 percent to 91 percent. In addition, targeted nutrition programmes in key countries in the region, aimed at young children, pregnant women and women of reproductive age, likely contributed to a rapidly declining CU5. Examples include, among others, the Integrated Child Development Scheme, implemented in India since 1975, and the Bangladesh Integrated Nutrition Programme, funded by the World Bank. Despite the rapid decrease in the CU5, the indicator was still much higher compared with those of all other Asian subregions. This suggests that much more progress can be achieved in the future by combining policy interventions that enhance both food availability and utilization.

South-Eastern Asia

South-Eastern Asia is among the regions that showed faster progress across the first seven MDGs. This also holds for the hunger target as measured by both the PoU and CU5. Undernourishment and underweight in children were both above 30 percent at the beginning of the monitoring period, but the PoU declined more quickly throughout the 2000s. This would be in line with the view that policy interventions to improve hygiene conditions—for instance water and sanitation infrastructure—typically require higher investments compared with those aimed at enhancing food availability. The CU5 has declined rapidly in the region, but is still above 20 percent in more than one country. Rapid progress has been made in improving hygiene conditions, with 71 percent of the population having access to better sanitation.[23] In view of the good growth prospects in the region, this also means that more progress will be possible, provided that interventions improve the diets of poor population groups and ensure wider access to clean water and sanitation facilities.

Western Asia

Western Asia shows a unique pattern of change. While the PoU has increased since the early 1990s, reflecting political instability in a number of countries, the CU5 has continued to decline. Underweight in children is at a low level virtually everywhere, while the sparse data available indicate high proportions in Yemen—well beyond 20 percent—and to a lower extent in other countries, such as Iraq and Syria, where data for the 2000s point to shares not far from 10 percent. Hygiene conditions in the region are generally good, with more than 90 percent of the population having access to clean water sources, and 88 percent of the population having access to improved sanitation facilities in 2012. The rise in the PoU, as shown in the previous section, reflects political and social problems together with war and civil strife in a limited number of countries in the region, which generated large migrant and refugee populations.

Latin America and the Caribbean

In the region as a whole, the two hunger indicators have converged over time, at a faster rate after the year 2000, when progress in reducing the PoU accelerated. The PoU, estimated at 14.7 percent in 1990–92, dropped to 5.5 percent by 2014–16, while the CU5 has decreased from 7.0 percent to 2.7 percent over the same period. The CU5 is generally low, with few exceptions. Within the region, Central America remains the most problematic area, with almost no improvement recorded over the MDG monitoring period. The PoU and the CU5 in Central America were close to each other in the early 1990s (at about 11 percent of the population) and both indicators have seen little progress since then. Shares higher than 10 percent have been reported for Haiti in recent periods; in this country the indicator has decreased since the early 1990s, when it exceeded 20 percent. Relatively high values have been reported, in recent years, also for Guatemala, Honduras and Guyana, although not exceeding 15 percent.

Progress for both indicators stems from economic growth combined with a stronger commitment to social protection, especially over the last decade. Many countries have made hunger and malnutrition eradication a high political priority. At the continental level, important commitments started in 2005 with the Hunger-Free Latin America and the Caribbean Initiative, and eventually led, through various other initiatives, to the *Santiago Declaration of the Community of Latin American and Caribbean States* in January 2013. Despite progress, major challenges remain. Many countries are witnessing growing overweight and obesity rates and, as a result, the increasing prevalence of non-communicable diseases.

Oceania

This region is characterized by high rates of underweight among children. Without progress over 25 years, the CU5 is now not far from levels prevalent in many parts of sub-Saharan Africa. Slow progress is also observed for the PoU. The common trends for the two indicators suggest related underlying drivers, especially low food availability and dietary diversity. In many small island developing states in the region, the variety of nutrients available and acquired is somewhat limited.

Slow progress in increasing access to safe drinking water and improved sanitation facilities has also contributed to lack of progress in reducing food insecurity. Only 55 percent of households in the region have access to safe water, while only 35 percent have access to improved sanitation facilities. Several indicators for the underlying drivers even suggest some deterioration of the situation. While access to safe water has improved by just 12 percent since the early 1990s, access to sanitation facilities has declined by about 1 percent per year over the same period.

Moreover, the region suffers from a malnutrition problem not well captured by the PoU and CU5, namely the growing coexistence of undernutrition and overnutrition. One contributing factor to overnutrition has been the "westernization" of food consumption patterns, which is associated with a rising prevalence of overweight and obesity.

Food Security and Nutrition: The Drivers of Change

In 2000, world leaders met and adopted the United Nations Millennium Declaration. Later, eight Millennium Development Goals (MDGs) were set out, including the first one on halving hunger and extreme poverty rates, reflecting the world's commitment to improving the lives of billions of people.

Half a year remains before the end of 2015, the deadline for achieving most of the MDG targets, including the hunger target, MDG 1c, traditionally measured using the prevalence of undernourishment (PoU) indicator. As this report shows, since 1990–92, over 216 million people have been rescued from a life of hunger—to date, 72 countries have already reached the MDG 1c hunger target, with another nine just short by a small margin. Of these, 12 developing countries already had undernourishment rates below 5 percent in 1990–92. Meanwhile, twenty-nine countries have accomplished the more ambitious 1996 World Food Summit (WFS) goal of halving the number of chronically underfed people.

Progress towards food security and nutrition targets requires that food is available, accessible and of sufficient quantity and quality to ensure good nutritional outcomes. Proper nutrition contributes to human development; it helps people realize their full potential and take advantage of opportunities offered by the development process. As past editions of this report (2010, 2012 and 2014) have shown, good governance, political stability and the rule of law, and the absence of conflict and civil strife, weather-related shocks or excessive food price volatility, are conducive to all dimensions of food security.

This section looks at a range of factors that enable progress towards food security and nutrition goals. The list of factors—economic growth, agricultural productivity growth, markets (including international trade) and social protection—is by no means exhaustive. The section also shows how being in a protracted crisis has deleterious effects on progress in hunger reduction. Preliminary

quantitative analysis, using data from the period 1992–2013, has helped identify these drivers of change and their relative importance in shaping progress against hunger.[24]

Economic growth is central to the fight against hunger—countries that become richer are less susceptible to food insecurity. Policy-makers in rapidly growing economies have increased capacity and resources to dedicate to improving food security and nutrition. But this is not always the case. Economic growth, while a necessary condition for progress in poverty and hunger reduction, especially in the face of an expanding population, is not sufficient. It is *inclusive* growth that matters—growth that promotes equitable access to food, assets and resources, particularly for poor people and women, so that individuals can develop their potential.[25]

Across the developing world, the majority of the poor and most of the hungry live in rural areas, where family farming and smallholder agriculture is a prevailing—albeit not universal—mode of farm organization. Although the ability of family farming and smallholder agriculture to spur growth through productivity increases varies considerably, its role in reducing poverty and hunger is key. Growth in family farming and smallholder agriculture, through labour and land productivity increases, has significant positive effects on the livelihoods of the poor through increases in food availability and incomes.

The linkages between food security and international trade are complex and context-specific. Policies that affect exports and imports of food contribute to determining relative prices, wages and incomes in the domestic market, and hence shape the ability of poor people to access food. Trade, in itself, is neither a threat nor a panacea when it comes to food security. The opportunities and risks to food security associated with trade openness should be carefully assessed and addressed through an expanded set of policy instruments.

Social protection systems have become an important tool in the fight against hunger. More than one hundred countries implement conditional or unconditional cash transfer programmes that focus on promoting food security and nutrition, health and education, particularly for children. Food distribution schemes and employment guarantee programmes are also important. The expansion of social protection across the developing world has been critical for progress towards the MDG 1c hunger target. Providing regular and predictable cash transfers to poor households often plays a critical role in terms of filling immediate food gaps, but can also help improve the lives and livelihoods of the poor by alleviating constraints to their productive capacity. Combining social protection with complementary agricultural development measures, such as the Purchase from Africans for Africa programme, which links family farmers and smallholders to school-feeding programmes, can maximize the poverty- reducing impact of these programmes.

In 1990, only 12 countries in Africa were facing food crises, of which only four were in protracted crises.[26] Just 20 years later, a total of 24 countries were experiencing food crises, with 19 in crisis for eight or more of the previous ten years. Food insecurity can be both a cause and effect of protracted crises and can be instrumental in triggering or deepening conflict and civil strife—it increasingly lies at the root of protracted crisis situations. The impact of conflict on food security can be more dramatic than the direct impact of war, and mortality caused by conflict through food insecurity, and famine can far exceed deaths directly caused by violence.[27]

Economic Growth and Progress Towards Food Security and Nutrition Targets

Economic growth is necessary for alleviating poverty and reducing hunger and malnutrition; it is critical for sustainably increasing employment and incomes, especially in low-income countries. Since the beginning of the 1990s and up to 2013 (most of the MDG monitoring period) global output per capita has increased by 1.3 percent per year, on average. The economies of low- and middle-income countries—including all developing countries—grew more rapidly, by 3.4 percent per year. Nevertheless, these numbers mask considerable variation in economic growth performance across regions and countries.

The relationship between economic growth and hunger is complex. Economic growth increases household incomes, through higher wages, increased employment opportunities, or both, due to stronger demand for labour. In a growing economy, more household members are able to find work and earn incomes. This is essential for improving food security and nutrition and contributes to a virtuous circle as better nutrition strengthens human capacities and productivity, thus leading to better economic performance. However, the question here is whether or not those people who are living in extreme poverty and are most affected by hunger will be given the opportunity to participate in the benefits of growth and, if they are, whether they will be able to take advantage of it.

On average, and across the developing world since 1990–92, economic growth has brought strong and persistent hunger reduction. This is evident when GDP growth per capita is plotted against the PoU. Increases in the incomes of the poor are associated with higher intake of dietary energy and other nutrients. But in the longer term,

as economies grow and countries become richer, this relationship weakens—increases in GDP growth may bring relatively fewer people out of hunger. . . . Among the early success stories is Ghana, which has experienced average annual growth rates of over 3 percent, and has witnessed impressive hunger reduction rates—the PoU in the country fell from 47 percent in 1990–92 to below 5 percent in 2012–14.

In several cases, the positive effects of economic growth on food security and nutrition are related to greater participation of women in the labour force. In Brazil, for example, labour force participation of women rose from 45 percent in 1990–94 to 60 percent in 2013. In Costa Rica, the proportion of women workers increased by 23 percent between 2000 and 2008. Spending by women typically involves more household investments in food and nutrition, but also in health, sanitation and education, compared with the case when resources are controlled by men.[28]

But not all countries that experienced strong economic growth performed well in terms of hunger reduction. Some countries progressed well towards the international hunger targets, while others experienced setbacks. In general, there has been uneven progress in translating economic growth into improvements in food security.

Inclusive Economic Growth and Poverty Reduction

On the whole, progress in alleviating poverty has been faster than in fighting hunger. This is because the hungry are the poorest of the poor; they have limited or no access to physical and financial assets, little or no education, and often suffer from ill health. Poor agricultural households lack access to sufficient, high-quality land and other natural resources or to remunerative sources of income (self-employment, wage labour). At the same time, hunger creates a trap from which people cannot easily escape. Hunger and undernutrition mean less-productive individuals, who are more prone to disease and thus often unable to earn more and improve their livelihoods. This, in turn, hinders progress in alleviating extreme poverty and fighting hunger—particularly as labour is the principal asset held by the poor.

Not all types of growth are effective in reducing hunger and malnutrition. Very poor people cannot participate in growth processes that require capital or generate employment for the educated and skilled. For example, economic growth generated by capital-intensive exploitation of resources, such as minerals and oil, is likely to have very few or weak direct linkages to the poor. The greater the inequality in the distribution of assets, such as land, water, capital, education and health, the more difficult it is for the poor to improve their situation and the slower the progress in reducing undernourishment.[29]

Inclusive economic growth improves the incomes of the poor. If these incomes grow more rapidly than the growth rate of the economy, income distribution also improves. What matters for effectively improving food security is for economic growth to reach those in extreme poverty—the bottom quintile of the income distribution. Approximately three-quarters of the world's poor live in rural areas, with the share even higher in low-income countries.[30] In most developing regions, the risk of working poverty (workers who live on less than US$1.25 a day) is highest for employment in agriculture—about eight out of ten working poor are engaged in vulnerable employment in the informal economy, particularly in agriculture.[31]

Agriculture on its own can trigger growth in countries with a high share of agriculture in GDP. But even if other sectors of the economy, such as mining or services, were to grow, agriculture, through targeted investments, can become an avenue through which the poor participate in the growth process. Empirical evidence suggests that agricultural growth in low-income countries is three times more effective in reducing extreme poverty compared with growth in other sectors. In sub-Saharan Africa, agricultural growth can be 11 times more effective in reducing poverty than growth in non-agricultural sectors.[32] Investments and policies that promote increased agricultural labour productivity lead to increases in rural income. Countries that have invested in their agriculture sectors—and especially in improving productivity of smallholders and family farming—have made significant progress towards the MDG 1c hunger target.

The accommodation of gender considerations is crucial for economic growth in countries with agriculture-dependent economies. Women play important roles as producers, managers of productive resources and income earners, and they are key providers of unpaid care work in rural households and communities. However, despite decades of efforts to address gender inequalities, many rural women continue to face gender-based constraints that limit their capacity to contribute to growth and take advantage of new opportunities arising from the changes shaping national economies. This has serious consequences for well-being—not only for women themselves, but also for their families and societies at large—and it represents one of the main reasons for the economic underperformance of agriculture in poorer countries.[33] While it is sometimes argued that economic growth inevitably leads to gender equality, the empirical evidence is weak and inconsistent. Much seems to depend on policies and strategies aimed at shaping inclusive markets and reducing poverty.[34] Agriculture-based solutions need to be complemented by interventions that promote the productive potential of the rural space. In addition, direct support to rural livelihoods through social protection programmes provides immediate relief to the most vulnerable. Such programmes also have long-term benefits—they enable broad participation of the poor in the growth process through better access to education, health and proper nutrition, all of which expand and strengthen human potential.

Social protection can establish a virtuous circle of progress involving the poor with increased incomes, employment and wages. For example, the Zero Hunger Programme and the *Bolsa Família* in Brazil were crucial for achieving inclusive growth in the country. *Bolsa Família* reached almost a quarter of the population, mainly women, transferring above US$100 every month to each family, as long as they sent their children to school.[35] With the Brazilian economy growing at 3 percent per year since 2000, thus providing the necessary public revenues, these programmes have significantly reduced income inequality—between 2000 and 2012, the average incomes of the poorest quintile of the population grew three times as fast as those of the wealthiest 20 percent.[36]

The Contribution of Family Farming and Smallholder Agriculture to Food Security and Nutrition

More than 90 percent of the 570 million farms worldwide are managed by an individual or a family, relying predominately on family labour. These farms produce more than 80 percent of the world's food, in terms of value. Globally, 84 percent of family farms are smaller than 2 hectares and manage only 12 percent of all agricultural land. While small farms tend to have higher yields than larger farms, labour productivity is less and most small family farmers are poor and food-insecure.[37] The sustainability and future food security of these farms may be threatened by intensive resource use. Public policies that recognize the diversity and complexity of the challenges faced by family farms throughout the value chain are necessary for ensuring food security.

Improved productivity of agricultural resources through sustainable intensification plays a key role in increasing food availability and improving food security and nutrition. At the global level, productivity and food availability have been increasing, contributing significantly to reductions in undernourishment worldwide. Higher agricultural labour productivity is generally associated with lower levels of undernourishment.

Public policies should provide incentives for the adoption of sustainable agricultural intensification practices and techniques—sustainable land management, soil conservation, improved water management, diversified agricultural systems and agroforestry—in order to produce more outputs from the same area of land while reducing negative environmental impacts. More conventional yield-enhancing technologies, such as improved seed varieties and mineral fertilizers, are also valuable options, especially when combined with greater attention to using these inputs efficiently.

With increased productivity, farmers grow more food, become more competitive and receive higher incomes. Productivity growth in small family farms contributes to more inclusive growth, not only by reducing the prices of staple foods but also by improving access to food. With well-functioning rural labour markets, such productivity growth increases the demand for labour in rural areas, generating jobs for the poor and raising the unskilled labour wage rate. Rural household members diversify their income sources by obtaining better-paid off-farm work, which helps poverty and hunger to decline.

In spite of overall progress, marked regional differences persist. In the early 1990s, average value added per worker in agriculture was lowest in sub-Saharan Africa, approximately US$700 in 2005 prices, compared with other regions, such as Eastern Asia and Latin America, where it amounted to US$4 600 and US$4 400, respectively. By 2010–13, average value added per worker in agriculture in sub-Saharan Africa amounted to US$1 199, whereas in Eastern Asia and Latin America, it had risen to US$15 300 and US$6 000, respectively. Gains in labour productivity have also been slower in sub-Saharan Africa, and so have been the reductions in the PoU, with current levels systematically higher than in other regions.

The evidence suggests that agricultural productivity gains have helped countries reduce undernourishment. For example, over the period 1990–92 to 2012–14 in sub-Saharan Africa, where agriculture is dominated by small family farms, countries that made little progress towards achieving the MDG 1c hunger target, such as Botswana, Côte d'Ivoire, Liberia, Namibia, Swaziland, Uganda, the United Republic of Tanzania and Zambia, experienced average gains in agricultural value added per worker of only 25 percent. These gains were significantly lower than those experienced in Angola, Benin, Ethiopia, Gabon, Ghana, and Mali, countries that have met the MDG 1c hunger target. On average, labour productivity in agriculture in these countries increased by 69 percent between 1990–92 and 2012–14. Over the same period, in sub-Saharan African countries that have made progress towards the

target but not yet achieved it, average agricultural value added per worker increased by 42 percent.

Similar patterns are observed when looking at a more traditional measure of agricultural productivity—output per hectare. Significant yield gaps—the difference between farmers' yields and technical potential yields achieved using the latest varieties and under the best of conditions—still persist, particularly in sub-Saharan Africa. Such yield gaps reflect a largely suboptimal use of inputs and insufficient adoption of the most productive technologies. In Mali (an MDG 1c achiever), for example, the yield gap for rainfed maize—in 2008–10—was 75 percent, a significantly high value but lower than those observed in Uganda (83 percent) and the United Republic of Tanzania (88 percent), suggesting a linkage between agricultural productivity and progress towards food security.[38]

In the recent past, in many sub-Saharan African countries, agricultural growth has been mostly driven by more extensive use of land and reallocation of productive factors—not necessarily oriented to supplying local markets and reducing food insecurity—rather than the support of public policies to expand access to agricultural credit and insurance, advisory services and sustainable technologies.

Other constraining factors that compromise agricultural productivity gains and the generation of stable incomes for family farmers include weather-related shocks; poor transport, storage and communications infrastructure; and missing or inefficient markets. Weak institutions and inadequate public agricultural and rural development policies are major causes of such failures.

Inclusive markets for smallholders and family farmers are an important ingredient in promoting food security and nutrition. Markets not only facilitate the flow of food from surplus to deficit areas, ensuring food availability; they also transmit price signals to farmers to adjust their production and input use.[39] Well-functioning markets that foster price stability and predictability are crucial—on the one hand, a significant share of farmers rely on markets for generating part of their cash income, while, on the other hand, many family farmers are net-food buyers relying on markets to purchase part of their food needs. Smallholder and family farming productivity and access to markets are interlinked and contribute to both food availability and access to food. Improving access to marketing opportunities can also help boost productivity.

One relevant approach to increasing family farmers' access to markets is local food procurement by different levels of government (local, regional and national). Not only can public purchase schemes guarantee food security for vulnerable populations and income for smallholders and family farmers, but they may also enhance collective action to strengthen their marketing capacities and ensure greater effectiveness.

To accelerate progress in improving access to food by the poor, lagging regions, particularly sub-Saharan Africa, will increasingly have to transform their agricultural policies to significantly improve agricultural productivity and increase the quantity of food supplied by family farmers. The importance of family farming and smallholder agriculture is best reflected by the Comprehensive Africa Agriculture Development Programme (CAADP), which has established a goal of 6 percent annual agricultural growth. The expected impacts are primarily to improve food security and nutrition, reduce poverty and increase employment.

International Trade and Food Security Linkages

International trade and trade policies affect the domestic availability and prices of goods and those of the factors of production such as labour, with implications for food access. International trade can also affect market structure, productivity, sustainability of resource use, nutrition and various population groups in different ways. Assessing its impact on food security is thus highly complex. For example, banning grain exports can boost domestic supplies and reduce prices in the short run. This benefits consumers, but has negative implications for farmers producing for export. Import or export restrictions by major players affect global supplies and exacerbate price volatility at the global level. Lowering import duties reduces food prices paid by consumers, but can put pressure on the incomes of import-competing farmers, whose own food security may be negatively affected. . . . In practice, the picture is further complicated by market imperfections in national local markets, which prevent the transmission of global price changes to those markets.

Lessons from Trade Policy Reforms
Policies to increase openness to international trade have generally taken place in the context of wider economic reforms, and it is therefore difficult to disentangle their effects. A number of case studies have attempted to analyse the impact of trade on food security, and, not surprisingly, the results have been mixed.[40] In China, economic reforms have generated positive results for growth, poverty reduction and food security. Trade, which has continued to grow rapidly, has played a part, although domestic reforms appear to have been more important in stimulating growth. Also in Nigeria, domestic reforms improved

incentives for agricultural commodity producers, and per capita calorie intake increased substantially after the implementation of trade reforms, pointing to a possible positive impact on food security.

Similarly, in Chile, trade openness and the elimination of policy distortions stimulated both agricultural and overall economic growth and the transition from traditional crops to more profitable products for export. Research has shown that the reforms have contributed substantially to poverty reduction and food security. Peru is another example of the positive food security outcomes of institutional and economic transformations aimed at strengthening private-sector initiatives, including trade openness. However, the country implemented social protection policies and programmes, to address uneven growth across sectors and income inequality and mitigate the negative effects of the reforms on vulnerable parts of the population.

Conversely, in Guatemala, Kenya, Senegal and the United Republic of Tanzania, the food security outcomes of economic and trade reforms appear to have been disappointing. In Guatemala, while the reforms resulted in diversified production of more profitable crops, external factors (such as lower coffee prices) have undermined the potential to improve food security. In Kenya, limited coordination in policy sequencing seems to have slowed progress against hunger. The reforms in Senegal have shown mixed results; although the PoU declined on aggregate, female-headed households became less food-secure.

Indeed, the constraints faced by rural women, in terms of lack of access to productive factors, such as land, credit, inputs, storage and technology, may undermine their capacity to adopt new technologies and/or take advantage of economies of scale to improve their competitiveness. In several developing countries, female small farmers who are unable to compete with cheaper agricultural imports have been forced to abandon or sell their farms, which in turn can contribute to their food insecurity.[41]

While trade in itself is not intrinsically detrimental to food security, for many countries, particularly those at earlier levels of development, trade reforms can have negative effects on food security in the short-to-medium term. Recent research shows that countries supporting the primary sector tend to be better off on most dimensions of food security (food availability, access, and utilization), while taxation of this sector is detrimental to food security.[42] However, the evidence also shows that excessive support can also lead to poor performances on all dimensions of food security.

As countries become more open to international trade in agricultural products, they become more exposed and potentially more vulnerable to sudden changes in global agricultural markets. For example, import surges—sudden increases in the volume of imports from one year to the next—can hinder the development of agriculture in developing countries.

Food sectors in developing countries that are characterized by low productivity and lack of competitiveness are especially vulnerable to import surges. A sudden disruption of domestic production can have disastrous impacts on domestic farmers and workers—loss of jobs and reduced incomes, with potentially negative consequences for food security. During the period 1984–2013, China, Ecuador, India, Kenya, Nigeria, Pakistan, Uganda, the United Republic of Tanzania and Zimbabwe were prone to sudden increases in imports (defined as imports exceeding the average of the previous three years by more than 30 percent), registering more than a hundred surges.[43]

The factors that lead to an import surge may originate in the importing country itself as a result of domestic supply shortfalls or rapid increases in demand. Other factors are exogenous, for example when countries providing significant support to the production and/or export of food products channel production surpluses to the international markets. Surges resulting from external factors can be difficult for the affected countries to manage.

Serious disruptions in domestic markets and negative food security outcomes have been used to support arguments for a more cautious approach to greater openness to agricultural trade and for the establishment of effective safeguards in new trade agreements. In circumstances where the agriculture sector has yet to play out its potential growth-enhancing role, trade policy, including trade remedies, and incentives to boost domestic production can have potentially important roles to play. At the same time, complementary policies (as in the case of Peru) can protect the most vulnerable groups from the possible negative effects of openness to trade.

Trade in the New Agricultural Markets Context
The international agricultural market context has changed from one characterized by depressed and stable prices to one where market reactions to climatic and economic shocks can give rise to sudden price increases or falls. Such changes have prompted reassessment of the role of trade and trade policies in promoting food security.

As food import bills have risen significantly following increases in food prices in 2008, confidence in global markets as reliable sources of affordable food has waned, and attention has turned to support for domestic food production. As a result, some developing countries have adopted policies designed to influence domestic prices directly through border measures and price controls, or to create incentives for increasing domestic supply. Among the available trade policy instruments, export restrictions and the elimination of import tariffs have been the preferred policies to address food security concerns during periods of high and volatile prices.

Trade, in itself, is neither a threat nor a panacea when it comes to food security, but it can pose challenges and even risks that need to be considered in policy decision-making. To ensure that countries' food security and development needs are addressed in a consistent and systematic manner, they need to have a better overview of all the policy instruments available to them and the flexibility to apply the most effective policy mix for achieving their goals.

The Relevance of Social Protection for Hunger Trends Between 1990 and 2015

Social protection has directly contributed to hunger reduction over the MDG monitoring period. Since the late 1990s, there has been a global trend towards the extension of cash transfers and other social assistance programmes, triggered in part by the financial crises in emerging market economies during that time.[44] Social protection has since been progressively anchored in national legislation, increasing its coverage to support vulnerable groups.

Coverage has increased for many reasons, including the recognition that social protection can be instrumental in promoting sustainable and inclusive growth. Social protection is a crucial part of the policy spectrum that addresses high and persistent levels of poverty and economic insecurity, high and growing levels of inequality, insufficient investments in human resources and capabilities, and weak automatic stabilizers of aggregate demand in the face of economic shocks.

With sufficient coverage and proper implementation, social protection policies can promote both economic and social development in the short and longer term, by ensuring that people enjoy income security, have effective access to health care and other social services, are able to manage risk and are empowered to take advantage of economic opportunities. Such policies play a crucial role in fostering inclusive and sustainable growth, strengthening domestic demand, facilitating the structural transformation of national economies, and promoting decent work.[45]

Between 1990 and 2015, social protection programmes have grown exponentially. Although much of this increase occurred in high- and middle-income countries, significant progress in social protection coverage has also been made in the developing regions, as for example in Africa, through innovative cash transfer and health-care programmes.[46] Today, every country in the world has at least one social assistance programme in place. School-feeding programmes—the most widespread type of social protection programme—are implemented in 130 countries. Unconditional cash transfers are also common, and are now implemented in 118 countries globally. Likewise, conditional cash transfer and public works/community asset programmes continue to expand rapidly.[47] Global and regional efforts have also been instrumental, including the push for national social protection floors endorsed by International Labour Organization (ILO) Recommendation 202.[48] Yet, despite the proliferation of programmes around the world, the ILO estimates that 70 percent of the world's poor still do not have access to adequate social protection. [49]

International organizations, such as FAO and WFP, play important roles in designing and implementing efficient and effective safety net programmes and social protection systems in the countries with a focus on food security and nutrition. Social protection systems often meet immediate food gap needs and, if designed accordingly, they can help improve lives and livelihoods—a key factor for reducing the number of hungry people in the world.

Recent research concludes that approximately 150 million people worldwide are prevented from falling into extreme poverty thanks to social protection.[50] However, the impact of social assistance programmes such as cash transfers on well-being extends beyond the direct effects of transfers. Transfers can help households manage risk and mitigate the impact of shocks that keep households mired in poverty.

Social assistance programmes such as cash transfer programmes can influence the productive capacity of beneficiaries, in particular those with limited access to financial services for investment and risk mitigation. The provision of regular and predictable cash transfers brings significant benefits when markets are missing or do not function well. When transfers are of sufficient size and combined with additional support to beneficiaries, they can often be saved and/or invested in productive assets and can improve social inclusion for even greater returns over the participants' lifetimes.[51] In combination with savings and credit, environmental rehabilitation and agricultural insurance, transfers can encourage prudent risk-taking and increase productive outcomes—even for the poorest households.[52]

Social assistance programmes, particularly when combined with additional interventions in the areas of drinking water supply, health and/or education, have been shown to enhance nutritional outcomes and promote human capital. The integration of nutrition objectives into social assistance programmes also has the potential to significantly accelerate progress in reducing undernutrition and raising economic productivity.[53] Furthermore, women are direct beneficiaries of many social assistance programmes, such as cash transfers. With more control of resources, this has empowered them with positive impacts on food security and nutritional status, especially of children.[54] However, such positive outcomes depend on other contextual factors and require complementary interventions.

Over the past twenty five years, evidence has emerged that social protection programmes can play a significant role in achieving food security and nutrition targets. The evidence suggests that increasing spending for strengthened social protection programmes can be a highly cost-effective way to promote rural poverty reduction and improved food security and nutrition, and, hence, to achieve development goals.[55] The fact that, despite the rapid growth of social protection programmes in recent decades, about 70 percent of the world population still lacks access to more adequate, formal forms of social security, indicates there is still considerable need for expanded coverage and, hence, scope for accelerating the eradication of hunger. However, just expanding social protection programmes will not suffice. The most effective social protection policies for improved food security and rural poverty reduction have been those that are well integrated with agriculture sector policies and fully aligned with the priorities and vision set out in broader strategies aimed at creating viable and sustainable livelihoods for the poor.

Protracted Crises and Hunger

Countries and areas in protracted crisis are "environments in which a significant proportion of the population is acutely vulnerable to death, disease and disruption of livelihoods over a prolonged period of time. Governance in these environments is usually very weak, with the state having limited capacity to respond to, and mitigate, threats to the population, or to provide adequate levels of protection."[56] Based on the criteria set out in *The State of Food Insecurity in the World 2010*,[57] the list of countries considered to be in protracted crisis situations was updated in 2012 to encompass 20 countries.[58] However, it should be noted that some protracted crisis situations are limited to specific geographic areas, and may not affect the entire country, let alone the entire population.

Although protracted crises are diverse in both their causes and effects, food insecurity and malnutrition are common prevalent features.[59] Food insecurity and malnutrition are particularly severe, persistent and widespread in protracted crisis contexts. The approximate combined population in protracted crisis situations in 2012 was 366 million people, of whom approximately 129 million were undernourished between 2010 and 2012 (including conservative estimates for countries lacking data). This accounted for approximately 19 percent of the global total of food-insecure people. In 2012, the mean prevalence of undernourishment in protracted crisis situations was 39 percent, compared with 15 percent, on average, in the rest of the developing world.

Achieving the MDG 1c target of halving the proportion of undernourished population in these countries poses an enormous challenge. Of the 20 countries in protracted crisis identified above, only one, Ethiopia, has reached the MDG 1c target. All the others report either insufficient progress or even deterioration.

The Typology of Crises

Over the past 30 years, the typology of crises has gradually evolved from catastrophic, short-term, acute and highly visible events to more structural, longer-term and protracted situations resulting from a combination of multiple contributing factors, especially natural disasters and conflicts, with climate change, financial and price crises increasingly frequent among the exacerbating factors. In other words, protracted crises have become the new norm, while acute short-term crises are now the exception. Indeed, more crises are considered protracted today than in the past.[60]

From a food security and nutrition perspective, in 1990, only 12 countries in Africa were facing food crises, of which only four were in protracted crisis. Just 20 years later, a total of 24 countries were facing food crises, with 19 of these having been in crisis for eight or more of the previous ten years.[61] Moreover, the growing imperative of dealing with the long-term contexts of these emergencies is becoming evident. For instance, the Bosphorus Compact[62] reported that global humanitarian appeals between 2004 and 2013 increased by 446 percent overall—rising from US$3 billion to US$16.4 billion. Similarly, the number of displaced people at the end of 2013 was 51.2 million, more than at any point since the end of World War Two. The average length of displacement in major refugee situations is now 20 years. And nine out of ten humanitarian appeals continue for more than three years, with 78 percent of spending by the Organisation for Economic Co-operation and Develop-

ment's Development Assistance Committee donors allocated to protracted emergencies.

Over the past three decades, the causes of crises have become more interconnected, displaying an evolving trend of triggers for protracted crises due to natural causes, either human-induced, or stemming from a combination of human and natural causes.[63] Conflicts are increasingly the main underlying cause, with the prevalence of human-induced conflicts higher than previously. As such, conflicts are now a common feature of crises. The complex relation between conflict and food security and nutrition is still to be fully explored.

Ways in Which Crises Impact Food Security

Protracted crises undermine food security and nutrition in multiple ways, affecting the availability, access and utilization of food. Disruptions to crop production, livestock rearing and trade can have a negative impact on food availability. People's access to food is frequently affected in crises because of displacement, disruptions to livelihoods, or when land is taken. For example, when state and customary institutions are unable or unwilling to protect and promote individuals' legal rights, attempts to take land from women, orphans and other vulnerable individuals go unchecked.[64] Finally, the utilization of food can be impacted by changes in intra-household and community relations and power dynamics and by inequitable service delivery.

Food insecurity can be further deepened and self-perpetuating as people use up their reserves of food, finance and other assets, and turn to unsustainable coping mechanisms, such as selling off productive assets and taking up activities that lead to land degradation to meet immediate food needs.

Gender and age are two powerful determinants of the impact of protracted crises on individuals. Women are more likely than men to be affected and their access to aid can be undermined by gender-based discrimination. Pre-existing gender-based disparities in access to assets such as land, property or credit mean that women have often fewer financial resources than men to cope with impacts such as loss of productive capacity, leaving them unable to afford the increased prices of food in crisis-affected areas.[65] Protracted crises have also been found to put additional care burdens on women post-crisis, while limited mobility and work opportunities outside the home reduce their range of coping strategies. Often, with male household members absent, due to death, migration or recruitment into armed forces, women are not always able to claim family assets, such as land, livestock, tools and machinery, previously owned by their husbands, especially if they are illiterate or insufficiently aware of their legal rights with significant negative implications for food security.

Why Is It So Difficult to Deal with Food Insecurity and Malnutrition in Protracted Crises?

Addressing food insecurity and malnutrition in protracted crises is particularly challenging. Evidence shows that stakeholders need to address the critical manifestations of protracted crises, such as hunger and malnutrition, and disruption to and depletion of livelihoods, while simultaneously addressing underlying causes such as poor governance, inadequate capacities, limited access to scarce natural resources and conflict.

In addition, policies and actions have to consider the specific features and complex challenges presented by protracted crises, including their longevity; the particular need to protect marginalized and vulnerable groups, and to respect basic human rights; the mismatch between short-term funding mechanisms and long-term needs, and how best to integrate humanitarian and development assistance; the often poor coordination of responses; and inadequate ownership by national stakeholders of response-related processes. Finally, the context specificity of protracted crises makes it difficult, and undesirable, to adopt "one-size-fits-all" approaches.

Nonetheless, examples exist of good practices in addressing some of the issues at the root of protracted crises, ranging from innovative funding mechanisms such as crisis modifiers, to more comprehensive country owned processes. In addition, rural women should be seen as partners in the rehabilitation process rather than simply "victims." Indeed, evidence indicates that relief programmes that adopt a gender perspective can avert widespread malnutrition and lead to quick and more extensive recovery in food production and other aspects of livelihoods.[66]

Protracted crises are becoming an increasingly important global problem, negatively impacting on people's food security and nutrition, and often the result of instability and conflict. Successful experiences exist, but need to be scaled up, which requires a high degree of political commitment at all levels. Current efforts by the Committee on World Food Security (CFS) to finalize a Framework for Action for Food Security and Nutrition in Protracted Crises could be an important first step to mobilize political commitment and guide action.

Notes

1. The proportion of undernourished people in the total population is the indicator known as prevalence of undernourishment (PoU).

2. Reference is made here to the developing regions as defined by the M49 country classification of the United Nations (see http://unstats.un.org/unsd/methods/m49/ m49regin.htm).

3. If China and India are excluded from the aggregate of the developing regions, the reduction in undernourishment follows a more stable, continuous downward trend. China and India alone account for 81 percent of the total reduction of the number of undernourished people in the developing regions between 1990–92 and 2014–16, and China alone accounts for almost two-thirds.

4. Rome Declaration on World Food Security, adopted at the World Food Summit, Rome, 13–17 November 1996.

5. This is known as target 1c of Millennium Development Goal 1 (MDG 1) (see http://www.un.org/millenniumgoals/).

6. The assessment of progress towards these targets, started by FAO at the end of the 1990s, took 1990–92 as the base period. Both the WFS and MDG hunger targets are to be reached by the end of 2015. To maintain consistency, progress has been assessed with reference to a three-year average centred on 2015, that is, 2014–16. Achievement of the MDGs are meant to be assessed for the 25-year period, from 1990 to 2015, but, as observations are only available for the 24-year period from 1990–92 to 2014–16, the 50 percent change required for reaching the targets has had to be adjusted by a factor of 24/25. This corresponds to a 48 percent reduction of the PoU with respect to 1990–92.

7. The share of sub-Saharan Africa increased from 45 percent to over 60 percent.

8. This is the case if the region is considered without Sudan, which was recently added to the Northern Africa subregion after the partition of the country when South Sudan became an independent state in 2011.

9. See note 6 for details of the assessment of countries that reached the MDG 1c and WFS targets.

10. This is the region called "Middle Africa" in the M49 country classification adopted by the United Nations (http://unstats.un.org/unsd/methods/m49/m49regin.htm for the full listing).

11. Current annual growth rates are, for instance, 2.5 percent in the Gambia and Ghana; 2.6 percent in Mauritania and Togo; 2.7 percent in Benin and Cameroon; 2.9 percent in Malawi, Mali, Mozambique, Nigeria and Sao Tome and Principe; and 3.2 percent in Angola. See Population Reference Bureau. 2014. *World Population*

Data Sheet 2014 (available at http://www.prb.org /Publications/Datasheets/2014/2014-world-population -data-sheet/data-sheet.aspx).

12. Following the split of former Sudan into two countries in 2011, South Sudan was classified as part of sub-Saharan Africa, while Sudan was added to Northern Africa. In order to allow appropriate assessments of progress between 1990–92 and 2014–16, Sudan is not considered in the Northern Africa region.

13. See, for example, the case study on Tajikistan in the 2013 edition of this report.

14. See, for example, the case study on Yemen in the 2014 issue of this report.

15. FAO/ECLAC/ALADI. 2015. *The CELAC Plan for Food and Nutrition Security and the Eradication of Hunger 2025*. Executive summary (available at http://www.fao.org/fileadmin/user_upload/rlc/docs/celac /ENG_Plan_CELAC_2025.pdf).

16. See, for example, the case study on Haiti in the 2014 issue of this report.

17. Cyclone Pam with 270 km/hour winds, hit Vanuatu as a category 5 cyclone, the second strongest ever to have formed in the Southern Pacific region.

18. One obvious methodological difference between the two indicators is the population coverage: underweight is measured only for children below five years of age, while undernourishment is measured for the entire population. Other differences relate to the way indicators are compiled. The height and weight of children are directly measured in household surveys, while the availability of and access to sufficient food are estimated using a statistical model that draws from multiple data sources.

19. The starting point for monitoring the CU5 was the year 1990, whereas it was 1990–92 for the PoU. The last available data point for CU5 is 2013, whereas for the PoU it is 2014–16. Information for the PoU and the CU5 is not available for the same sets of countries. All comparisons are therefore limited to regional aggregates.

20. The Human Development Index was 0.399 in subSaharan Africa in 1990, compared with a world average of 0.597. See UNDP. 2014. *Human Development Report 2014. Sustaining human progress: reducing vulnerabilities and building resilience*. New York, USA, Table 2 (available at http://hdr.undp.org/en/content /table-2-human-development-index-trends-1980-2013).

21. The share of GDP devoted to health expenditure in sub-Saharan Africa was three percentage points lower than for the world (6 percent versus 9 percent).

22. For a summary of the debate on this point, see N. Alexandratos and J. Bruinsma. 2012. *World agriculture towards 2030/2050: the 2012 revision*. ESA Working paper No. 12-03. Rome, FAO.

23. See, FAO. 2015. *Food security indicators*. Web page (available at http://www.fao.org/economic/ess/ess-fs/ess -fadata/it/#.VRuyjOEZbqc).

24. P. Karfakis, G. Rapsomanikis and E. Scambelloni. 2015 (forthcoming). *The drivers of hunger reduction*. ESA Working Paper. Rome, FAO.

25. Commission on Growth and Development. 2008. *The growth report: strategies for sustained growth and inclusive development*. Washington, DC. World Bank.

26. For a definition of protracted crises, see FAO and WFP. 2010. *The State of Food Insecurity in the World 2010. Addressing food security in protracted crises*. Rome, FAO.

27. See The Geneva Declaration on Armed Violence and Development. 2011. *Global Burden of Armed Violence 2011: lethal encounters*. Geneva, Switzerland (http://www.genevadeclaration.org/measurability/global -burden-of-armed-violence/global-burden-of-armed -violence-2011.html); and FAO. 2013. *Study suggests 258 000 Somalis died due to severe food insecurity and famine*. News release (available at http://www.fao.org /somalia/news/detail-events/en/c/247642/).

28. J.P. Azevedo, G. Inchauste and V. Sanfelice. 2013. *Decomposing the recent inequality decline in Latin America*. Policy Research Working Paper 6715. Washington, DC, World Bank.

29. FAO, IFAD and WFP. 2012. *The State of Food Insecurity in the World 2012. Economic growth is necessary but not sufficient to accelerate reduction of hunger and malnutrition*. Rome, FAO.

30. International Labour Organization (ILO). 2012. *Global Employment Trends 2012. Preventing a deeper job crisis*. Geneva, Switzerland.

31. FAO. 2012. *Decent rural employment for food security: a case for action*. Rome.

32. FAO, IFAD and WFP, 2012 (see note 29) and L. Christiaensen, L. Demery and J. Kuhl. 2011. The (evolving) role of agriculture in poverty reduction: an empirical perspective. *Journal of Development Economics*, 96: 239–254.

33. FAO. 2011. *State of Food and Agriculture 2010–11. Women in agriculture: closing the gender gap for development*. Rome.

34. N. Kabeer. 2014. *Gender equality and economic growth: a view from below*. Paper prepared for UN Women Expert Group Meeting "Envisioning women's rights in the post-2015 context", New York, 3–5 November 2014.

35. International Policy Centre for Inclusive Growth. 2009. *What explains the decline in Brazil's inequality?* One Pager No. 89. Brasilia, International Policy Centre for Inclusive Growth, Poverty Practice, Bureau for Development Policy, United Nations Development Programme and the Government of Brazil.

36. Government of Brazil. 2014. *Indicadores de Desenvolvimento Brasileiro 2001–2012.* Brasilia.

37. FAO. 2014. *The State of Food and Agriculture 2014. Innovation in family farming.* Rome.

38. The calculations are based on data collected from the Global Yield Gap Atlas, an initiative by the University of Nebraska-Lincoln, Wageningen University and Water for Food (see http://www .yieldgap.org/).

39. World Bank. 2008. *World Development Report 2008. Agriculture for development.* Washington, DC; and IFAD. 2011. *Rural Poverty Report 2011. New realities, new challenges: new opportunities for tomorrow's generation.* Rome.

40. H. Thomas, ed. 2006. *Trade reforms and food security: country case studies and synthesis.* Rome, FAO.

41. WomenWatch. 2011. *Gender equality and trade policy.* Resource paper (available at http://www .un.org/womenwatch/feature/trade/gender_equality _and_trade_policy.pdf).

42. E. Magrini, P. Montalbano, S. Nenci and L. Salvatici. 2014. *Agricultural trade policies and food security: is there a causal relationship?* FOOD-SECURE Working Paper No. 25 (available at http://www3.lei.wur.nl/FoodSecurePublications/25 _Salvatici_et_al_Agtrade-policies-FNS.pdf).

43. FAO. 2014. *Policy responses to high food prices in Latin America and the Caribbean: country case studies,* edited by D.Dawe and E. Krivonos. Rome.

44. ILO. 2014. *World Social Protection Report 2014/15. Building economic recovery, inclusive development and social justice.* Geneva, Switzerland.

45. *Ibid.*

46. International Social Security Association. 2011. *Africa: a new balance for social security.* Geneva, Switzerland.

47. U. Gentilini, M. Honorati, and R. Yemtsov. 2014. *The State of Social Safety Nets 2014.* Washington, DC, World Bank.

48. International Labour Conference. 2012. Recommendation no. 202 concerning national floors for social protection (available at http://www.ilo .org/brussels/WCMS_183640/lang--en/index.htm).

49. ILO, 2014 (see note 51).

50. A. Fiszbein, R. Kanbur and R. Yemtsov. 2014. Social protection and poverty reduction: global patterns and some targets. *World Development,* 61(1): 167–177.

51. WFP. 2012. *Bangladesh food security for the ultra poor: lessons learned report 2012.* Rome.

52. M. Madajewicz, A.H. Tsegay and M. Norton. 2013. *Managing risks to agricultural livelihoods: impact evaluation of the HARITA Program in Tigray, Ethiopia, 2009–2012.* Boston, USA, Oxfam America; and FAO. 2014. *The economic impacts of cash transfer programmes in sub-Saharan Africa.* From Protection to Production Policy Brief (available at http://www.fao.org/3/a-i4194e.pdf).

53. *The Lancet.* 2013. Maternal and Child Nutrition series. *The Lancet,* 382(9890); and The Transfer Project. 2015. *The broad range of cash transfer impacts in sub-Saharan Africa: consumption, human capital and productive activity.* Research brief (available at http://ovcsupport.net/wp-content /uploads/2015/03/TP-Broad-Impacts-of-SCT-in -SSA_NOV-2014.pdf).

54. See, for instance, M. Van den Bold, A. Quisumbing and S. Gillespie. *Women's empowerment and nutrition.* IFPRI Discussion Paper No. 01294. Washington, DC, International Food Policy Research Institute.

55. H. Alderman and M. Mustafa. 2013. *Social protection and nutrition.* Note prepared for the Technical Preparatory Meeting for the International Conference on Nutrition (ICN2), Rome, 13–15 November 2013. Rome, FAO and World Health Organization.

56. A. Harmer and J. Macrae, eds. 2004. *Beyond the continuum: aid policy in protracted crises.* HPG Report No.18, p. 1. London, Overseas Development Institute.

57. Criteria for identifying countries in protracted crises: (i) longevity of crisis—at least eight of the past ten years on the Global Information and Early Warning System (GIEWS) list; (ii) aid flows—at least 10 percent of total official development assistance received in the form of humanitarian assistance (between 2000 and 2010); (iii) economic and food security status—countries appear on the list of low-income food-deficit countries. It should be recognized that the methodology employed in *The State of Food Security in the World 2010* (see note 26) used three of a number of possible criteria, and that the list therein is not definitive.

58. The updated list of countries in protracted crisis includes: Afghanistan, Burundi, Central African Republic, Chad, Congo, Cote d'Ivoire, Democratic People's Republic of Korea, Democratic Republic of the Congo, Eritrea, Ethiopia, Guinea, Haiti, Iraq, Kenya, Liberia, Sierra Leone, Somalia, Sudan, Uganda and Zimbabwe.

59. P. Pingali, L. Alinovi and J. Sutton. 2005. Food security in complex emergencies: enhancing food system resilience. *Disasters*, 29(51): S5–S24.

60. High-Level Expert Forum. 2012. *Food insecurity in protracted crises—an overview*. Brief prepared for the High Level Expert Forum on Food Insecurity in Protracted Crises, Rome, 13–14 September 2012.

61. Global Information and Early Warning System (GIEWS) list of Countries Requiring External Assistance (available at http://www.fao.org/Giews/english/hotspots/index.htm).

62. Work is ongoing to agree a new compact on how to more effectively manage risk in recurrent and protracted crises, known as the Bosphorus Compact. The compact is expected to be launched in May 2016 at the World Humanitarian Summit.

63. GIEWS list (see note 61).

64. J. Adoko and S. Levine. 2004. *Land matters in displacement: the importance of land rights in Acholiland and what threatens them*. Kampala, Civil Society Organizations for Peace in Northern Uganda.

65. United Nations Develoment Programme (UNDP). 2012. *Africa Human Development Report 2012. Towards a food secure future*. New York, USA.

66. FAO and WFP, 2010 (see note 26).

THE FOOD AND AGRICULTURE ORGANIZATION OF THE UNITED NATIONS, headquartered in Rome, Italy, is a governmental organization with 195 nation members and present in 130 countries. Among its functions is to help eliminate hunger, food insecurity, and malnourishment, and to make agriculture, forestry, and fisheries more productive and sustainable.

The British Government Office for Science → **NO**

The Future of Food and Farming: Challenges and Choices for Global Sustainability

Executive Summary—Key Conclusions for Policy Makers

1 Introduction[1]

The global food system will experience an unprecedented confluence of pressures over the next 40 years. On the demand side, global population size will increase from nearly seven billion today to eight billion by 2030, and probably to over nine billion by 2050; many people are likely to be wealthier, creating demand for a more varied, high-quality diet requiring additional resources to produce. On the production side, competition for land, water and energy will intensify, while the effects of climate change will become increasingly apparent. The need to reduce greenhouse gas emissions and adapt to a changing climate will become imperative. Over this period globalisation will continue, exposing the food system to novel economic and political pressures.

Any one of these pressures ("drivers of change") would present substantial challenges to food security; together they constitute a major threat that requires a strategic reappraisal of how the world is fed. Overall, the Project has identified and analysed five key challenges for the future. Addressing these in a pragmatic way that promotes resilience to shocks and future uncertainties will be vital if major stresses to the food system are to be anticipated and managed. The five challenges are:

A. Balancing future demand and supply sustainably— to ensure that food supplies are affordable.
B. Ensuring that there is adequate stability in food supplies—and protecting the most vulnerable from the volatility that does occur.
C. Achieving global access to food and ending hunger. This recognises that producing enough food in the world so that everyone can *potentially* be fed is not the same thing as ensuring food security for all.
D. Managing the contribution of the food system to the mitigation of climate change.

E. Maintaining biodiversity and ecosystem services while feeding the world.

These last two challenges recognise that food production already dominates much of the global land surface and water bodies, and has a major impact on all the Earth's environmental systems.

In recognising the need for urgent action to address these future challenges, policy-makers should not lose sight of major failings in the food system that exist today.

Although there has been marked volatility in food prices over the last two years, the food system continues to provide plentiful and affordable food for the majority of the world's population. Yet it is failing in two major ways which demand decisive action:

- **Hunger remains widespread.** 925 million people experience hunger: they lack access to sufficient of the major macronutrients (carbohydrates, fats and protein). Perhaps another billion are thought to suffer from "hidden hunger," in which important micronutrients (such as vitamins and minerals) are missing from their diet, with consequent risks of physical and mental impairment. In contrast, a billion people are substantially over-consuming, spawning a new public health epidemic involving chronic conditions such as type 2 diabetes and cardiovascular disease. Much of the responsibility for these three billion people having suboptimal diets lies within the global food system.
- **Many systems of food production are unsustainable.** Without change, the global food system will continue to degrade the environment and compromise the world's capacity to produce food in the future, as well as contributing to climate change and the destruction of biodiversity. There are widespread problems with soil loss due to erosion, loss of soil fertility, salination and other forms of degradation;

Foresight. The Future of Food and Farming (2011). Final Project Report. The Government Office for Science, London.

rates of water extraction for irrigation are exceeding rates of replenishment in many places; overfishing is a widespread concern; and there is heavy reliance on fossil fuel-derived energy for synthesis of nitrogen fertilisers and pesticides. In addition, food production systems frequently emit significant quantities of greenhouse gases and release other pollutants that accumulate in the environment.

In view of the current failings in the food system and the considerable challenges ahead, this Report argues for decisive action that needs to take place now.

The response of the many different actors involved will affect the quality of life of everyone now living, and will have major repercussions for future generations. Much can be achieved immediately with current technologies and knowledge given sufficient will and investment. But coping with future challenges will require more radical changes to the food system and investment in research to provide new solutions to novel problems. This Report looks across all of these options to draw out priorities for policy-makers.[2]

The analysis of the Project has demonstrated the need for policy-makers to take a much broader perspective than hitherto when making the choices before them— they need to consider the *global food system* from production to plate.

The food system is not a single designed entity, but rather a partially self-organised collection of interacting parts. For example, the food systems of different countries are now linked at all levels, from trade in raw materials through to processed products. Besides on-farm production, capture fisheries and aquaculture are also important, in terms of both nutrition and providing livelihoods, especially for the poor—about a billion people rely on fish as their main source of animal protein. Many vulnerable communities obtain a significant amount of food from the wild ("wild foods"), which increases resilience to food shocks.

Most of the economic value of food, particularly in high-income countries, is added beyond the farm gate in food processing and in retail, which together constitute a significant fraction of world economic activity. At the end of the food chain, the consumer exerts choices and preferences that have a profound influence on food production and supply, while companies in the food system have great political and societal influence and can shape consumer preferences. All of the above imply the need to give careful consideration to the complex ramifications of possible future developments and policy changes in the global food system.

Policy-makers also need to recognise food as a unique class of commodity and adopt a broad view of food that goes far beyond narrow perspectives of nutrition, economics and food security.

Food is essential for survival and for mental and physical development—nutritional deficiencies during pregnancy and in early growth (especially the first two years) can have lifelong effects. For the very poor, obtaining a minimum amount of calories becomes a dominant survival activity. However, issues of culture, status and religion also strongly affect both food production and demand, and hence shape the basic economics of the food system. Also, food production, cooking and sharing are major social and recreational activities for many in middle- and high-income countries.

Box 1.1
High-Level Conclusions

A major conclusion of this Report is the critical importance of interconnected policy-making. Other studies have stated that policy in all areas of the food system should consider the implications for volatility, sustainability, climate change and hunger. Here it is argued that policy in other sectors outside the food system also needs to be developed in much closer conjunction with that for food. These areas include energy, water supply, land use, the sea, ecosystem services and biodiversity. Achieving much closer coordination with all of these wider areas is a major challenge for policy-makers.

There are three reasons why broad coordination is needed. First, these other areas will crucially affect the food system and therefore food security. Secondly, food is such a critical necessity for human existence, with broad implications for poverty, physical and mental development, well-being, economic migration and conflict, that if supply is threatened, it will come to dominate policy agendas and prevent progress in other areas. And, thirdly, as the food system grows, it will place increasing demands on areas such as energy, water supply and land—which in turn are closely linked with economic development and global sustainability. Progress in such areas would be made much more difficult or impossible if food security were to be threatened.

However, there is a tension between the Report's identification of five key challenges to the food system and its stress on the importance of considering policy development in the round. The following highlight a number of key themes and conclusions that both summarise the findings and cut across the different challenges, with an emphasis on what needs to be done immediately.

1. Substantial changes will be required throughout the different elements of the food system and beyond if food security is to be provided for a predicted nine billion people. Action has to occur on all of the following four fronts simultaneously:
   - More food must be produced sustainably through the spread and implementation of existing knowledge, technology and best practice, and by investment in new science and innovation and the social infrastructure that enables food producers to benefit from all of these.
   - Demand for the most resource-intensive types of food must be contained.
   - Waste in all areas of the food system must be minimised.
   - The political and economic governance of the food system must be improved to increase food system productivity and sustainability.

The solution is not just to produce more food, or change diets, or eliminate waste. The potential threats are so great that they cannot be met by making changes piecemeal to parts of the food system. It is essential that policy-makers address all areas at the same time.

2. Addressing climate change and achieving sustainability in the global food system need to be recognised as dual imperatives. Nothing less is required than a redesign of the whole food system to bring sustainability to the fore.

The food system makes extensive use of non-renewable resources and consumes many renewable resources at rates far exceeding replenishment without investing in their eventual replacement. It releases greenhouse gases, nitrates and other contaminants into the environment. Directly, and indirectly through land conversion, it contributes to the destruction of biodiversity. Unless the footprint of the food system on the environment is reduced, the capacity of the earth to produce food for humankind will be compromised with grave implications for future food security. Consideration of sustainability must be introduced to all sectors of the food system, from production to consumption, and in education, governance and research.

3. It is necessary to revitalise moves to end hunger. Greater priority should be given to rural develop-

ment and agriculture as a driver of broad-based income growth, and more incentives provided to the agricultural sector to address issues such as malnutrition and gender inequalities. It is also important to reduce subsidies and trade barriers that disadvantage low-income countries. Leadership in hunger reduction must be fostered in both high-, middle- and low-income countries.

Though the proportion of the world's population suffering from hunger has declined over the last 50 years, there are worrying signs that progress is stalling and it is very unlikely that the Millennium Development Goals for hunger in 2015 will be achieved. Ending hunger requires a well-functioning global food system that is sensitive to the needs of low-income countries, although it also requires concerted actions that come from within low-income countries.

4. Policy options should not be closed off. Throughout, the Project's Final Report has argued the importance of, within reason, excluding as few as possible different policy options on a *priori* grounds. Instead, it is important to develop a strong evidence base upon which to make informed decisions.

Food is so integral to human wellbeing that discussions of policy options frequently involve issues of ethics, values and politics. For example, there are very different views on the acceptability of certain new technologies, or on how best to help people out of hunger in low-income countries. . . . Achieving a strong evidence base in controversial areas is not enough to obtain public acceptance and approval—genuine public engagement and discussion needs to play a critical role.

5. This Report rejects food self-sufficiency as a viable option for nations to contribute to global food security, but stresses the importance of crafting food system governance to maximise the benefits of globalisation and to ensure that they are distributed fairly. For example, it is important to avoid the introduction of export bans at time of food stress, something that almost certainly exacerbated the 2007–2008 food price spike.

The food system is globalised and interconnected. This has both advantages and disadvantages. For example, economic disruptions in one geographical region can quickly be transmitted to others, but supply shocks in one region can be compensated for by producers elsewhere. A globalised food system also improves the global efficiency of food production by allowing breadbasket regions to export food to less favoured regions.

2   Important Drivers of Change Affecting the Food System

This is a unique time in history—decisions made now and over the next few decades will disproportionately influence the future:

- For the first time, there is now a high likelihood that growth in the global population will cease, with the number of people levelling in the range of eight to ten billion towards the middle of the century or in the two decades that follow.
- Human activities have now become a dominant driver of the Earth system: decisions made now to mitigate their detrimental effects will have a very great influence on the environment experienced by future generations, as well as the diversity of plant and animal species with which they will share the planet.
- There is now a developing global consensus, embodied in the Millennium Development Goals, that there is a duty on everyone to try to end poverty and hunger, whether in low-income countries or among the poor in more wealthy nations.

Threats from interacting drivers of change will converge in the food system over the next 40 years. Careful assessment of the implications of these drivers is essential if major pressures are to be anticipated, and future risks managed. Six particularly important drivers are outlined here. This Project has considered the combined effect of such drivers on the food system to explore interactions, feedbacks and non-linear effects.

I. **Global population increases.** Policy-makers should assume that today's population of about seven billion is most likely to rise to around eight billion by 2030 and probably to over nine billion by 2050. Most of these increases will occur in low- and middle-income countries; for example, Africa's population is projected to double from one billion to approximately two billion by 2050. However, population projections are uncertain and will need to be kept under review. Factors affecting population size include GDP growth, educational attainment, access to contraception and gender equality; possibly the single most important factor is the extent of female education. Population growth will also combine with other transformational changes, particularly in low- and middle-income countries as rising numbers of people move from rural areas to cities that will need to be serviced with food, water and energy.

II. **Changes in the size and nature of per capita demand.** Dietary changes are very significant for the future food system because, per calorie, some food items (such as grain-fed meat) require considerably more resources to produce than others. However, predicting patterns of dietary change is complex because of the way pervasive cultural, social and religious influences interact with economic drivers.
- *Meat:* different studies have predicted increases in per capita consumption (kg/capita/annum) from 32 kg today to 52 kg by the middle of the century. In high-income countries, consumption is nearing a plateau. Whether consumption of meat in major economies such as Brazil and China will stabilise at levels similar to countries such as the UK, or whether they will rise further to reach levels more similar to the USA is highly uncertain. However, major increases in the consumption of meat, particularly grain-fed meat, would have serious implications for competition for land, water and other inputs, and will also affect the sustainability of food production.
- *Fish:* demand is expected to increase substantially, at least in line with other protein foods, and particularly in parts of east and south Asia. The majority of this extra demand will need to be met by further expansion of aquaculture, which will have significant consequences for the management of aquatic habitats and for the supply of feed resources.

Major uncertainties around future per capita consumption include:

- the degree to which consumption will rise in Africa
- the degree to which diets will converge on those typical of high-income countries today
- whether regional differences in diet (particularly in India) persist into the future
- the extent to which increased GDP is correlated with reduced population growth and increased per capita demand—the precise nature of how these different trade-offs develop will have a major effect on gross demand.

III. **Future governance of the food system at both national and international levels.** Many aspects of governance have a significant impact on the workings of the food system:
- The globalisation of markets has been a major factor shaping the food system over recent decades and the extent to which this continues will have a substantial effect on food security.

- The emergence and continued growth of new food superpowers, notably Brazil, China and India. Russia is already significant in global export markets, and likely to become even more so, with a large supply of underutilised agricultural land.
- The trend for consolidation in the private sector, with the emergence of a limited number of very large transnational companies in agribusiness, in the fisheries sector, and in the food processing, distribution and retail sectors. There is some evidence that this trend may be reversing, with the entry into international markets of new companies from emerging economies.
- Production subsidies, trade restrictions and other market interventions already have a major effect on the global food system. How they develop in the future will be crucial.
- The extent to which governments act collectively or individually to face future challenges, particularly in shared resources, trade and volatility in agricultural markets. The inadequate governance of international fisheries, despite severe resource and market pressures, illustrates in microcosm many of the political and institutional obstacles to effective collective action.
- The adequacy of the current international institutional architecture to respond to future threats to the global food system, and the political will to allow them to function effectively, is unclear.
- The control of increasing areas of land for food production (such as in Africa) will be influenced by both past and future land-purchase and leasing agreements—involving both sovereign wealth funds and business.

IV. **Climate change.** This will interact with the global food system in two important ways:
- Growing demand for food must be met against a backdrop of rising global temperatures and changing patterns of precipitation. These changing climatic conditions will affect crop growth and livestock performance, the availability of water, fisheries and aquaculture yields and the functioning of ecosystem services in all regions. Extreme weather events will very likely become both more severe and more frequent, thereby increasing volatility in production and prices. Crop production will also be indirectly affected by changes in sea level and river flows, although new land at high latitudes may become suitable for cultivation and some degree of increased carbon dioxide fertilisation is likely to take place (due to elevated atmospheric carbon dioxide concentrations). The extent to which adaptation occurs (for example, through the development of crops and production methods adapted to new conditions) will critically influence how climate change affects the food system.
- Policies for climate change mitigation will also have a very significant effect on the food system—the challenge of feeding a larger global population must be met while delivering a steep reduction in greenhouse gas emissions.

V. **Competition for key resources.** Several critical resources on which food production relies will come under more pressure in the future. Conversely, growth in the food system will itself exacerbate these pressures:
- *Land for food production:* Overall, relatively little new land has been brought into agriculture in recent decades. Although global crop yields grew by 115% between 1967 and 2007, the area of land in agriculture increased by only 8% and the total currently stands at approximately 4,600 million hectares. While substantial additional land could in principle be suitable for food production, in practice land will come under growing pressure for other uses. For example, land will be lost to urbanisation, desertification, salinisation and sea level rise, although some options may arise for salt-tolerant crops or aquaculture. Also, while it has been estimated that the quality of around 16% of total land area including cropland, rangeland and forests is improving, the International Soil Reference and Information Centre has estimated (2009) that of the 11.5 billion hectares of vegetated land on earth, about 24% has undergone human-induced soil degradation, in particular through erosion. In addition, with an expanding population, there will be more pressure for land to be used for other purposes. And while some forms of biofuels can play an important role in the mitigation of climate change, they may lead to a reduction in land available for agriculture.

  There are strong environmental grounds for limiting any significant expansion of agricultural land in the future (although restoration of derelict, degraded or degrading land will be important). In particular, further conversion of rainforest to agricultural land should be avoided as it will increase greenhouse gas emissions very significantly and accelerate the loss of biodiversity.
- *Global energy demand:* This is projected to increase by 45% between 2006 and 2030 and could double between now and 2050. Energy

prices are projected to rise and become more volatile, although precise projections are very difficult to make. Several parts of the food system are particularly vulnerable to higher energy costs—for example, the production of nitrogen fertilisers is highly energy intensive: the roughly fivefold increase in fertiliser price between 2005 and 2008 was strongly influenced by the soaring oil price during this period.The financial viability of fishing (particularly capture fisheries) is also strongly affected by fuel price.

- *Global water demand:* Agriculture already currently consumes 70% of total global 'blue water' withdrawals from rivers and aquifers available to humankind. Demand for water for agriculture could rise by over 30% by 2030, while total global water demand could rise by 35–60% between 2000 and 2025, and could double by 2050 owing to pressures from industry, domestic use and the need to maintain environmental flows. In some arid regions of the world, several major non-renewable fossil aquifers are increasingly being depleted and cannot be replenished, for example in the Punjab, Egypt, Libya and Australia. Estimates suggest that exported foods account for between 16% and 26% of the total water used for food production worldwide, suggesting significant potential for more efficient global use of water via trade, although there is the risk of wealthy countries exploiting water reserves in low-income countries.

VI. **Changes in values and ethical stances of consumers.** These will have a major influence on politicians and policy makers, as well as on patterns of consumption in individuals. In turn, food security and the governance of the food system will be affected. Examples include issues of national interest and food sovereignty, the acceptability of modern technology (for example genetic modification, nanotechnology, cloning of livestock, synthetic biology), the importance accorded to particular regulated and highly specified production methods such as organic and related management systems, the value placed on animal welfare, the relative importance of environmental sustainability and biodiversity protection, and issues of equity and fair trade.

Notes

1. The contents of this Executive Summary closely follow the findings of the Foresight Project's Final Report, although the emphasis here is on the high-level conclusions and priority actions. All the supporting references for the analysis and figures contained in this Executive Summary are provided in the Final Report.

2. See Box 1.1 for a list of the Project's high-level conclusions.

THE BRITISH GOVERNMENT OFFICE FOR SCIENCE is an organization of about 80 permanent staff whose main function is to ensure that government policies and decisions are informed by the best scientific evidence and strategic long-term thinking.

EXPLORING THE ISSUE

Will the World Be Able to Feed Itself in the Foreseeable Future?

Critical Thinking and Reflection

1. Do the data from the past decade regarding possible progress in addressing global undernourishment and hunger show the glass to be half-full or half-empty?
2. Is the real problem of undernourishment and hunger due to factors unrelated to the actual growing of food?
3. Is the wide range of pressures adversely affecting food production simply too daunting to allow for dramatic increases in production?
4. What effect will economic development in the poorer regions of the globe have on the capacity of the future world to feed itself?
5. Is it realistic to think that people will unilaterally choose to become self-sufficient (grow their own food) and practice better food anti-waste patterns of behavior?

Is There Common Ground?

Analysts are in agreement that increased population and increased affluence among the currently less-affluent peoples of the world will increase the demand for food in the next 50 years. There is also significant agreement that enough food is currently produced but that the distribution system is flawed for whatever reason. There is also agreement that real progress has been made over the past few decades in lowering the percentage of undernourishment and hunger in the world.

Additional Resources

Bread for the World Institute, Hunger 2009: *Global Development Charting a New Course, 19th Annual Report on the State of the World* (Bread for the World Institute, 2009)

This report presents a somewhat optimistic viewpoint about future food prospects.

Clarke, Kevin, "Starving for Attention," *U.S. Catholic* (Vol. 79, October 2014)

The article focuses on issues related to food production and global hunger.

Cribb, Julian, *The Coming Famine: The Global Food Crisis and What We Can Do to Avoid It* (University of California Press, 2010)

The author describes the confluence of a number of shortages—water, land, energy, technology, and knowledge—that will constrain the world's ability to address future increased demand for food.

Fraser, Evan D.G. and Rimas, Andrew, *Empires of Food: Feast, Famine, and the Rise and Fall of Civilizations* (Free Press, 2010)

This book describes the history of food's role in the rise and fall of civilizations. It emphasizes today's negative impact of climate change, rising fuel costs, and a shrinking agricultural frontier on food availability, and argues the importance of global food systems.

Headley, Derek, *Reflections on the Global Food Crisis: How Did It Happen? How Has It Hurt? And How Can We Prevent the Next One?* (International Food Policy Research Institute, 2010)

This book examines the 2008 food crisis, blaming it on rising energy prices, growing demand for biofuels, the U.S. dollar depreciation, and various trade stocks.

International Food Policy Research Institute, *2012 Global Food Policy Report* (International Food Policy Research Institute, 2013)

This is the second in an annual series that examines major food policy developments and events.

Margulis, Matias E., "The Regime Complex for Food Security: Implications for the Global Hunger Challenge," *Global Governance* (Vol. 19, March 2013)

The article focuses on reform efforts for the global governance of food security.

Organization for Economic Co-Operation and Development, *Global Food Security: Challenges for the Food and Agricultural System* (OECD Publishing, 2013)

This report focuses on the challenges facing global efforts to ensure enough food for the world's population.

Rieff, David, *The Reproach of Hunger* (Simon & Schuster, 2015)

This book is an objective analysis of whether ending extreme poverty and widespread hunger is a possibility by 2050.

Rosin, Christopher et al., *Food Systems Failure: The Global Food Crisis and the Future of Agriculture* (Earthscan, 2011)

This book describes the current food crisis, focusing on contradictions in policy and practice that prevent solutions.

Saad, Majda Bne, *The Global Hunger Crisis: Tracking Food Insecurity in Developing Countries* (Pluto Press, 2013)

The author identifies the causes of global hunger in the current global political and economic system and identifies challenges facing low-income food-deficit countries.

United Nations, Food and Agriculture Organization, *Food Outlook: Biannual Report on Global Food Markets* (United Nations, October 2015)

This report describes food prices in the near future.

Internet References . . .

Food First

www.foodfirst.org

The Hunger Project

www.thp.org

International Food Policy Research Institute

www.ifpri.org

UN Food and Agriculture Organization

www.fao.org

UN World Food Programme

www.wfp.org

Selected, Edited, and with Issue Framing Material by:
James E. Harf, *Maryville University*
and
Mark Owen Lombardi, *Maryville University*

ISSUE

Can the Global Community Successfully Confront the Global Water Shortage?

YES: UNICEF and World Health Organization, from "Progress on Sanitation and Drinking Water: 2015 Update and MDG Assessment," *United Nations Publications* (2015)

NO: WWAP (United Nations World Water Assessment Programme), from "Water for a Sustainable World," The United Nations World Development Report 2015, *UNESCO* (2015)

Learning Outcomes

After reading this issue, you will be able to:

- Describe recent advances made in improving access to safe drinking water.
- Understand how the developed and developing worlds differ with respect to access to improved drinking water.
- Understand how urban and rural areas differ with respect to access to improved drinking water.
- Understand the factors associated with increased global water demand.
- Understand the relationship between water and the dimensions of sustainable development.
- Understand why experts predict that the global water crisis threatens the stability of nations and the health of billions of people.

ISSUE SUMMARY

YES: The report by UNICEF and WHO details how far the global community has come in the past 15 years in providing safe drinking water for its citizens, suggesting that over 90 percent of the world's population now has access to improved sources of drinking water.

NO: The WWAP report predicts, however, that by 2025, the global demand for water is projected to increase by 55 percent, which will tax the planet's ability to meet that increased demand.

In March 2011, high-level experts from around the globe met at the invitation of the InterAction Council, an international organization of former heads of government, to discuss the status of the world's freshwater supply as it related to global security. The main conclusion of the group was that the global water crisis is both real and urgent, as one billion people were without reliable water sources and two billion were without adequate sanitation. Their conclusions mirrored those of others. As the World Water Council suggests, the times of "easy water" are long gone. Water shortages and other water problems are occurring with greater frequency, particularly in large cities. Some observers have speculated that the situation is reminiscent of the fate that befell ancient glorious cities like Rome.

A 2015 White Paper Prepared by the United Nations Food and Agricultural Organization, *Towards a Water and Food Secure Future: Critical Perspectives for Policy-Makers,* built on these views, suggesting that while the outlook for 2050 "is encouraging, […] much work is needed to achieve sustainable water use […]." While the report suggested that at the global level sufficient water will exist for food production, an increasing number of regions will "face growing water scarcity." Urban demands for water will increase in the developing world because of increased population in the coming decades. By 2050 agriculture will continue to be the biggest user of water globally. And climate change is expected, according to the white paper, to bring greater challenges to water management.

A much more stark assessment appeared in February 2016 in the journal *Science Advances*, with the claim that global water shortages are worse than previously thought.

Recognition that the supply of water is a growing problem is not new. As early as 1964, the United Nations Environmental Programme (UNEP) revealed that close to a billion people were at risk from desertification. At the Earth Summit in Rio de Janeiro in 1992, world leaders reaffirmed that desertification was of serious concern.

Moreover, since these early warnings about global water, in conference after conference and study after study, increasing population growth and declining water supplies and quality are being linked together, as is the relationship between the planet's ability to meet its growing food needs and available water. Lester R. Brown, in "Water Deficits Growing in Many Countries: Water Shortages May Cause Food Shortages," *Eco-Economy Update 2002–11* (August 6, 2002), sums up the problem this way: "The world is incurring a vast water deficit. It is largely invisible, historically recent, and growing fast." The World Water Council's study, "World Water Actions Report, Third Draft" (October 31, 2002), describes the problem in much the same way: "Water is no longer taken for granted as a plentiful resource that will always be available when we need it." The report continued with the observation that increasing numbers of people in more and more countries are discovering that water is a "limited resource that must be managed for the benefit of people and the environment, in the present and in the future." In short, water is fast becoming both a food-related issue and a health-related problem. Some scholars are now arguing that water shortage is likely to become the twenty-first century's analog to the oil crisis of the last half of the previous century. The one major difference, as scholars are quick to point out, is that water is not like oil; there is no substitute.

Proclamations of impending water problems abound. Peter Gleick, in *The World's Water 1998–99: The Biennial Report on Freshwater Resources* (Island Press, 1998), reports that the demand for freshwater increased sixfold between 1900 and 1995, twice the rate of population growth. The UN study *United Nations Comprehensive Assessment of Freshwater Resources of the World* (1997) suggested that one-third of the world's population lives in countries having medium to high water stress. One 2001 headline reporting the release of a new study proclaimed that "Global thirst 'will turn millions into water refugees'" (*The Independent*, 1999). News reports released by the UN Food and Agricultural Organization in conjunction with World Food Day 2002 asserted that water scarcity could result in millions of people having inadequate access to clean water or sufficient food. And the World Meteorological Organization predicts that two out of every three people will live in water-stressed conditions by 2050 if consumption patterns remain the same.

Sandra Postel, in Pillar *of Sand: Can the Irrigation Miracle Last?* (W. W. Norton, 1999), suggests another variant of the water problem. For her, the time-tested method of maximizing water usage in the past, irrigation, may not be feasible as world population marches toward 7 billion. She points to the inadequacy of surface water supplies, increasing depletion of groundwater supplies, the salinization of the land, and the conversion of traditional agricultural land to other uses as reasons for the likely inability of irrigation to be a continuing panacea. Yet the 1997 UN study concluded that annual irrigation use would need to increase 30 percent for annual food production to double, necessary for meeting food demands of 2025.

The issue of water quality is also in the news. The World Health Organization reports that in some parts of the world, up to 80 percent of all transmittable diseases are attributable to the consumption of contaminated water. Also, a UNEP-sponsored study, *Global Environment Out-look 2000,* reported that 200 scientists from 50 countries pointed to the shortage of clean water as one of the most pressing global issues.

More recent studies bring the same message. The UN *World Water Development Report* 2009 lays out the problems and the challenges of global water, and suggests that current decision-making processes are not up to the challenge of addressing these problems. The report outlines major issues in the areas of access to drinking water infrastructure, sanitation infrastructure, effects of population growth, agriculture and livestock, energy, sanitation treatment of waste water, climate change, and migration issues related to water scarcity.

In 2010, the World Water Council, an international water think tank sponsored by the World Bank and the UN, also sounded the alarm, suggesting that more than one out of six humans lack access to safe drinking water, while more than two out of six lack adequate sanitation. The Council suggests that during the next 50 years, population growth coupled with industrialization and urbanization will dramatically increase the demand for safe water.

Water.org cites a variety of official UN and other international governmental organization studies to reveal bullet points relating to the statistics of the global water problems. A sampling of its information includes the following:

- 3.575 million people die each year from water-related diseases.
- People living in slums pay 5 to 10 times more per liter of water than the rich in the same city.
- An American taking a 5-minute shower uses more water than a developing world person uses in a day.
- Lack of sanitation is the world's biggest cause of infection.
- Diarrhea remains the second leading cause of death for children under five.
- Every 20 seconds, a child dies from a water-related disease.
- Children in poor environments often carry 1000 parasitic worms in their bodies at any given moment.
- In any one day, more than 200 million hours are spent collecting water for domestic use.
- At any given moment, half of the world's hospital beds are occupied by people with water-related diseases.
- Less than 1 percent of the world's fresh water is readily accessible for direct human usage.
- By 2025, 48 countries will face freshwater stress or scarcity.

The World Water Council recently suggested a comprehensive plan of action that focused on four major initiatives: supporting political action to improve water and sanitation services and water management, deepening the involvement of major water users, strengthening regional cooperation to achieve water security and economic development, and mobilizing citizens and consumers to address the global water crisis.

Although most allude to a current water shortage that will only worsen, a few analysts suggest otherwise. Bjørn Lomborg, the "skeptical scientist," has taken this issue in *The Skeptical Environmentalist: Measuring the Real State of the World* (2002) with the opposing view in the global water debate. His argument can be summed up in his simple quote: "Basically we have sufficient water." Lomborg maintains that water supplies rose during the twentieth century and that we have gained access to more water through technology. Benjamin Radford suggests that when one distills the evidence behind the dire headlines, there is "one little fact. There is no water shortage."

Although most of the discussion and debate center on either the dwindling fresh water supply or unequal access thereto, an alternative view suggests that the real issue may be corporate control of water. Maude Barlow, in *Blue Covenant* (2007), makes the case that it is privatization of fresh water and other corporate behavior that contribute significantly to the global water crisis. From putting massive amounts of water into bottles for sale to individuals to controlling large amounts of water used in industry to controlling the world water trade, for-profit private corporations have used supply/demand and other economic principles to undercut the global community's ability to address the global water crisis.

There is hope, according to the InterAction Council, but action is needed. Twelve specific steps were recommended for the policy community by the Council following the March 2011 meeting:

- Continue the dialogue on the water crisis.
- Endorse the human right to water.
- Support ratification of the UN Watercourses Convention.
- Encourage the UN Security Council to focus on water security.
- Support increased universal sanitation coverage and safe water supply.
- Facilitate links between national and global water, agricultural, and energy policies.
- Support necessary hydro-climatic monitoring.
- Support protection of ecological sustainability boundaries and investment in ecosystem restoration.
- Encourage cooperation and act as a mediator in water conflicts.
- Call on national governments to strengthen water education programs.
- Involve the private sector.
- Create a white paper supporting the above recommendations.

In the YES selection from UNICEF, the authors conclude that much progress in achieving safe drinking water has been made with the phrase "the report shows how far we have come." It asserts that over 90 percent of the world's population has access to "improved sources of drinking water." In the NO selection from UN-Water, the authors conclude that unless "the balance between demand and finite supplies is restored, the world will face an increasingly severe global water deficit."

YES ↵

UNICEF and World Health Organization

Progress on Sanitation and Drinking Water—2015 Update and MDG Assessment

Introduction

In 2000 the Member States of the United Nations signed the Millennium Declaration, which later gave rise to the Millennium Development Goals (MDGs). Goal 7, to ensure environmental sustainability, included a target that challenged the global community to halve, by 2015, the proportion of people without sustainable access to safe drinking water and basic sanitation. The WHO/UNICEF Joint Monitoring Programme for Water Supply and Sanitation (JMP), which began monitoring the sector in 1990, has provided regular estimates of progress towards the MDG targets, tracking changes over the 25 years to 2015.

In 1990, global coverage of the use of improved drinking water sources and sanitation facilities stood at 76 per cent and 54 per cent, with respective MDG targets of 88 per cent and 77 per cent by 2015. The challenges were huge, as the global figures hid vast disparities in coverage between countries, many of which were battling poverty, instability and rapid population growth.

The JMP has monitored the changes in national, regional and global coverage, establishing a large and robust database and presenting analysis not only of the indicators detailed in the original framework for the MDGs, but also many other parameters. The analysis has helped shed light on the nature of progress and the extent to which the ambition and vision of the MDGs have been achieved. It has also helped to identify future priorities to be addressed in the post-2015 Sustainable Development Goals.

. . .

The MDG target called for the proportion of the population without sustainable access to safe drinking water to be halved between 1990 and 2015. During the MDG period it is estimated that, globally, use of improved drinking water sources[1] rose from 76 per cent to 91 per cent.

The MDG target of 88 per cent was surpassed in 2010, and in 2015, 6.6 billion people use an improved drinking water source. There are now only three countries with less than 50 per cent coverage, compared with 23 in 1990.

Despite the achievements of the MDG period, a great deal remains to be done. Behind the global headline figures, huge disparities in access remain. While many developed regions have now achieved universal access, coverage with improved drinking water sources varies widely in developing regions. The lowest levels of coverage are found in the 48 countries designated as the least developed countries by the United Nations, particularly those in sub-Saharan Africa.

Globally, 2.6 billion people have gained access to an improved drinking water source since 1990. In most regions, over one third of the 2015 population gained access during the MDG period. Developing regions with low baselines and those experiencing rapid population growth have had to work much harder to maintain and extend coverage. Although sub-Saharan Africa missed the MDG target, over 40 per cent of the current population gained access since 1990.

In 2015, it is estimated that 663 million people worldwide still use unimproved drinking water sources, including unprotected wells and springs and surface water. The majority of them now live in two developing regions. Nearly half of all people using unimproved drinking water sources live in sub-Saharan Africa, while one fifth live in Southern Asia.

Use of improved drinking water has increased in all regions of the world since 1990, but rates of progress have varied during the MDG period. Coverage in Eastern Asia increased dramatically—by 27 percentage points—and exceeded the MDG target, with over half a billion people gaining access in China alone. Access in Southern Asia and South-eastern Asia also rose steeply, by 20 and 19 per cent respectively, and these regions also met the target.

Sub-Saharan Africa fell short of the MDG target but still achieved a 20 percentage point increase in the use of improved sources of drinking water. This means 427 million people gained access during the MDG period—an average of 47,000 people per day for 25 years. Over the same period, the Caucasus and Central Asia and Oceania[2] achieved increases of just 2 per cent and 5 per cent, respectively, and also missed the target.

The least developed countries have faced the greatest challenges in meeting the MDG target, given low coverage and high population growth. Half of these are classified by the World Bank as 'fragile situations',[3] and many have been affected by conflict during the MDG period, but have nevertheless made progress. Between 1990 and 2015, the proportion of people in least developed countries using improved drinking water sources increased from 51 per cent to 69 per cent, but use of piped water on premises only increased from 7 per cent to 12 per cent.

Significant proportions of the population in sub-Saharan Africa and Oceania continue to use rivers, lakes, ponds and irrigation canals as their main source of drinking water. Since 1990 the proportion of the population using surface water has been more than halved in sub-Saharan Africa, but remains largely unchanged in Oceania.

The MDG water and sanitation targets called for reporting on progress in both rural and urban areas. In 1990 the majority of the global population (57 per cent) lived in rural areas, but since then the situation has reversed, and in 2015 the proportion living in urban areas is 54 per cent.

It is estimated that 96 per cent of the urban population now uses improved drinking water sources, compared with 84 per cent of the rural population.[4] The gap in coverage between rural and urban areas has steadily decreased since 1990. But while rural coverage has increased rapidly, urban coverage has stagnated. The number of people without access in rural areas has decreased by over half a billion, but the number without access in urban areas has not changed significantly.

Urban coverage with piped water on premises has also remained largely unchanged since 1990, whereas rural coverage has almost doubled. However, the gap between access to piped water on premises in urban and rural areas remains large. Four out of five people living in urban areas now have access to piped drinking water on premises, compared with just one in three people living in rural areas.

In 2015, the vast majority of those who do not have access to improved drinking water sources live in rural areas. It is estimated that 79 per cent of the people

using unimproved sources and 93 per cent of people using surface water live in rural areas.

Nearly three quarters of the 2.6 billion people gaining access to an improved drinking water source over the MDG period use piped water on premises. Over half of the 951 million people gaining access to improved drinking water sources in rural areas and over three quarters of the 1.6 billion people gaining access in urban areas are using piped water.

However, the balance between increases in piped water on premises and increases in other improved sources has varied widely between regions. In most developing regions, coverage gains over the MDG period have been driven by gains in access to piped water on premises. This is particularly striking in Eastern Asia, which contributed 723 million new users of piped water on premises, with 694 million gaining access in China alone. Piped water on premises also dominated in Latin America and the Caribbean, Western Asia and Northern Africa. In these regions the number of users of other improved sources declined over the MDG period.

By contrast, in Southern Asia, South-eastern Asia and sub-Saharan Africa, coverage gains over the MDG period have been mainly driven by other improved sources. Since 1990, 471 million people in Southern Asia and 348 million people in sub-Saharan Africa gained access to other improved sources, such as wells and springs.

The rate of increase in piped water coverage has generally been higher in rural than in urban areas. As a result, during the MDG period the rural-urban gap in access to piped water on premises has narrowed in all regions, except for Oceania, where neither urban nor rural coverage has changed. In sub-Saharan Africa, urban coverage has declined by ten percentage points.

While most regions recorded an overall increase in access to piped water on premises, a small number of countries have increased coverage of this higher level of service by more than 25 percentage points. By increasing access from 28 per cent to 73 per cent, China significantly boosted the regional average for Eastern Asia. Despite slow regional progress in sub-Saharan Africa, Botswana, and Senegal all increased coverage by more than a third, as did Belize, El Salvador, Guatemala and Paraguay in Latin America and the Caribbean.

Of the 663 million people still using unimproved drinking water sources, those who use surface water face the greatest risks to their health and well-being. Those with no service at all, who have not benefited from any investment, are increasingly concentrated in three regions. Rural populations are particularly disadvantaged, accounting for 93 per cent of the people using surface water. Seven out of ten of the 159 million people relying on water taken directly from rivers, lakes and other surface waters live in sub-Saharan Africa, eight times more than any other region.

The MDG target called for halving the proportion of the population without sustainable access to basic sanitation between 1990 and 2015. During the MDG period, it is estimated that use of improved sanitation facilities rose from 54 per cent to 68 per cent globally. The global MDG target of 77 per cent has therefore been missed by nine percentage points and almost 700 million people.

Despite encouraging progress on sanitation, much unfinished business remains from the MDG period. In addition to the shortfall against the global target, large disparities in access still exist. Almost all developed countries have achieved universal access, but sanitation coverage varies widely in developing countries. Since 1990 the number of countries with less than 50 per cent of the population using an improved sanitation facility has declined slightly, from 54 to 47, and countries with the lowest coverage are now concentrated in sub-Saharan Africa and Southern Asia.

Nearly one third of the current global population has gained access to an improved sanitation facility since 1990, a total of 2.1 billion people. However, the proportion gaining access varies across developing regions. Western Asia and Northern Africa have provided access to 50 per cent and 41 per cent of the current population since 1990. By contrast, sub-Saharan Africa has provided access to less than 20 per cent of the current population. Despite failing to meet the target of halving the proportion of the population without access, Southern Asia nevertheless managed to provide access to 32 per cent of the current population.

In 2015 it is estimated that 2.4 billion people globally still use unimproved sanitation facilities. The vast majority live in just three regions, with 40 percent in Southern Asia. There are now twice as many people using unimproved sanitation facilities in sub-Saharan Africa than in Eastern Asia. The nearly 700 million people who would have been served if the MDG target for sanitation had been met is equal to the number of unserved people in sub-Saharan Africa.

Use of improved sanitation facilities increased in all regions, except for Oceania, but rates of progress varied widely. The Caucasus and Central Asia, Eastern Asia, Northern Africa and Western Asia were the only developing regions to meet the MDG target. Eastern Asia dramatically increased coverage—by 28 percentage points—to meet the target. South-eastern Asia also achieved a significant increase, of 24 per percentage points, but narrowly missed the target. While Southern Asia and sub-Saharan Africa had similarly low levels of coverage in 1990 (22 per cent and 24 per cent, respectively), the former increased coverage by 25 percentage points, while the latter only achieved a 6 per cent increase.

In Southern Asia, which had the lowest baseline coverage in 1990, 576 million people gained access to improved sanitation facilities during the MDG period—an average of 63,000 people per day for 25 years. Over the same period, improved sanitation coverage in sub-Saharan Africa and Oceania has stagnated. However, the vast majority (64 per cent) of those without access to improved sanitation in Southern Asia still practise open defecation, compared with 33 per cent in sub-Saharan Africa and just 18 per cent in Oceania.

Between 1990 and 2015, open defecation declined in all regions, with the most dramatic reductions seen in the least developed countries (from 45 per cent in 1990 to 20 per cent in 2015), representing an important first step on the sanitation ladder.

Globally, it is estimated that 82 per cent of the urban population now uses improved sanitation facilities, compared with 51 per cent of the rural population. Inequalities in access to improved sanitation between rural and urban areas have decreased during the MDG period. The number of people without access to improved sanitation in rural areas has decreased by 15 per cent, and open defecation rates have decreased from 38 per cent to 25 per cent.

Notes

1. In the absence of nationally representative data on the safety of drinking water for the majority of countries, the agreed proxy indicator for monitoring 'sustainable access to safe drinking water' during the MDG period has been 'use of an improved drinking water source'.

2. It should be kept in mind throughout this report that data from Oceania are limited. Estimates for each of the small island states in the region draw upon a very small number of data points, many of which date back several years, making it difficult to produce robust estimates for 2015.

3. The World Bank, 'Harmonised List of Fragile Situations', 2015, http://sit-eresources.worldbank.org/EXTLICUS/Resources/511777-1269623894864/FY15FragileSituationList.pdf, accessed May 10, 2015.

4. JMP estimates are based on national surveys and censuses. Official definitions of urban and rural

vary across countries and may not be directly comparable. While all surveys are representative of total urban and rural populations, samples may not be representative of all population subgroups or those living in specific geographic locations, including informal settlements or remote rural areas.

UNICEF is the UN Children's Emergency Fund headquartered in New York. It provides long-term humanitarian and developmental assistance to children and mothers in developing countries. WHO, the World Health Organization, is a specialized agency of the United Nations and is headquartered in Geneva, Switzerland. It focuses on world health issues.

WWAP   ➡ **NO**

Water for a Sustainable World

Water is at the core of sustainable development. Water resources, and the range of services they provide, underpin poverty reduction, economic growth and environmental sustainability. From food and energy security to human and environmental health, water contributes to improvements in social well-being and inclusive growth, affecting the livelihoods of billions.

. . .

The Consequences of Unsustainable Growth

Unsustainable development pathways and governance failures have affected the quality and availability of water resources, compromising their capacity to generate social and economic benefits. Demand for freshwater is growing. Unless the balance between demand and finite supplies is restored, the world will face an increasingly severe global water deficit.

Global water demand is largely influenced by population growth, urbanization, food and energy security policies, and macro-economic processes such as trade globalization, changing diets and increasing consumption. By 2050, global water demand is projected to increase by 55%, mainly due to growing demands from manufacturing, thermal electricity generation and domestic use.

Competing demands impose difficult allocation decisions and limit the expansion of sectors critical to sustainable development, in particular food production and energy. The competition for water—between water 'uses' and water 'users'—increases the risk of localized conflicts and continued inequities in access to services, with significant impacts on local economies and human well-being.

Over-abstraction is often the result of out-dated models of natural resource use and governance, where the use of resources for economic growth is under-regulated and undertaken without appropriate controls. Groundwater supplies are diminishing, with an estimated 20% of the world's aquifers currently over-exploited. Disruption of ecosystems through unabated urbanization, inappropriate agricultural practices, deforestation and pollution are among the factors undermining the environment's capacity to provide ecosystem services, including clean water.

Persistent poverty, inequitable access to water supply and sanitation services, inadequate financing, and deficient information about the state of water resources, their use and management impose further constraints on water resources management and its ability to help achieve sustainable development objectives.

Water and the Three Dimensions of Sustainable Development

Progress in each of the three dimensions of sustainable development—social, economic and environmental—is bound by the limits imposed by finite and often vulnerable water resources and the way these resources are managed to provide services and benefits.

Poverty and Social Equity

While access to household water supplies is critical for a family's health and social dignity, access to water for productive uses such as agriculture and family-run businesses is vital to realize livelihood opportunities, generate income and contribute to economic productivity. Investing in improved water management and services can help reduce poverty and sustain economic growth. Poverty-oriented water interventions can make a difference for billions of poor people who receive very direct benefits from improved water and sanitation services through better health, reduced health costs, increased productivity and time-savings.

Economic growth itself is not a guarantee for wider social progress. In most countries, there is a wide—and often widening—gap between rich and poor, and between those who can and cannot exploit new opportunities.

WWAP (United Nations World Water Assessment Programme). 2015. The United Nations World Water Development Report 2015: Water for a Sustainable World. Paris, UNESCO.

Access to safe drinking water and sanitation is a human right, yet its limited realization throughout the world often has disproportionate impacts on the poor and on women and children in particular.

Economic Development

Water is an essential resource in the production of most types of goods and services including food, energy and manufacturing. Water supply (quantity and quality) at the place where the user needs it must be reliable and predictable to support financially sustainable investments in economic activities. Wise investment in both hard and soft infrastructure that is adequately financed, operated and maintained facilitates the structural changes necessary to foster advances in many productive areas of the economy. This often means more income opportunities to enhance expenditure in health and education, reinforcing a self-sustained dynamic of economic development.

Many benefits may be gained by promoting and facilitating use of the best available technologies and management systems in water provision, productivity and efficiency, and by improving water allocation mechanisms. These types of interventions and investments reconcile the continuous increase in water use with the need to preserve the critical environmental assets on which the provision of water and the economy depends.

Environmental Protection and Ecosystem Services

Most economic models do not value the essential services provided by freshwater ecosystems, often leading to unsustainable use of water resources and ecosystem degradation. Pollution from untreated residential and industrial wastewater and agricultural run-off also weakens the capacity of ecosystem to provide water-related services.

Ecosystems across the world, particularly wetlands, are in decline. Ecosystem services remain under-valued, under-recognized and under-utilized within most current economic and resource management approaches. A more holistic focus on ecosystems for water and development that maintains a beneficial mix between built and natural infrastructure can ensure that benefits are maximized.

Economic arguments can make the preservation of ecosystems relevant to decision-makers and planners. Ecosystem valuation demonstrates that benefits far exceed costs of water-related investments in ecosystem conservation. Valuation is also important in assessing trade-offs in ecosystem conservation, and can be used to better inform development plans. Adoption of 'ecosystem-based management' is key to ensuring water long-term sustainability.

Water's Role in Addressing Critical Developmental Challenges

Interlinkages between water and sustainable development reach far beyond its social, economic and environmental dimensions. Human health, food and energy security, urbanization and industrial growth, as well as climate change are critical challenge areas where policies and actions at the core of sustainable development can be strengthened (or weakened) through water.

Lack of **water supply, sanitation and hygiene** (WASH) takes a huge toll on health and well-being and comes at a large financial cost, including a sizable loss of economic activity. In order to achieve universal access, there is a need for accelerated progress in disadvantaged groups and to ensure non-discrimination in WASH service provision. Investments in water and sanitation services result in substantial economic gains; in developing regions the return on investment has been estimated at US$5 to US$28 per dollar. An estimated US$53 billion a year over a five-year period would be needed to achieve universal coverage—a small sum given this represented less than 0.1% of the 2010 global GDP.

The increase in the number of people without access to water and sanitation in **urban areas** is directly related to the rapid growth of slum populations in the developing world and the inability (or unwillingness) of local and national governments to provide adequate water and sanitation facilities in these communities. The world's slum population, which is expected to reach nearly 900 million by 2020, is also more vulnerable to the impacts of extreme weather events. It is however possible to improve performance of urban water supply systems while continuing to expand the system and addressing the needs of the poor.

By 2050, **agriculture** will need to produce 60% more food globally, and 100% more in developing countries. As the current growth rates of global agricultural water demand are unsustainable, the sector will need to increase its water use efficiency by reducing water losses and, most importantly, increase crop productivity with respect to water. Agricultural water pollution, which may worsen with increased intensive agriculture, can be reduced through a combination of instruments, including more stringent regulation, enforcement and well-targeted subsidies.

Energy production is generally water-intensive. Meeting ever-growing demands for energy will generate increasing stress on freshwater resources with repercussions on other users, such as agriculture and industry. Since these sectors also require energy, there is room to create synergies as they develop together. Maximizing the water efficiency of power plant cooling systems and increasing the capacity of wind, solar PV and geothermal energy will be a key determinant in achieving a sustainable water future.

Global water demand for the **manufacturing industry** is expected to increase by 400% from 2000 to 2050, leading all other sectors, with the bulk of this increase occurring in emerging economies and developing countries. Many large corporations have made considerable progress in evaluating and reducing their water use and that of their supply chains. Small and medium-sized enterprises are faced with similar water challenges on a smaller scale, but have fewer means and less ability to meet them.

The negative impacts of **climate change** on freshwater systems will most likely outweigh its benefits. Current projections show that crucial changes in the temporal and spatial distributing of water resources and the frequency and intensity of water-related disasters rise significantly with increasing greenhouse gas emissions. Exploitation of new data sources, better models and more powerful data analysis methods, as well as the design of adaptive management strategies can help respond effectively to changing and uncertain conditions.

Regional Perspectives

The challenges at the interface of water and sustainable development vary from one region to another.

Increasing resource use efficiency, reducing waste and pollution, influencing consumption patterns and choosing appropriate technologies are the main challenges facing **Europe and North America**. Reconciling different water uses at the basin level and improving policy coherence nationally and across borders will be priorities for many years to come.

Sustainability in the **Asia and the Pacific** region is intimately linked with progress in access to safe water and sanitation; meeting water demands across multiple uses and mitigating the concurrent pollution loads; improving groundwater management; and increasing resilience to water-related disasters.

Water scarcity stands at the forefront when considering water-related challenges that impede progress towards sustainable development in the **Arab** region, where unsustainable consumption and over-abstraction of surface and groundwater resources contribute to water shortages and threaten long-term sustainable development. Options being adopted to enhance water supplies include water harvesting, wastewater reuse and solar energy desalination.

A major priority for the **Latin America and the Caribbean** region is to build the formal institutional capacity to manage water resources and bring sustainable integration of water resources management and use into socio-economic development and poverty reduction. Another priority is to ensure the full realization of the human right to water and sanitation in the context of the post-2015 development agenda.

The fundamental aim for **Africa** is to achieve durable and vibrant participation in the global economy while developing its natural and human resources without repeating the negatives experienced on the development paths of some other regions. Currently only 5% of the Africa's potential water resources are developed and average per capita storage is 200 m^3 (compared to 6,000 m^3 in North America). Only 5% of Africa's cultivated land is irrigated and less than 10% of hydropower potential is utilized for electricity generation.

. . .

Unsustainable Growth

Unsustainable development pathways and governance failures have generated immense pressures on water resources, affecting its quality and availability, and in turn compromising its ability to generate social and economic benefits. The planet's capacity to sustain the growing demands for freshwater is being challenged, and there can be no sustainable development unless the balance between demand and supply is restored.

Global gross domestic product (GDP) rose at an average of 3.5% per year from 1960 to 2012 (World Economics, 2014), and much of this economic growth has come at a significant social and environmental cost. During this same period, population growth, urbanization, migration and industrialization, along with increases in production and consumption, have generated ever-increasing demands for freshwater resources. These same processes have also contributed to the polluting of water resources, further reducing their immediate accessibility and thus compromising the capacity of ecosystems and the natural water cycle to satisfy the world's growing demand for water.

1.1 Increasing Global Water Demand

Global water demand is largely influenced by population growth, urbanization, food and energy security policies, and macro-economic processes such as trade globalization and changing consumption patterns.

Over the past century, the development of water resources has been largely driven by the demands of expanding populations for food, fibre and energy. Strong income growth and rising living standards of a growing middle class have led to sharp increases in water use, which can be unsustainable, especially where supplies are vulnerable or scarce and where its use, distribution, price, consumption and management are poorly managed or regulated. Changing consumption patterns, such as increasing meat consumption, building larger homes, and using more motor vehicles, appliances and other energy-consuming devices, typically involves increased water consumption for both production and use.

Demand for water is expected to increase in all sectors of production (WWAP, 2012). By 2030, the world is projected to face a 40% global water deficit under the business-as-usual (BAU) scenario.

Population growth is another factor, but the relationship is not linear: over the last decades, the rate of demand for water has doubled the rate of population growth. The world's population is growing by about 80 million people per year. It is predicted to reach 9.1 billion by 2050, with 2.4 billion people living in Sub-Saharan Africa, the region with the most heterogeneously distributed water resources.

Increasing urbanization (see Chapter 6) is causing specific and often highly localized pressures on freshwater resource availability, especially in drought-prone areas. More than 50% of people on the planet now live in cities, with 30% of all city dwellers residing in slums. Urban populations are projected to increase to a total of 6.3 billion by 2050. Developing countries account for 93% of urbanization globally, 40% of which is the expansion of slums. By 2030, the urban population in Africa and Asia will double.

Excessive water withdrawals for agriculture and energy (see Chapters 7 and 8) can further exacerbate water scarcity. Freshwater withdrawals for energy production, which currently account for 15% of the world's total, are expected to increase by 20% through 2035. The agricultural sector is already the largest user of water resources, accounting for roughly 70% of all freshwater withdrawals globally, and over 90% in most of the world's least-developed countries. Practices like efficient irrigation techniques can have a dramatic impact on reducing water demand, especially in rural areas.

Many of the pressures that impact water sustainability occur at local and national levels, and are influenced by rules and processes established at those levels. Increasingly, however, the rules and processes that govern global economics—investment of capital, trade, financial markets, as well as international aid and development assistance—influence local and national economies, which in turn dictate local water demand and the sustainability of water resources at the basin level.

1.2 Potential Impacts of Increasing Demand

Competing demands will lead to increasingly difficult allocation decisions and limit the expansion of sectors critical to sustainable development, in particular food production and energy. Inter-sectoral competition and the delicate trade-offs between energy and agricultural production can already be seen in the debate about biofuels. Production of biofuel from food crops, such as corn, wheat and palm

. . .

oil, has induced additional competition for land and water even within the agricultural sector, especially in regions already under water stress, and has also been associated with increased food prices. Growing food crops for biofuel has spurred debate over ethical considerations regarding future food security as well as existing efforts to combat malnutrition.

Increasing industrial production will also lead to increased water use, with potential impacts on water quality. In certain areas where water use for industrial production is not well-regulated or enforced, pollution could increase dramatically, closely linking increasing economic activity with the degradation of environmental services.

Competition for water between water 'uses' and water 'users' increases the risk of localized conflicts and continued inequities in access to services. In this competition, the need to maintain water and ecosystem integrity in order to sustain life and economic development is often ignored. Frequently, the environment, as well as marginalized and vulnerable people, are the biggest losers in the competition for water.

Inter-state and regional conflicts may also emerge due to water scarcity and poor management structures. It is noteworthy that 158 of the world's 263 transboundary water basins lack any type of cooperative management framework. Of the 105 water basins with water institutions, approximately two-thirds include three or more riparian states (adjacent to rivers and streams); yet, less than 20% of the accompanying agreements are multilateral. This indicates that the mechanisms, political will and/or resources to manage shared water resources bilaterally or multilaterally and share potential benefits efficiently and effectively are missing.

Competition over water highlights the difficult policy choices that are posed by the water-food-energy-nexus and the trade-offs involved in managing each sector, either separately or together. These three pillars of any functioning society are closely interlinked, and choices made in one area will inevitably impact the choices and hence resources available in the others.

Over-abstraction is often the result of out-dated models of natural resource use and governance, where the use of resources for economic growth is under-regulated and undertaken without appropriate controls. Unsustainable withdrawals of surface and groundwater can severely affect water availability for ecosystems and the services they provide, with significant impacts on local economies and human well-being. Inadequate assessment of water resources, especially groundwater, and in some cases a disregard of information when it is available, have contributed to water resources management failures in many parts of the world.

If institutional mechanisms within governments and other governance structures continue to follow narrow objectives along sector-specific mandates, fundamental disconnects will continue to occur. This situation has already led to negative impacts for the most vulnerable and marginalized people; accelerated ecosystem degradation; depleted natural resources; and slowed progress towards development goals, poverty reduction and conflict mitigation.

1.3 Water Resources: Status and Availability

The distribution and availability of freshwater resources, through precipitation and runoff, can be erratic, with different areas of the globe receiving different quantities of water over any given year. There can be considerable variability between arid and humid climates and over wet and dry seasons. However, compounded yearly averages show significant variations in per capita water availability between countries.

Climate change will exacerbate the risks associated with variations in the distribution and availability of water resources. Increasing variability in precipitation patterns, which many countries have already begun to experience, are leading to direct and indirect effects on the whole of the hydrological cycle, with changes in runoff and aquifer recharge, and in water quality. In addition, higher water temperatures due both to warmer climates and increasing discharges of waste heat, are expected to exacerbate many forms of pollution. These include sediments, nutrients, dissolved organic carbon, pathogens, pesticides and salt, as well as thermal pollution, with possible negative impacts on ecosystems, human health, and water system reliability and operating costs.

Groundwater plays a substantial role in water supply, in ecosystem functioning and human well-being. Worldwide, 2.5 billion people depend solely on groundwater resources to satisfy their basic daily water needs, and hundreds of millions of farmers rely on groundwater to sustain their livelihoods and contribute to the food security of so many others. Groundwater reportedly provides drinking water to at least 50% of the global population and accounts for 43% of all water used for irrigation (Groundwater Governance, n.d.). Groundwater also sustains the baseflows of rivers and important aquatic ecosystems. Uncertainty over the availability of groundwater resources and their replenishment rates pose a serious challenge to their management and in particular to their ability to serve as a buffer to offset periods of surface water scarcity.

Groundwater supplies are diminishing, with an estimated 20% of the world's aquifers being over-exploited, leading to serious consequences such as land subsidence and saltwater intrusion in coastal areas. Groundwater levels are declining in several of the world's intensely used agricultural areas and around numerous mega-cities (Groundwater Governance, n.d.). In the Arabian Peninsula, freshwater withdrawal, as a percentage of internal renewable water resources, was estimated at 505% in 2011 (FAO AQUASTAT), with significant volumes of groundwater reserves being transboundary in nature.

Water availability is also affected by pollution. Most problems related to water quality are caused by intensive agriculture, industrial production, mining and untreated urban runoff and wastewater. Expansion of industrial agriculture has led to increases in fertilizer applications. These and other industrial water pollutants create environmental and health risks. Excessive loads of nitrogen and phosphate, the most common chemical contaminants in the world's freshwater resources, contribute to the eutrophication of freshwater and coastal marine ecosystems, creating 'dead zones' and erosion of natural habitats.

The human interference with phosphorus and nitrogen cycles is well beyond safe thresholds. Eutrophication of surface water and coastal zones is expected to increase almost everywhere until 2030. Thereafter, it may stabilize in developed countries, but is likely to continue to worsen in developing countries. Globally, the number of lakes with harmful algal blooms will increase by at least 20% until 2050. Phosphorus discharges will increase more rapidly than those of nitrogen and silicon, exacerbated by the rapid growth in the number of dams.

The disruption of ecosystems through unabated urbanization, inappropriate agricultural practices, deforestation and pollution is undermining the environment's capacity to provide basic water-related services (e.g. purification, storage). Degraded ecosystems can no longer regulate and restore themselves; they lose their resilience, further accelerating the decline in water quality and availability.

Global environmental degradation, including climate change, has reached a critical level with major ecosystems approaching thresholds that could trigger their massive collapse. This is a result of past failures to design decision-making mechanisms that would appropriately govern the global and national commons and the earth's shared natural resources. Despite efforts to create cooperation around environmental treaties and agreements, decisions directly affecting environmental issues are often taken outside of environmental policy circles. Any predominance of economic logic without the integration of social and environmental considerations, as it currently exists in many development approaches, means that long-term environmental objectives may be set aside in favour of short-term economic goals.

1.4 Constraints on Water Resources Management

The previous sections of this chapter have provided a summary view of the processes that drive increasing demands for water, their potential consequences and what these could mean for water resources. However, there are additional constraints that pose significant challenges to improving water resources management. These transcend any type of pressure-state-response analysis, yet they are tangible and merit a critical level of consideration when addressing water-related issues in the context of sustainable development.

1.4.1 Persistent Poverty

Persistent poverty is usually the result of a vicious cycle in which limited income converges with limited access to resources. Safe water and sanitation are precursors to health care, education and jobs (see Chapter 2). For the last 15 years, eradication of extreme poverty and hunger has been the number one priority under the MDGs. Nevertheless, as of 2012, 1.2 billon people still lived in extreme poverty, the majority located in slums, often lacking adequate drinking water and sanitation services. The global limit of ecological sustainability of water available for abstraction is reported to have been reached. Regionally, this limit has already been exceeded for about one-third of the human population and it will rise to about half by 2030. Apart from access to sanitation and clean drinking

water, the world's 850 million rural poor also lack access to water for agricultural production, which is usually their primary source of income. Without access to improved agricultural water management, poverty in these regions will persist.

Women and youth are disproportionately impacted both by water scarcity and the lack of safe drinking water, increasing the vulnerability associated with persistent poverty. Water policies are often based on generalized perspectives that lack gender perspectives and local knowledge. By failing to integrate gender considerations in water resources management and also in sectors such as agriculture, urban water supply, energy and industry, gender inequities will persist, preventing the adoption of innovative solutions that may be put forth by women.

. . .

1.4.2 Discrimination and Inequalities in Access to Drinking Water and Sanitation Services

Socio-economic inequalities, and the lack of policies to effectively address them, were among the main obstacles to the achievement of MDGs in general and improved access to sanitation and safe drinking water in particular. Many people around the globe including women, children, the elderly, indigenous peoples and people with disabilities have lower levels of access to safe drinking water, hygiene or sanitation facilities than other groups (see Chapter 5). While access to safe drinking water and sanitation is recognized as a basic human right, discrimination based on ethnicity, religion, economic class, social status, gender, age or physical abilities often restricts people from accessing land and water resources and related services. Such exclusion has long-term social and economic effects, as the disadvantaged are more likely to remain poor, lacking opportunities for education, employment and social engagement.

Population dynamics also affect access. High urbanization rates in many countries have not been matched by governments' ability to provide adequate drinking water and sanitation infrastructure and improved service delivery. Human migration from rural to urban areas is posing a continuous challenge to the provision of basic drinking water and sanitation services, especially in poor peri-urban and slum areas, as well as to public health, in particular to prevent outbreaks of cholera and other water- related diseases.

In the rural context, which require different systems to those generally found in urban settings, providing adequate drinking water and sanitation is also challenging. The lack of infrastructure and services means that many people do not have access to adequate sanitation and must rely on unsafe water supplies. The lack of access to safe drinking water coupled with other shortages of basic services, scarce resources and limited income-generating possibilities, can further entrench vulnerability.

1.4.3 Insufficient and Unsustainable Financing for Water Resources Management and Services

Water services remain rather low on the scale of policy priorities, despite well-documented contributions to human and economic development. When compared with other development sectors, particularly education and health, sanitation and drinking water services receive a relatively low priority for both official development assistance (ODA) and national expenditure. This under-prioritization of water directly contravenes a State's obligation to expend maximum available resources to promote the progressive realization of the human right to water and sanitation for all persons, without discrimination. Financing for water resources management is also usually a low priority, in spite of it being a cornerstone of economic growth.

In most countries, funding for water infrastructure comes from government allocations, although many developing countries still depend on external assistance to fund water resources management and utilities. This is neither adequate, nor sustainable. Most countries report that information required for adequate financial planning in the water services sector, such as information on users and their potential contributions, is insufficient. Costs of infrastructure operation and maintenance are often neglected or not well factored into water mobilization projects. As a result, many water systems are inadequately maintained, leading to damages, losses, unreliability, and decreasing quality and quantity of service to users. Financing is reported to be particularly inadequate for sanitation, with drinking water absorbing the majority of funding available particularly in developing countries. Financing for wastewater treatment is chronically neglected.

Despite persistent management obstacles relating to financing in the water sector, over 50% of countries low in the Human Development Index (HDI) have reported that financing for water resources development and management from government budgets and official development assistance have been increasing over the last 20 years.

1.4.4 Data and Information

Monitoring water availability, use and the related impacts, represents a massive and persistent challenge. Reliable and objective information about the state of water resources, their use and management is often poor, lacking or otherwise unavailable. Worldwide, water observation networks

provide incomplete and incompatible data on surface and groundwater quality and quantity, and no comprehensive information exists on wastewater generation and treatment. Various studies and assessments provide a snapshot of the state and use of water resources at a given time and place, but generally do not provide a broader, more complete picture of how different dimensions of water are changing over time in different parts of the world.

In the context of sustainable development, where water is often a key driver—and a potential limiting factor—for economic growth, human well-being and environmental health, this lack of information and knowledge creates barriers to cohesive policy formulation and sound decision-making on developmental objectives. For example, there are often too few reliable metrics on which to track the outcomes of water productivity improvement measures.

From an economic perspective, there is a need to couple data and information on water resources and their use with indicators of growth in various economic sectors in order to assess its role and contribution in terms of economic development, and to garner a better understanding of its consequences on the resource and different users. Similarly, quantifying water's role in maintaining healthy ecosystems is often limited to the determination of 'environmental flows' (i.e. the quantity and timing of water flows required to sustain freshwater ecosystems). Although an important and useful tool for managing freshwater ecosystems, environmental flows are often based on the requirements of certain indicator species and may not take full account of the interconnections between ecological systems and their impacts on economic and social development.

In terms of human well-being, much of the focus has been on monitoring access to safe water supply and sanitation services, which has in large part been driven by MDG target 7.C. Here too, data and information challenges persist, including the difficulty in translating the term 'safe' into a measurable metric. Water quality testing is still not available in most countries, making it difficult to determine if improved sources actually deliver 'safe' water to the user or whether the risks of contracting water-related diseases remain. Furthermore, most countries do not report on other aspects of access to 'safe' water such as the quantity available, possible security threats including as risks on the journey to fetch water, frequency and duration of access or supply, and water's potentially prohibitive cost.

Although real progress has been made, full compliance with the human right to access safe drinking water clearly requires something that is significantly better than many of the improved sources that people have to use. This situation also highlights the need to select target indicators based on good (and readily available) data, and to craft and implement good monitoring mechanisms. The MDG indicators focused on aggregate outcomes, which masked the fact that access improvements often did not reach the most vulnerable groups such as the elderly poor, people with disabilities, women and children. Indicators that disaggregate data by gender, age and social group pose both a challenge and an opportunity for the SDGs in the post-2015 development agenda.

While data availability and quality remain a concern, a limited number of core indicators pertaining to various aspects of water resources, services, uses and management in the context of sustainable development are available. Several of these indicators are used in this Report. They are compiled and presented in *Case studies and Indicators report*.

UNESCO or UN Educational, Scientific and Cultural Organization is a specialized agency of the United Nations. Headquartered in Paris, it is the "intellectual agency" of the United Nations and aims to increase collaboration among its 195 member states.

EXPLORING THE ISSUE

Can the Global Community Successfully Confront the Global Water Shortage?

Critical Thinking and Reflection

1. Do the key findings presented in the YES article represent real progress in increasing improved drinking water throughout the globe?
2. Do you believe that the root cause of insufficient safe drinking water is due to factors such as unsustainable development strategies and governing failures rather than simply increased population, urbanization, and increased food requirements globally?
3. Do you buy into the prediction that the world will need 60 percent more food and the developing world 100 percent more food by the year 2050?
4. Do you believe that "simply" increasing agricultural efficiency in the developing world will avert a future global water crisis?

Is There Common Ground?

Although many analysts differ over whether a global water *crisis* is already here, there is an overwhelming consensus that given population growth and the growth of affluence, future generations will face a water crisis. There is also an emerging belief among many that if the international community is willing to act, technologies are or will be there that will allow the world to do so. The lessons from another diminishing resource, oil, should be instructive here.

Additional Resources

Bigas, Harriet et al. (eds.), *The Global Water Crisis: Addressing an Urgent Security Issue* (United Nations University, 2012)

Caldecott, Julian, *Water: The Causes, Costs, and Failure of a Global Crisis* (Virgin Books, 2010)

The author spells out the history, science, economics, and politics behind the global water crisis.

Dellapenna, et al., "Thinking about the Future of Global Water Governance," *Ecology & Society* (vol. 18, September 2013)

The article focuses on the increasing severity of global water problems.

Salaam-Blyther, Tiaji, *Global Access to Clean Drinking Water and Sanitation: U.S. and International Programs* (Congressional Research Service, 2012)

This report describes current programs to create access to clean water and sanitation around the globe.

Solomon, Steven, *Water: The Epic Struggle for Wealth, Power, and Civilization* (Harper Perennial, 2011)

The book describes how water has replaced oil as the primary cause of global conflicts.

Tolentino Jr., Amado S., "Water Scarcity Comes of Age," *Environmental Policy & Law* (vol. 46, 2016)

The article discusses the global water problem in the context of global warming.

UN-Water, *The United Nations World Water Development Report 2016* (United Nations, 2016)

This report describes the nexus between water and the economy both at the global level and at regional levels.

United States Senate, *Domestic and Global Water Supply Issues* (United States Senate, 2012)

This report stresses that in 20 years world demand for fresh water will exceed supply by 40 percent.

World Water Council, *7th World Water Forum 2015: Final Report* (World Water Council, 2015)

This report describes the activities of the forum attended by all principal countries of the world.

Vörösmarty, C.J., et al., "What Scale for Water Governance?" *Science* (vol. 349, July 31, 2015)

The article focuses on replacing localized views of water governance with more global perspectives.

Zetland, David, *The End of Abundance: Economic Solutions to Water Scarcity* (Aguanomics Press, 2011)

The book provides an overview of the many aspects of the world of water.

Internet References . . .

International Association for Environmental Hydrology

www.hydroweb.com

Pacific Institute

www.worldwater.org

UN-Water

www.unwater.org

Water Conservancy Worldwide

www.lmtf.org

World Water Council

www.worldwtaercoincil.org

Selected, Edited, and with Issue Framing Material by:
James E. Harf, *Maryville University*
and
Mark Owen Lombardi, *Maryville University*

ISSUE

Is the Global Oil Crisis of the Last Half-Century Over?

YES: **Marc Lallanilla**, from "Peak Oil: Theory or Myth?" Livescience.com (2015)

NO: **Michael T. Klare**, from "Peak Oil Is Dead. Long Live Peak Oil!," TomDispatch.com (2014)

Learning Outcomes
After reading this issue, you will be able to: <ul><li>Describe the evolution of the term "peak oil."</li><li>Describe the view that the cost of oil affects the amount of its availability.</li><li>Understand how the demand for fossil fuels in many developing parts of the world affects global demand.</li><li>Understand the argument that the supply of oil is growing worldwide and could eventually outpace consumption</li><li>Understand a new related theory called techno-dynamism and how it relates to peak oil.</li><li>Describe the major challenges to the theory of techno-dynamism as it relates to peak oil</li></ul>

ISSUE SUMMARY

YES: In this Internet post, Marc Lallanilla suggests that oil production will not reach a peak but rather a plateau, which will continue for some time before slowly declining.

NO: In this Internet post, Michael T. Klare argues that taking into account both political and physical constraints, peak oil "remains in our future," although there will be a "gradual disappearance" of what the author calls "easy" oil.

Oil prices reached their lowest level in 13 years in January 2016 when its price of $26 a barrel was more than $100 lower than its 2014 price. By February the price of a gallon of gas at the pump had dropped to well under $2, although by June it had inched up to around $2.25. U.S. oil stockpiles reached unprecedented high levels, with the Saudis stepping in to ensure such attainable levels. Yet just a few years ago, gas prices had passed the $3 a gallon level, and within a year, the unthinkable had occurred, as prices edged toward and then passed $4 a gallon. Had the oil crisis peaked? Is the crisis simply artificial, that is, created by supply and demand forces conspiring to make it so?

The sharp increase in the cost of oil a few years ago had coincided with a wave of political unrest sweeping the Middle East and North Africa. Observed first in Tunisia, it quickly spread to Yemen, Egypt, Bahrain, and eventually to Libya. Although these countries represent a small percentage of oil-exporting nations and although Saudi Arabia promised to make up the difference in oil production, the cost of a gallon of gas spiraled upward in conjunction with political turmoil. The cost of a gallon at the pump was also correlated with the rise in the price of a barrel of oil at the source. In early March 2010, the price of a barrel of oil was close to $120, for example, about $25 more than it should have been based on simple supply and demand. Was politics now playing an important role in the volatility of oil prices?

This was not the first oil crisis in recent times, however, as the beginning of the new millennium had witnessed an oil crisis, the third major crisis in the last 30 years (1972–2002 and 1979 were the dates of earlier problems), and the fourth crisis appeared in 2008 as prices passed $3 a gallon for a while. The crisis of 2000 manifested itself in the United States via much higher gasoline prices and in Europe via both rising prices and shortages at the pump. Both were caused by the inability of national distribution systems to adjust to the Organization of Petroleum Exporting Countries' (OPEC's) changing production levels. The 2000 panic eventually subsided but reappeared in 2005 in the wake of the uncertainty surrounding the Iraqi war and the war on terrorism. Four major crises and a minor one in less than 40 years thus have characterized the oil issue.

These earlier major fuel crises are discrete episodes in a much larger problem facing the human race, particularly the industrial world. That is, oil, the earth's current principal source of energy, is a finite resource that ultimately will be totally exhausted. And unlike earlier energy transitions, where a more attractive source invited a change (such as from wood to coal and from coal to oil), the next energy transition is being forced upon the human race in the absence of an attractive alternative. In short, we are being pushed out of our almost total reliance on oil toward a new system with a host of unknowns. What will the new fuel be? Will it be from a single source or some combination? Will it be a more attractive source? Will the source be readily available at a reasonable price, or will a new cartel emerge that controls much of the supply? Will its production and consumption lead to major new environmental consequences? Will it require major changes to our lifestyles and standards of living? When will we be forced to jump into using this new source?

The 2010 crisis was different from the earlier crises in the 1970s in one way in that the earlier crises had the imprint of OPEC vindictive behavior to the West on it. Today, the Middle East is dominated by Arab Spring, the phrase coined to capture the series of rebellions and revolutions that have been spreading from Tunisia to Bahrain, and affecting many countries in between. Also within the region, one major oil player, Iraq, struggles to create a stable government while reentering the global oil market. But oil prices rose nonetheless as oil speculators cornered the market, forcing prices upward. The recent spike in the price at the pump is one example of the effect of such speculation.

But the 2010 and 2014 crises were also similar to other oil crises in that price increases do not reflect the simple supply and demand equation but, instead, are in part also a function of external unrelated forces, be they vindictive behavior of a cartel of nations or greedy behavior of individual investors. And for the latter, high oil prices is the goal, not certainty of supply. This is unfortunate for large Western consumers of oil as well as the largest producers such as Saudi Arabia. As long as consuming countries must rely on autocratic and sometimes unfriendly national regimes for their oil, fear, volatility, and uncertainly will characterize the scene.

Since these consumers cannot dramatically increase their own internal supply of oil under present technology, the answer may lie in alternative sources of energy. Before considering new sources of fuel, other questions need to be asked. Are the calls for a viable alternative to oil premature? Are we simply running scared without cause? Did we learn the wrong lessons from the earlier energy crises? More specifically, were these crises artificially created or a consequence of the actual physical unavailability of the energy source? Have these crises really been about running out of oil globally, or were they due to other phenomena at work, such as poor distribution planning by oil companies or the use of oil as a political weapon by oil-exporting countries?

For well over half a century now, Western oil-consuming countries have been predicting the end of oil. Using a model known as Hubbert's Curve (named after a U.S. geologist who designed it in the 1930s), policymakers have predicted that the world would run out of oil at various times; the most recent prediction is that oil will run out a couple of decades from now. Simply put, the model visualizes all known available resources and the patterns of consumption on a timeline until the wells run dry. Despite such prognostication, it was not until the crisis of the early 1970s that national governments began to consider ways of both prolonging the oil system and finding a suitable replacement. Prior to that time, governments as well as the private sector encouraged energy consumption. "The more, the merrier" was an oft-heard refrain. Increases in energy consumption were associated with economic growth. After Europe recovered from the devastation of World War II, for example, every 1 percent increase in energy consumption brought a similar growth in economic output. To the extent that governments engaged in energy policy making, it was designed solely to encourage increased production and consumption. Prices were kept low and the energy was readily available. Policies making energy distribution systems more efficient and consumption patterns both more efficient and lowered were almost nonexistent.

Yet, when one reads the UN assessment of foreseeable world energy supplies (Hisham Khatib et al., World Energy Assessment: Energy and the Challenge of Sustainability

(United Nations Development Programme, 2002), a sobering message appears. Do not panic just yet. The study revealed no serious energy shortage during the first half of the twenty-first century. In fact, the report suggested that oil supply conditions had actually improved since the crises of the 1970s and early 1980s. The report went further in its assessment, concluding that fossil fuel reserves are "sufficient to cover global requirements throughout this century, even with a high-growth scenario." Another source suggesting no shortages for some time is "The Myth of the Oil Crisis: Overcoming the Challenges of depletion, Geopolitics, and Global Warming" (Robin M. Mills, Praeger, 2008). Francis R. Stabler argued in "The Pump Will Never Run Dry" (The Futurist, November 1998), that technology and free enterprise will combine to allow the human race to continue its reliance on oil far into the future. For Stabler, the title of his article tells the reader everything. The pump will not run dry!

More recently, the optimists have become bolder, empowered by adjustments in future estimates and the pushing back of the date on which the world will reach "peak oil." One such observer is Robin Mills, CEO of Qamar Energy. He points to the return of Iran as a global supplier of oil as a key reason for such optimism.

These modern-day optimists had an earlier supporter in Julian L. Simon, who argued in his The Ultimate Resource 2 (1996) that even God may not know exactly how much oil and gas are "out there." Chapter 11 of Simon's book is titled "When Will We Run Out of Oil? Never!" Another supporter of Stabler has been Bjørn Lomborg in The Skeptical Environmentalist: Measuring the Real State of the World (Cambridge University Press, 2001). Arguing that the world seems to find more fossil energy than it consumes, he concludes that "we have oil for at least 40 years at present consumption, at least 60 years' worth of gas, and 230 years' worth of coal." Simon and Lomborg are joined by Michael C. Lynch in a published article on the Web under global oil supply (msn.com) titled "Crying Wolf: Warnings about Oil Supply."

Today, the search for an alternative to oil still continues despite more optimistic predictions about future oil brought on by lowed prices. Nuclear energy, once thought to be the answer, may play a future role, but at a reduced level. Both water power and wind power remain possibilities, as do biomass, geo-thermal energy, and solar energy. Many also believe that the developed world is about to enter the hydrogen age to meet future energy needs. The question before us, therefore, is whether the international community has the luxury of some time before all deposits of oil are exhausted.

The two selections for this issue suggest different answers to the question of should we continue to rely on oil. In the YES selection, Marc Lallanilla provides arguments, albeit cautionary, against the idea of peak oil, suggesting that determining such a point in time is dependent upon "an ever-changing set of assumptions and variables." In the NO selection, Michael T. Klare argues not so fast when predicting the demise of peak oil. Increases in unconventional production will likely offset decreases in conventional production. When this will eventually affect peak oil depends on which scenario laid out by Klare will hold true.

YES ↵

Marc Lallanilla

Peak Oil: Theory or Myth?

Peak oil—the point in time when domestic or global oil production peaks and begins to forever decline—has been looming on the horizon for decades. Countless research reports, government studies and oil industry analyses have tried to pin down the exact year when peak oil will occur, to no avail.

The stakes are undeniably high: Much of human civilization is now inextricably linked to a readily available supply of inexpensive oil and petroleum products. From heating, electricity production and transportation to cosmetics, medicines and plastic bags, modern life runs on oil.

Peak Oil Theory: The Early Years

In October 1973, the world was roiled by the OPEC oil embargo. Members of the Organization of Petroleum Exporting Countries agreed to stop exporting oil to the United States, much of Western Europe, Japan and several other nations.

Though the oil embargo only lasted five months (until March 1974), it sent shock waves throughout the industrialized world and underscored our utter dependence on petroleum. Many government leaders and academic institutions realized, even after the embargo ended, that the global oil economy couldn't last forever.

Years earlier, in 1956, geologist M. King Hubbert at Shell Oil Company (and later at the U.S. Geological Survey) predicted that oil production in the lower 48 U.S. states would peak sometime around 1970.

Though his comments generated much controversy, he was later vindicated when institutions such as the National Academy of Sciences and the Energy Information Agency (EIA) confirmed that his now-famous bell curve predicting the 1970 peak was correct, despite much rosier predictions made by industry and government analysts.

'Hubbert got a lot of notoriety in his lifetime for correctly predicting U.S. oil would peak in 1970," said Alan Carroll, a geologist at the University of Wisconsin-Madison and author of "Geofuels: Energy and Earth" (Cambridge University Press, 2015). "That same logic was extended to

world oil production, and there have been many predictions that global production will reach a peak, none of which have happened yet," Carroll said.

When Hubbert turned his sights to global oil production in 1974, his report was equally disturbing, especially in light of the OPEC oil embargo: He predicted that the world's peak oil production would occur in 1995, assuming that current production and use trends continue.

In 1988, Hubbert said in an interview, "We are in a crisis in the evolution of human society. It's unique to both human and geologic history. It has never happened before and it can't possibly happen again. You can only use oil once."

Does Peak Oil Even Exist?

Since Hubbert introduced the concept of peak oil, countless forecasters from every corner of the industrial, governmental and academic worlds have tried to substantiate or refute Hubbert's prediction.

Geoscientist Kenneth S. Deffeyes, author of "When Oil Peaked" (Hill and Wang, 2010), asserted that peak oil happened on Thanksgiving Day 2005. Meanwhile, petroleum geologist Colin Campbell, a founder of the Association for the Study of Peak Oil (ASPO), once estimated that peak oil had occurred around 2010, but his views have shifted somewhat as new data have become available.

The trouble is, determining when peak oil will occur, if it already has occurred, or if it will happen at all, are all dependent on an ever-changing set of assumptions and variables.

"The basic assumption of peak oil analysis is that you have prior knowledge of what the available reserves are, and in fact we do not," Carroll said.

Reserves are the known amount of oil that can be extracted given present-day prices and present-day technology, Carroll explained. But peak oil also depends on oil prices and available technology. For example, hydraulic fracturing, aka fracking, has opened up numerous oil fields in areas that were once considered played out or too costly to develop.

As a result of expanded fracking production, places like North Dakota—home of the Bakken formation of oil-bearing shale rock—are now experiencing an oil boom, and are likely to shift the global energy picture in dramatic ways over the next decade.

Thanks to fracking, instead of resembling a bell curve, U.S. oil production is back on the rise. Through the first half of 2014, the United States produced an average 8.3 million barrels a day. "We may have a second peak [of oil production] in the U.S.," Carroll said. "Maybe Hubbert wasn't right."

Oil Supply and Oil Demand

Demand for fossil fuels is another critical factor in the debate over peak oil. Developing countries like China, India and Brazil have become big markets for oil (and other fossil fuels such as coal). As these enormous markets expand—and as the global population continues to increase beyond the 7 billion mark—the demand for oil increases.

And as the demand for fossil fuels like oil increases, the supply of these resources dwindles, or so some have argued. But the amount of available oil is not uniform. For instance, reserve estimates may be inaccurate. In California's San Joaquin Valley, production has well exceeded its initial 800 million barrel estimate, with 2.5 billion barrels already drilled and production continuing to grow through secondary recovery efforts, Carroll said.

Oil industry analysts often describe oil resources in terms of conventional and unconventional oil. Conventional oil describes oil that's available through more traditional, less expensive technologies like the oil wells that dot landscapes from West Texas to Saudi Arabia.

Unconventional oil, however, isn't readily or cheaply available. Sources like the tar oil sands of Canada, shale oils from the Bakken formation, coal oil (liquefied fuel from coal) and biofuels (ethanol, biodiesel and other liquid fuels from plants like switchgrass) are expected to form an increasingly important resource in the 21st century.

"We might hit a peak in terms of conventional oil, but coming in behind that are the oil sands, oil shales, the methane hydrates, and they will prevent consumption from simply dropping in a peak fashion," Carroll said.

Costs and Benefits of Unconventional Oil

EIA administrator Adam Sieminski points out a crucial issue in what makes oil available—its cost. When the price of oil reaches a certain point, it becomes profitable to drill in areas and in ways that would not be profitable if oil were too cheap.

"The question is not when you will run out of oil, but when you will run out of money to get the oil," Carroll said.

Deepwater drilling, for example, is an expensive and risky drilling procedure that usually takes place miles offshore in waters more than 500 feet (152 meters) deep. Roughly 80 percent of the oil produced in the Gulf of Mexico comes from deepwater wells, according to the U.S. Energy Information Administration.

The risks of deepwater drilling—and all unconventional oil development—were thrown into sharp relief in 2010, when BP's Deepwater Horizon well exploded, killing 11 people and spilling an estimated 205 million gallons (776 million liters) of oil into the Gulf of Mexico. It was the largest oil spill in U.S. history, eclipsing even the Exxon Valdez oil spill of 1989.

Despite the high costs and the risks, unconventional oil exploration and drilling makes sense when the price of oil is high—and according to energy consultants Wood Mackenzie, spending on deepwater drilling should grow from $43 billion in 2012 to $114 billion in 2022.

Thus, the amount of oil that's available for refining isn't fixed, even though the overall quantity of oil on Earth is finite.

A Peak, or a Plateau?

In a much-quoted (and much-criticized) report from 2006, Cambridge Energy Research Associates (CERA) presented an analysis that found 3.74 trillion barrels of oil available—far more than the 1.2 trillion barrels estimated by some earlier analyses.

Their research suggested that oil production won't simply reach a peak, followed by a precipitous decline. Instead, "global production will eventually follow an 'undulating plateau' for one or more decades before declining slowly."

From their research, CERA also determined that "the global production profile will not be a simple logistic or bell curve postulated by geologist M. King Hubbert, but it will be asymmetrical—with the slope of decline more gradual ... it will be an undulating plateau that may well last for decades."

Their analysis calls into question the very idea of "peak oil" as a useful model for energy forecasting or governmental policy: "The 'peak oil' theory causes confusion and can lead to inappropriate actions and turn attention away from the real issues," CERA director Peter M. Jackson said. "Oil is too critical to the global economy to allow fear to replace careful analysis about the very real challenges with delivering liquid fuels to meet the needs of growing economies."

Whether oil production peaks or plateaus, one underlying fact drives the issue: "World production of conventional oil will reach a maximum and decline thereafter," according to a 2005 in-depth analysis co-authored by Robert L. Hirsch and commissioned by the U.S. Department of Energy (widely referred to as "the Hirsch report").

"Prediction of the peaking is extremely difficult because of geological complexities, measurement problems, pricing variations, demand elasticity, and political influences," the report concludes. "Peaking will happen, but the timing is uncertain."

Other scientists, such as Carroll, question whether a true peak will ever be reached, given the remarkable quantities of carbon stored in the planet's crust. "There's potential for an enormous increase in quantity if one is willing to go for lower quality," he said.

Regardless of when or how oil production begins to decline, according to the Hirsch report, its effects will be global and will be accompanied by dramatic social, political, economic and environmental upheaval.

Mitigation of these effects—through conservation and development of alternative energy sources—will require advance planning and "an intense effort over decades," according to the report. "There will be no quick fixes. Even crash [mitigation] programs will require more than a decade to yield substantial relief."

The final word on peak oil may belong to Campbell, who was among the first to foresee its arrival: "The Stone Age did not end because we ran out of stone, but because bronze and iron proved to be better substitutes," he wrote in 2001. "Firewood gave way to coal; and coal to oil and gas, not because they ran out or went into short supply but because the substitutes were cheaper and more efficient. But now, oil production does reach a peak without sight of a preferred substitute."

MARC LALLANILLA is a journalist, editor and online producer. Previously he worked as a producer at ABC News. He has a long list of clients.

Michael T. Klare

NO

Peak Oil Is Dead: Long Live Peak Oil!

Among the big energy stories of 2013, "peak oil"—the once-popular notion that worldwide oil production would soon reach a maximum level and begin an irreversible decline—was thoroughly discredited. The explosive development of shale oil and other unconventional fuels in the United States helped put it in its grave.

As the year went on, the eulogies came in fast and furious. "Today, it is probably safe to say we have slayed 'peak oil' once and for all, thanks to the combination of new shale oil and gas production techniques," declared Rob Wile, an energy and economics reporter for Business Insider. Similar comments from energy experts were commonplace, prompting an R.I.P. headline at Time.com announcing, "Peak Oil is Dead."

Not so fast, though. The present round of eulogies brings to mind Mark Twain's famous line: "The reports of my death have been greatly exaggerated." Before obits for peak oil theory pile up too high, let's take a careful look at these assertions. Fortunately, the International Energy Agency (IEA), the Paris-based research arm of the major industrialized powers, recently did just that—and the results were unexpected. While not exactly reinstalling peak oil on its throne, it did make clear that much of the talk of a perpetual gusher of American shale oil is greatly exaggerated. The exploitation of those shale reserves may delay the onset of peak oil for a year or so, the agency's experts noted, but the long-term picture "has not changed much with the arrival of [shale oil]."

The IEA's take on this subject is especially noteworthy because its assertion only a year earlier that the U.S. would overtake Saudi Arabia as the world's number one oil producer sparked the "peak oil is dead" deluge in the first place. Writing in the 2012 edition of its *World Energy Outlook*, the agency claimed not only that "the United States is projected to become the largest global oil producer" by around 2020, but also that with U.S. shale production and Canadian tar sands coming online, "North America becomes a net oil exporter around 2030."

That November 2012 report highlighted the use of advanced production technologies—notably horizontal drilling and hydraulic fracturing ("fracking")—to extract oil and natural gas from once inaccessible rock, especially shale. It also covered the accelerating exploitation of Canada's bitumen (tar sands or oil sands), another resource previously considered too forbidding to be economical to develop. With the output of these and other "unconventional" fuels set to explode in the years ahead, the report then suggested, the long awaited peak of world oil production could be pushed far into the future.

The release of the 2012 edition of *World Energy Outlook* triggered a global frenzy of speculative reporting, much of it announcing a new era of American energy abundance. "Saudi America" was the headline over one such hosanna in the *Wall Street Journal*. Citing the new IEA study, that paper heralded a coming "U.S. energy boom" driven by "technological innovation and risk-taking funded by private capital." From then on, American energy analysts spoke rapturously of the capabilities of a set of new extractive technologies, especially fracking, to unlock oil and natural gas from hitherto inaccessible shale formations. "This is a real energy revolution," the *Journal* crowed.

But that was then. The most recent edition of *World Energy Outlook*, published this past November, was a lot more circumspect. Yes, shale oil, tar sands, and other unconventional fuels will add to global supplies in the years ahead, and, yes, technology will help prolong the life of petroleum. Nonetheless, it's easy to forget that we are also witnessing the wholesale depletion of the world's existing oil fields and so all these increases in shale output must be balanced against declines in conventional production. Under ideal circumstances—high levels of investment, continuing technological progress, adequate demand and prices—it might be possible to avert an imminent peak in worldwide production, but as the latest IEA report makes clear, there is no guarantee whatsoever that this will occur.

Inching Toward the Peak

Before plunging deeper into the IEA's assessment, let's take a quick look at peak oil theory itself.

As developed in the 1950s by petroleum geologist M. King Hubbert, peak oil theory holds that any individual oil field (or oil-producing country) will experience a high rate of production growth during initial development, when drills are first inserted into a oil-bearing reservoir. Later, growth will slow, as the most readily accessible resources have been drained and a greater reliance has to be placed on less productive deposits. At this point—usually when about half the resources in the reservoir (or country) have been extracted—daily output reaches a maximum, or "peak," level and then begins to subside. Of course, the field or fields will continue to produce even after peaking, but ever more effort and expense will be required to extract what remains. Eventually, the cost of production will exceed the proceeds from sales, and extraction will be terminated.

For Hubbert and his followers, the rise and decline of oil fields is an inevitable consequence of natural forces: oil exists in pressurized underground reservoirs and so will be forced up to the surface when a drill is inserted into the ground. However, once a significant share of the resources in that reservoir has been extracted, the field's pressure will drop and artificial means—water, gas, or chemical insertion—will be needed to restore pressure and sustain production. Sooner or later, such means become prohibitively expensive.

Peak oil theory also holds that what is true of an individual field or set of fields is true of the world as a whole. Until about 2005, it did indeed appear that the globe was edging ever closer to a peak in daily oil output, as Hubbert's followers had long predicted. (He died in 1989.) Several recent developments have, however, raised questions about the accuracy of the theory. In particular, major private oil companies have taken to employing advanced technologies to increase the output of the reservoirs under their control, extending the lifetime of existing fields through the use of what's called "enhanced oil recovery," or EOR. They've also used new methods to exploit fields once considered inaccessible in places like the Arctic and deep oceanic waters, thereby opening up the possibility of a most un-Hubbertian future.

In developing these new technologies, the privately owned "international oil companies" (IOCs) were seeking to overcome their principal handicap: most of the world's "easy oil"—the stuff Hubbert focused on that comes gushing out of the ground whenever a drill is inserted—has already been consumed or is controlled by state-owned "national oil companies" (NOCs), including Saudi Aramco, the National Iranian Oil Company, and the Kuwait National Petroleum Company, among others. According to the IEA, such state companies control about 80% of the world's known petroleum reserves, leaving relatively little for the IOCs to exploit.

To increase output from the limited reserves still under their control—mostly located in North America, the Arctic, and adjacent waters—the private firms have been working hard to develop techniques to exploit "tough oil." In this, they have largely succeeded: they are now bringing new petroleum streams into the marketplace and, in doing so, have shaken the foundations of peak oil theory.

Those who say that "peak oil is dead" cite just this combination of factors. By extending the lifetime of existing fields through EOR and adding entire new sources of oil, the global supply can be expanded indefinitely. As a result, they claim, the world possesses a "relatively boundless supply" of oil (and natural gas). This, for instance, was the way Barry Smitherman of the Texas Railroad Commission (which regulates that state's oil industry) described the global situation at a recent meeting of the Society of Exploration Geophysicists.

Peak Technology

In place of peak oil, then, we have a new theory that as yet has no name but might be called techno-dynamism. There is, this theory holds, no physical limit to the global supply of oil so long as the energy industry is prepared to, and allowed to, apply its technological wizardry to the task of finding and producing more of it. Daniel Yergin, author of the industry classics, *The Prize* and *The Quest*, is a key proponent of this theory. He recently summed up the situation this way: "Advances in technology take resources that were not physically accessible and turn them into recoverable reserves." As a result, he added, "estimates of the total global stock of oil keep growing."

From this perspective, the world supply of petroleum is essentially boundless. In addition to "conventional" oil—the sort that comes gushing out of the ground—the IEA identifies six other potential streams of petroleum liquids: natural gas liquids; tar sands and extra-heavy oil; kerogen oil (petroleum solids derived from shale that must be melted to become usable); shale oil; coal-to-liquids (CTL); and gas-to-liquids (GTL). Together, these "unconventional" streams could theoretically add several trillion barrels of potentially recoverable petroleum to the global supply, conceivably extending the Oil Age hundreds of years into the future (and in the process, via climate change, turning the planet into an uninhabitable desert).

But just as peak oil had serious limitations, so, too, does techno-dynamism. At its core is a belief that rising world oil demand will continue to drive the increasingly costly investments in new technologies required to exploit the remaining hard-to-get petroleum resources. As suggested in the 2013 edition of the IEA's *World Energy Outlook*, however, this belief should be treated with considerable skepticism.

Among the principal challenges to the theory are these:

1. *Increasing Technology Costs*: While the costs of developing a resource normally decline over time as industry gains experience with the technologies involved, Hubbert's law of depletion doesn't go away. In other words, oil firms invariably develop the easiest "tough oil" resources first, leaving the toughest (and most costly) for later. For example, the exploitation of Canada's tar sands began with the strip-mining of deposits close to the surface. Because those are becoming exhausted, however, energy firms are now going after deep-underground reserves using far costlier technologies. Likewise, many of the most abundant shale oil deposits in North Dakota have now been depleted, requiring an increasing pace of drilling to maintain production levels. As a result, the IEA reports, the cost of developing new petroleum resources will continually increase: up to $80 per barrel for oil obtained using advanced EOR techniques, $90 per barrel for tar sands and extra-heavy oil, $100 or more for kerogen and Arctic oil, and $110 for CTL and GTL. The market may not, however, be able to sustain levels this high, putting such investments in doubt.

2. *Growing Political and Environmental Risk*: By definition, tough oil reserves are located in problematic areas. For example, an estimated 13% of the world's undiscovered oil lies in the Arctic, along with 30% of its untapped natural gas. The environmental risks associated with their exploitation under the worst of weather conditions imaginable will quickly become more evident—and so, faced with the rising potential for catastrophic spills in a melting Arctic, expect a commensurate increase in political opposition to such drilling. In fact, a recent increase has sparked protests in both Alaska and Russia, including the much-publicized September 2013 attempt by activists from Greenpeace to scale a Russian offshore oil platform—an action that led to their seizure and arrest by Russian commandos. Similarly, expanded fracking operations have provoked a steady increase in anti-fracking activism. In response to such protests and other factors, oil firms are being forced to adopt increasingly stringent environmental protections, pumping up the cost of production further.

3. *Climate-Related Demand Reduction*: The techno-optimist outlook assumes that oil demand will keep rising, prompting investors to provide the added funds needed to develop the technologies required. However, as the effects of rampant climate change accelerate, more and more polities are likely to try to impose curbs of one sort or another on oil consumption, suppressing demand—and so discouraging investment. This is already happening in the United States, where mandated increases in vehicle fuel-efficiency standards are expected to significantly reduce oil consumption. Future "demand destruction" of this sort is bound to impose a downward pressure on oil prices, diminishing the inclination of investors to finance costly new development projects.

Combine these three factors, and it is possible to conceive of a "technology peak" not unlike the peak in oil output originally envisioned by M. King Hubbert. Such a techno-peak is likely to occur when the "easy" sources of "tough" oil have been depleted, opponents of fracking and other objectionable forms of production have imposed strict (and costly) environmental regulations on drilling operations, and global demand has dropped below a level sufficient to justify investment in costly extractive operations. At that point, global oil production will decline even if supplies are "boundless" and technology is still capable of unlocking more oil every year.

Peak Oil Reconsidered

Peak oil theory, as originally conceived by Hubbert and his followers, was largely governed by natural forces. As we have seen, however, these can be overpowered by the application of increasingly sophisticated technology. Reservoirs of energy once considered inaccessible can be brought into production, and others once deemed exhausted can be returned to production; rather than being finite, the world's petroleum base now appears virtually inexhaustible.

Does this mean that global oil output will continue rising, year after year, without ever reaching a peak? That appears unlikely. What seems far more probable is that we will see a slow tapering of output over the next decade or two as costs of production rise and climate change—along with opposition to the path chosen by the energy giants—gains momentum. Eventually, the forces tending to reduce supply will overpower those favoring higher output, and a peak in production will indeed result, even if not due to natural forces alone.

Such an outcome is, in fact, envisioned in one of three possible energy scenarios the IEA's mainstream experts lay out in the latest edition of *World Energy Outlook*. The first assumes no change in government policies over the next 25 years and sees world oil supply rising from 87 to 110 million barrels per day by 2035; the second assumes some effort to curb carbon emissions and so projects output reaching "only" 101 million barrels per day by the end of the survey period.

It's the third trajectory, the "450 Scenario," that should raise eyebrows. It assumes that momentum develops for a global drive to keep greenhouse gas emissions below 450 parts per million—the maximum level at which it might be possible to prevent global average temperatures from rising above 2 degrees Celsius (and so cause catastrophic climate effects). As a result, it foresees a peak in global oil output occurring around 2020 at about 91 million barrels per day, with a decline to 78 million barrels by 2035.

It would be premature to suggest that the "450 Scenario" will be the immediate roadmap for humanity, since it's clear enough that, for the moment, we are on a highway to hell that combines the IEA's first two scenarios. Bear in mind, moreover, that many scientists believe a global temperature increase of even 2 degrees Celsius would be enough to produce catastrophic climate effects. But as the effects of climate change become more pronounced in our lives, count on one thing: the clamor for government action will grow more intense, and so eventually we're likely to see some variation of the 450 Scenario take shape. In the process, the world's demand for oil will be sharply constricted, eliminating the incentive to invest in costly new production schemes.

The bottom line: global peak oil remains in our future, even if not purely for the reasons given by Hubbert and his followers. With the gradual disappearance of "easy" oil, the major private firms are being forced to exploit increasingly tough, hard-to-reach reserves, thereby driving up the cost of production and potentially discouraging new investment at a time when climate change and environmental activism are on the rise.

Peak oil is dead! Long live peak oil!

MICHAEL T. KLARE is a professor of peace and world security studies at Hampshire College. He is the author of *The Race for What's Left* (Picador, 2012) and *Blood and Oil* (Holt Paperbacks, 2005).

EXPLORING THE ISSUE

Is the Global Oil Crisis of the Last Half-Century Over?

Critical Thinking and Reflection

1. How comfortable are you in the fact that there is disagreement over when and even whether the world will see peak oil?
2. Do you believe that the proper phrase to use when discussing the future level of available oil is an oil plateau rather than an oil peak?
3. Do you believe that lowered oil prices have as much of a negative effect on the world as many people believe?
4. Can you envisage your lives changing in any way 20 years from now as a consequence of the changing power structure in the oil market?
5. Does the theory of techno-dynamism seem plausible as it relates to peak oil, given how the world has observed technological advances in the past prolonging an adequate supply of oil?

Is There Common Ground?

Since oil is a finite resource, the planet will obviously run out of the resource someday. No responsible analyst denies this scenario. There is also recognition by many in the field that one great unknown factor is the extent to which new future technologies will extend the life of oil as the dominant resource, and if so, for how long? Modern technologies have trumped earlier predictions several times and there is little reason to believe that they may not play a role in the future. There is also strong consensus that no single alternative energy source is poised to replace oil yet. And the recent nuclear energy disaster in Japan only emphasizes this point.

Additional Resources

Crane, Hewitt, A Cubic Mile of Oil: Realities and Options for Averting the Looming Global Energy Crisis (Oxford University Press, 2010). This book conceptualizes future global energy use, introducing the concept of one cubic mile of oil (CMO) as a measure. It argues that annual consumption is currently 3.0 CMOs but by midcentury, it will be between 6 and 9 CMOs.

Goodman, Leah McGrath, The Asylum: The Renegades Who Hijacked the World's Oil Market (William Morrow, 2011). This book suggests that the price of oil is not controlled by OPEC but by a group of a few hundred speculators in Manhattan.

Gorelick, Steven M., Oil Panic and the Global Crisis: Predictions and Myths (Wiley-Blackwell, 2009). This book exposes the myths on both sides about global oil. It argues that although the supply of oil is not infinite, the ultimate size of oil reserves is poorly known.

Hartnady, Chris J.H., "Traversing the Global Oil Summit," South African Journal of Science (January/February 2012). The article shows the data for world oil production in mid-2012.

Koerth-Baker, Maggie, Before the Lights Go Out: Conquering the Energy Crisis Before It Conquers Us (Wiley, 2012). The author suggests that during the next 20 years we will be forced to radically change our energy systems that have shaped us for the past 100 years.

Library of Congress, "U.S. and Global Oil Markets," from "U.S. Offshore Oil and Gas Resources: Prospects and Processes," Congressional Research Service Report (April 26, 2010). This article reveals trends in the U.S. oil demand and production for the 2000s.

Lowery, John, Life Without Oil and Other Fossil Fuels (John Lowery, 2013). This book examines the alternatives to fossil fuels and presents an optimistic future.

Mills, Robin M., The Myth of the Oil Crisis: Overcoming the Challenges of Depletion, Geopolitics, and Global Warming (Praeger, 2008). This book by an oil insider purports to debunk myths about global oil, suggesting that enough oil exists for decades to come.

Rasizade, Alec, "The End of Cheap Oil," Contemporary Review (Autumn, 2008). This article examines the recent history of oil pricing, coming to the conclusion that the major oil and gas producers now dominate the global economic situation.

Rogoff, Kenneth, "What's Behind the Drop in Oil Prices?" World Economic Forum, (March 2, 2016). This short piece discusses why oil prices have dropped and its impact on the economy.

Ruppert, Michael, Confronting Collapse: The Crisis of Energy and Money in a Post Peak Oil World (Chelsea Green Publishing, 2009). This book describes the relationship of oil shortages and pricing. It suggests that the world is on the verge of collapse as a consequence. It then lays out a 25-point plan of action to avert it.

Sorrell, Steve, et al., "Shaping the Global Oil Peak: A Review of the Evidence on Field Sizes, Reserve Growth, Decline Rates and the Depletion Rates," Energy (January 2012). This article summarizes and evaluates the evidence on the four issues in the title of the article.

Strahan, Davis, "Still on a Slippery Slope," New Scientist (Vol 215, 2012). The article argues in favor of peak oil.

Whipple, Tom, "Peak Oil," Bulletin of the Atomic Scientists (November/ December, 2008). This article addresses the debate over when oil production will begin to fall, thus forcing transition to other energy sources.

Yergin, Daniel, *The Quest: Energy, Security, and the remaking of the Modern World* (Penguin Books, 2012)

Zervos, Sara, "Saudi Arabia, Shale & Iran: Everything You Need to Know about the Oil Crisis" (Forbes, January 26, 2016)

Internet References . . .

American Petroleum Institute

www.api.org

Hubbert Peak

www.hubbertpeak.com

International Energy Agency (IEA)

www.worldenergyoutlook.org

Plantforlife

www.plantforlife.com

World Energy Council

www.worldenergycouncil.org

World Oil

www.worldoil

Selected, Edited, and with Issue Framing Material by:
James E. Harf, *Maryville University*
and
Mark Owen Lombardi, *Maryville University*

ISSUE

Is the Paris Climate Change Agreement a Good Deal?

YES: European Commission, from "The Road from Paris: Assessing the Implications of the Paris Agreement and Accompanying the Proposal for a Council Decision on the Signing, on behalf of the European Union, of the Paris Agreement Adopted Under the United Nations Framework Convention on Climate Change," Communication from the Commission to the European Parliament and the Council (2016)

NO: Rupert Darwall, from "Paris: The Treaty That Dare Not Speak Its Name," *National Review* (2015)

Learning Outcomes
After reading this issue, you will be able to:
• Describe the main features of the Paris climate change agreement.
• Understand the potential historical significance of the agreement.
• Understand the arguments of those climate change activists that the agreement does not go far enough.
• Understand the arguments of those who believe that the agreement has major flaws.
• Gain an understanding of the extent to which scientists are divided on the issue of global warming.

ISSUE SUMMARY

YES: The communication from the European Commission of the European Union spells out the major details of the climate change agreement, emphasizing the positive aspects of the accord and discusses the implementation of the agreement.

NO: The *National Review*'s Rupert Darwall lays out several arguments against the treaty, such as the commitment of the West to paying billions of dollars to the developing world and to other binding obligations.

Earth Day 2016, Paris. The stage was set as leaders from 175 countries signed a historic agreement aimed to slow down the rise in greenhouse gas emissions harming the planet. The symbolic nature of the day was not lost on observers of the ceremony. *USA Today* saw the significance of the date, suggesting that the signing took place "using Earth Day as a backdrop." The United Nations Secretary-General, Ban Ki-moon, said that the world was in "a race against time" in efforts to combat the consequences of the rise in greenhouse gases. U.S. Secretary of State John Kerry signed the agreement on behalf of the U.S. government while holding his grand-daughter, who was joined

by 196 other children at the signing, symbolizing that tomorrow's generation of adults would pay a heavy price if today's generation of world leaders do not succeed in their quest to address a problem still viewed contentious in some quarters around the globe. And French President François Hollande exclaimed "History is here."

What were the major components of the agreement that it made it worthy of being called historic? Emily Gordon, energy editor of Great Britain's *Telegraph* newspaper, provided a succinct outline of the agreement's main provisions (*Telegraph*, December 12, 2015). First, the agreement sets a long-term goal to limit global warming to well below 2°C (Celsius) above pre-industrial levels, or even 1.5°C if possible. The 2°C level is the point at which "the worst extremes of global warming" will start to become evident. The expectation is that the developed world can reach this target faster than developing countries, so the emphasis should be on the former achieving this target sooner rather than later. The second goal focuses on more specific pledges to cut greenhouse gases in the 2020s. The Paris agreement commits countries to pursue policies that will allow them to reach targets during the decade of the 2020s. One-hundred-eighty-five countries covering 90 percent of global emissions had earlier submitted voluntary pledges to limit greenhouse gases. Gordon's third point focused on the Paris agreement's plan to "make countries pledge deeper emission cuts in the future," and improving their plan every five years. The reason for this inclusion, according to Gordon, is that the current pledges will result in only a 2.7°C warming by the end of this century. The non-binding part of the agreement calls for countries to revisit their pledges before 2020, while a binding part covers the same step for the post-2020 period. The fourth part of the Paris agreement calls on the richer countries of the globe to help poor countries financially with the costs "of going green." This was a particularly contentious issue during negotiations, as the developing world pushed the richer countries to provide funds. While the concept of financial help was part of the agreement, actual figures of aid were discussed as a nonbinding feature. Finally, the fifth component of the agreement creates a monitoring system to access progress every five years.

The December 2015 agreement signed on Earth Day 2016 was the culmination of a decades-long effort to reach agreement on a global plan to curtail greenhouse gas emissions, long thought to be at the root of global warming and climate change. And the recent action was the latest in a series of contentious steps to create a global assault against the presumed culprit. In December 2007, the Intergovernmental Panel on Climate Change (IPCC) and former U.S. Vice-President Al Gore were jointly awarded the Nobel Peace Prize for their work "to build up and disseminate greater knowledge about man-made climate change, and to lay the foundations for measures" to counteract such change. Critics of global warming were quick to respond, accusing the Nobel peace committee of having a liberal bias. This 2007 award, in turn, was the culmination of a story that had begun 15 years earlier at the UN-sponsored Earth Summit in Rio de Janeiro in 1992, when a Global Climate Treaty was signed. According to S. Fred Singer, in *Hot Talks, Cold Science: Global Warming's Unfinished Debate* (Independent Institute, 1998), the 1992 treaty rested on three basic assumptions. First, global warming has been detected in the records of climate of the last 100 years.

Second, a substantial warming in the future will produce catastrophic consequences—droughts, floods, storms, a rapid and significant rise in sea level, agricultural collapse, and the spread of tropical disease. And third, the scientific and policy-making communities knew (1) which atmospheric concentrations of greenhouse gases are dangerous and which ones are not, (2) that drastic reductions of carbon dioxide (CO_2) emissions as well as energy use in general by industrialized countries will stabilize CO_2 concentrations at close to current levels, and (3) that such economically damaging measures can be justified politically despite no significant scientific support for global warming as a threat.

In the years following the 1992 Earth Summit, it appeared that scientists opted for placement into one of three camps. The first camp has bought into the three assumptions outlined above. In late 1995, 2500 leading climate scientists announced in the first Intergovernmental Panel on Climatic Change (ICPP) report that the planet was warming due to coal and gas emissions. Scientists in a second camp suggested that while global warming was occurring and was continuing at that present moment, the source of such temperature rise could not be ascertained yet. The conclusions of the Earth Summit were misunderstood by many in the scientific community, the second camp would suggest. For these scientists, computer models, the basis of much evidence for the first group, have not yet linked global warming to human activities. A third group of scientists, representing a minority, have argued that we could not be certain that global warming was taking place, yet alone determine its cause. They presented a number of arguments in support of their position. Among them was the contention that pre-satellite data (pre-1979) showing a century-long pattern of warming was an illusion because satellite data (post-1979) revealed no such warming. Furthermore, when warming was present, it did not occur at the same time as a rise in greenhouse gases. Scientists in the third camp were also skeptical of studying global warming in the laboratory. They suggested, moreover, that most of the scientists who had opted for one of the first two camps had done so as a consequence of laboratory experiments, rather than of evidence from the real world.

Despite what some believe to be wide differences in scientific thinking about the existence of global warming and its origins, the global community moved forward with attempts to achieve consensus among the nations of the world for taking appropriate action to curtail human activities thought to affect warming. A 1997 international meeting in Kyoto, Japan, concluded with an agreement for reaching goals established at the earlier Earth Summit. Thirty-eight industrialized countries, including the United

States, agreed to reduction levels outlined in the treaty. However, the U.S. Senate never ratified the treaty, and the Bush Administration decided not to support it. Nonetheless, the two basic criteria for going into effect—the required number of countries (55) with the required levels of carbon dioxide's emissions (55 percent of carbon dioxide emissions from developed countries) must sign the treaty—were met when Russia ratified the treaty on November 18, 2004. The treaty went into effect on February 19, 2005. In the 2007 ICPP report (fourth in the series of IPCC reports), more than 2500 scientists reaffirmed the existence of global warming. The report suggested that among the risks are warming temperatures, heat waves, heavy rains, drought, stronger storms, decreased biodiversity, and sea-level rise.

And yet, there are loud voices in the scientific community and the media against the two basic premises about global warming or, more recently, climate change, as most are now beginning to call the phenomenon associated with changing global temperatures. On the one hand are scientists who eagerly seek various forums to articulate their dissenting minority voice against what appears to be a larger and growing consensus. The issue of global warming is to the current era what acid rain was to the 1970s. Just as the blighted trees and polluted lakes of Scandinavia captured the hearts of the then newly emerging group of environmentalists, the issue of global warming has been front page news for over two decades and fodder for environmentalists and policymakers everywhere. Library citations abound, making it the most often written about global issue today. Web sites pop up, public interest groups emerge, and scientists and nonscientists pick up the rallying cry for one side or another. Googling the words "global warming" or "climate change" on the Internet yields many millions in responses.

Both sides of the issue have found a substantial number of scientists measured in the thousands, although many more are on the pro-climate change side, to support their case that the Earth is or is not warming. Both sides have found hundreds of experts who will attest that the warming is either a cyclical phenomenon or the consequence of human behavior. Both sides have found a substantial number of policymakers and policy observers who will say that the Kyoto Treaty was humankind's best hope to reverse the global warming trend or that the treaty is seriously flawed with substantial negative consequences for the United States. It has been an issue whose debate heats up on occasion as the international community grapples with answers to the various disagreements summarized above. Finally, it is an issue whose potential solutions on the table will impact different sectors of the economy and different countries differently.

One particular scientific voice has been Richard Lindzen, MIT-endowed professor of meteorology. Contributing by printed word, in speech in public gatherings, and in testimony before the American government, Lindzen has painstakingly been providing evidence in support of his claims. A second type of voice has been found among conservative politicians who have argued against government expenditures for combating the climate phenomena. The continuing attempt of American conservative legislators to defund the Environmental Protection Agency is an example of this second type of activist. Finally, there have been the public print media who use their forum, Internet or print journalism, to argue against rushing to judgment on the global warming issue. The literature is replete with individuals who use public modes of communication to fight what they believe to be global warming alarmists.

This general fight over global warming and climate change took on an entirely new and more focused dimension with the signing of the recently concluded Paris Agreement on Earth Day 2016, as supporters and critics alike rushed to be heard. The Obama administration was quick to make its views known. The President began by describing his Administration's role in fighting climate change. Then turning to the agreement itself, the President suggested that it "sends a powerful signal that the world is firmly committed to a low-carbon future [. . .] The targets we've set are bold (but they represent) the best chance we've had to save the one planet that we've got." The President went on to exclaim that the moment could well "be a turning point for the world." *New York Magazine* called the agreement "President Obama's greatest accomplishment" (December 14, 2015). The head of Greenpeace International summed up the moment by saying that finally the "human race has found common cause." *The New Yorker*'s headline captured the spirit: "Good Reasons to Cheer the Paris Climate Deal" (December 14, 2015).

Critics too were quick to respond. Some suggested the agreement had not gone nearly far enough. Bill McKibben, a long-time advocate of the need to address greenhouse gas emissions, in an op-ed piece in *The New York Times* (December 13, 2015) under the headline "Falling Short on Climate in Paris," called the new pact "an ambitious agreement designed for about 1995." He argued that even if all voluntary actions take place, the planet will warm by 3.5C above pre-industrial levels. James Hansen, proclaimed by many as the father of climate change, called the agreement "a fraud really, a fake," suggesting that it is just promises and no action. He called for taxing greenhouse emissions across the board. Most criticism came from the right, however. In testimony before the U.S. House of Representatives committee on Science, Space, & Technology, Steven Groves of the Heritage Foundation, a conservative think tank, argued against ratification of the agreement and criticized the White House for ignoring Congress in the past and future on the issue. The Republican chairman of the Senate Environment and Public Works Committee immediately criticized many aspects of the deal, including the lack of a role for the U.S. Senate in the process.

In the YES selection, The Commission to the European Parliament and the Council said that the agreement provides a "last chance to hand over to future generations a world that is more stable, a healthier planet, fairer societies and more prosperous economies [. . .]." It called on the European Union to press forward both internally and internationally to secure ratification of the agreement and to create an "enabling environment" for its successful implementation, and thus a transition to a low-carbon economy. In the NO selection, Rupert Darwall, suggested that "one needs to be careful not to be taken in by the hoopla surrounding the treaty agreement." He calls the agreement much more reckless and irresponsible than previous agreements, particularly in not having reciprocal actions on the part of the developing world as part of its core structure.

YES ↩

European Commission

The Road from Paris: Assessing the Implications of the Paris Agreement and Accompanying the Proposal for a Council Decision on the Signing, on Behalf of the European Union, of the Paris Agreement Adopted Under the United Nations Framework Convention on Climate Change

1. Introduction

The 2015 Paris Agreement is a historically significant landmark in the global fight against climate change. The Agreement provides a lifeline, a last chance to hand over to future generations a world that is more stable, a healthier planet, fairer societies and more prosperous economies, also in the context of the 2030 Agenda on Sustainable Development. The Agreement will steer the world towards a global clean energy transition. This transition will require changes in business and investment behaviour and incentives across the entire policy spectrum. For the EU, this provides important opportunities, notably for jobs and growth. The transition will stimulate investment and innovation in renewable energy, thereby contributing to the EU's ambition of becoming the world leader in renewable energy, and increase the growth in markets for EU produced goods and services, for instance in the field of energy efficiency.

The Paris Agreement is the first multilateral agreement on climate change covering almost all of the world's emissions. The Paris Agreement is a success for the world and a confirmation of the EU's path to low carbon economy. The EU's negotiation strategy was decisive in reaching the Agreement. The EU has pushed for ambition, bringing its experience of effective climate policy and tradition of negotiations and rules-based international co-operation. The EU became the first major economy to present its climate plan (i.e. Intended Nationally Determined Contribution or "INDC") on March 6, 2015, reflecting the 2030 climate and energy policy framework set by the October 2014 European

Council[1] and the European Commission's blueprint for tackling global climate change beyond 2020.[2] The EU has set an ambitious economy-wide domestic target of at least 40% greenhouse gas emission reduction for 2030. The target is based on global projections that are in line with the medium term ambition of the Paris Agreement.

Throughout the Paris Conference, the EU maintained a high level of political coherence. All EU ministers in Paris showed willingness and determination to succeed. The EU acted as one, defending the EU position as agreed by the Environment Council. This allowed the EU to speak with a single and unified voice in all phases of the negotiations, a crucial element for the successful outcome in Paris. Most importantly as part of the EU's climate diplomacy outreach, the EU and it partners built a broad coalition of developed and developing countries in favour of the highest level of ambition. This High Ambition Coalition was instrumental in creating a positive dynamic during the negotiations and getting all big emitters on board the Paris Agreement.

Moreover, the global setting changed completely when compared to Copenhagen 2009 resulting in worldwide bottom up mobilisation of Governments and non-State actors, such as business, investors, cities and civil society. The French Presidency of the Climate Conference and the UN deserve credit for the positive dynamics ahead of Paris and during the conference.

The implementation of the commitments under the Paris Agreement requires maintaining the momentum and strong political determination to secure the transition to a climate resilient, climate neutral future, in a socially just manner. Climate change should remain on the political agendas of relevant international fora, including the G20 and G7 meetings. In this respect, the EU will continue its international leadership and its climate diplomacy.[3]

2. The Paris Agreement— A Global Deal

2.1 Key Features of the Paris Agreement

The Paris Agreement sets out a global action plan to put the world on track to avoid dangerous climate change acknowledging that this will require a global peaking of greenhouse gas emissions as soon as possible and achieving climate neutrality in the second half of this century. The Agreement has the following key features:

- It sets out a long-term goal to put the world on track to limit global warming to well below 2°C above pre-industrial levels—and pursue efforts to limit the temperature increase to 1.5°C; The aspirational goal of 1.5°C was agreed to drive greater ambition, and to highlight the concerns of the most vulnerable countries that are already experiencing the impacts of climate change.
- It sends a clear signal to all stakeholders, investors, businesses, civil society and policy-makers that the global transition to clean energy is here to stay and that resources have to shift away from fossil fuels; With 189 national climate plans covering some 98% of all emissions, tackling climate change is now become a truly global effort. With Paris, we are moving from action by a few to action by all.
- It provides a dynamic mechanism to take stock and strengthen ambition over time. Starting from 2023, Parties will come together every five years in a "global stocktake" to consider progress in emissions reductions, adaptation and support provided and received in view of the long-term goals of the Agreement.
- Parties have a legally binding obligation to pursue domestic mitigation measures, with the aim of achieving the objectives of their contributions.
- It sets up an enhanced transparency and accountability framework, including the biennial submission by all Parties of greenhouse gas inventories and the information necessary to track their progress, a technical expert review, a facilitative, multilateral consideration of Parties' progress and mechanism to facilitate implementation of and promote compliance.
- It provides an ambitious solidarity package with adequate provisions on climate finance and on addressing needs linked to adaptation and loss and damage associated with adverse effects of climate change. To promote individual and collective action on adaptation, the Paris Agreement establishes for the first time a global goal with the aim to enhance capacity, climate resilience and reduce climate vulnerability. Internationally, it encourages greater cooperation among Parties to share scientific knowledge on adaptation as well as information on practices and policies.

2.2 Ratification and Entry into Force of the Paris Agreement

Reaching the Paris Agreement has been a major achievement. The EU will remain proactive in the international climate negotiations to ensure that the ambition set by the Agreement is translated in all its implementing elements, such as detailed provisions on transparency and accountability, sustainable development mechanisms and technology mechanisms.

The immediate next step is the signature of the Paris Agreement. It will be opened for signature on the 22 April 2016 in New York, and enter into force when at least 55 Parties representing at least 55% of global emissions have ratified. Early ratification and entry into force is desirable as it would provide all countries with the legal certainty that the Agreement begins operating quickly. The EU should be in a position to ratify the Paris Agreement as soon as possible.

2.3 Medium-term Milestones Under the Paris Agreement

There are a number of medium-term milestones foreseen in the Paris Agreement. A clear understanding of the specific policy implications of a 1.5°C goal needs to be developed. The 5th Assessment Report of the Intergovernmental Panel on Climate Change (IPCC) was inconclusive on this aspect due to sparse scientific analysis. To address this, the IPCC has been requested to prepare a special report in 2018. The EU will provide input to the scientific work which will be carried out internationally for that purpose. The EU should participate in the first "facilitative dialogue", which will take place in 2018 to take stock of the collective ambition and progress in implementing commitments. In this respect, the EU will take part in the first global stocktake in 2023, which is relevant for considering progressively more ambitious action by all Parties for the period beyond 2030. In this sense, the EU, alongside the other parties, is invited to communicate, by 2020, their mid-century, long-term low greenhouse gas emission development strategies. To facilitate the preparation of these strategies, the Commission will prepare an in-depth analysis of the economic and social transformations in order to feed the political debate in the European Parliament, Council and with stakeholders.

3. How the EU Will Implement the Paris Agreement

The transition to a low carbon, resource-efficient economy demands a fundamental shift in technology, energy, economics, finance and ultimately society as a whole. The Paris Agreement is an opportunity for economic transformation, jobs and growth. It is a central element in achieving broader sustainable development goals, as well as the EU priorities of investment, competitiveness, circular economy, research, innovation and energy transition. Implementation of the Paris Agreement offers business opportunities for the EU to maintain and exploit its first mover advantage when fostering renewable energy, energy efficiency and competing on the development of other low carbon technology market globally. To reap those benefits, the EU will need to continue to lead by example and by action on regulatory policies to reduce emissions but also on enabling factors that accelerate public and private investment in innovation and modernisation in all key sectors, while ensuring other major economies press ahead with commitments. The transition to a low carbon economy needs to be properly managed, taking into account the differences in the energy mix and economic structure across the EU. That means also the need to anticipate and mitigate the societal impact of the transition in specific regions and socio-economic sectors.

3.1 Fostering the Enabling Environment for Low Carbon Transition

Energy Union Transition
EU's commitment to a clean energy transition is irreversible and non-negotiable. The Energy Union priority aims at *"moving away from an economy driven by fossil fuels, an economy where energy is based on a centralised, supply-side approach and which relies on old technologies and outdated business models, to empower consumers and (...) to move away from a fragmented system characterised by uncoordinated national policies, market barriers and energy-isolated areas"*.[4] The Energy Union Project, with all its dimensions, provides a broader framework within which the EU can provide the right enabling environment for the energy transition. According to the International Energy Agency, the full implementation of the climate plans will lead to investments of USD 13.5 trillion in energy efficiency and low-carbon technologies from 2015 to 2030, an annual average of USD 840 billion. The main impact of these climate plans is not only to scale up investment, but also to rebalance it across fuels and sectors, and across supply and demand. Among others, investments in renewables will be almost three times the investments into fossil-fuel power plants, while investment in energy efficiency (led by transport and the buildings sectors) is expected to equal in scale investments in other parts of the energy system.

Innovation and Competitiveness
The Paris Agreement gives a clear and ambitious direction of travel for low carbon innovation. At the margins of the Paris conference, 20 of the world leading economies launched "Mission Innovation" to reinvigorate public and private clean energy innovation, to develop and deploy breakthrough technologies and achieve cost reductions. The EU wishes to join this initiative given that the EU budget for low carbon related research under Horizon 2020 has already effectively been doubled for the period 2014-2020, and the EU has committed to invest at least 35% of Horizon 2020 into climate-related activities. Furthermore, the future research, innovation and competitiveness strategy for the Energy Union will tap into the synergies between energy, transport, circular economy, industrial and digital innovation. This should lead to greater competitiveness of present and future European low carbon and energy efficiency technologies.

Investment and Capital Markets
Shifting and rapidly scaling up private investment is essential to support the transition to a low emission and climate resilient economy, and for avoiding the "lock-in" of high emissions infrastructure and assets. EU funds will play an important role for mobilising the markets.[5] Investment support in the context of the Investment Plan for Europe, focussing on actions to remove barriers to investment in the European Union, as well as possible funding provided by the European Fund for Strategic Investments (EFSI) should promote emissions reduction and energy efficiency investments in the Single Market. The Investment Plan for Europe already has a promising track record in this area[6] and its full potential needs to be explored. The Commission has recently launched the European Investment Project Portal (EIPP) which will be fully operational shortly. Its purpose is to attract investors to viable and sound investment projects in Europe. Energy stakeholders are encouraged to send their projects to the EIPP in order to provide a comprehensive overview of projects for potential investors. As a matter of priority, the Commission will speed up technical assistance for stakeholders to establish, in 2016, schemes to aggregate smaller energy

efficiency projects, hence building a critical mass. These schemes should provide investors with better investment opportunities in energy efficiency and make capital better accessible for national, regional or local energy efficiency platforms and programmes. They will include strengthening technical and project development assistance in the context of the European Investment Advisory Hub (EIAH) set up by the Commission and the European Investment Bank to help public promoters to structure their projects and to promote financing schemes with standard terms and conditions, notably in the area of buildings.[7]

Financial institutions are key partners in this transition process. Well-functioning cross-border capital flows and integrated and sustainable capital markets are also important for this transition to happen. The measures already taken or under preparation in the context of the establishment of a Capital Markets Union[8] are essential in this context. To ensure such a transition—within the Single Market and beyond—the European Central Bank as well as national central banks, the European Investment Bank and the European Bank for Reconstruction and Development, the Green Climate Fund, and other international finance institutions like the World Bank but also national development banks can play a helpful role. In response to a request by the G20 in April 2015 to review how the financial sector can take account of climate-related issues, the Financial Stability Board (FSB) established a task force on climate-related financial disclosures, whose aim is to help market participants better understand climate-related risks and better manage those risks. The G20 has recently set up a study group to analyse green finance related issues (GFSG). At European level, the European Systemic Risk Board has published a report on the transition to a low-carbon economy and the potential risks for the financial sector.[9]

Carbon Pricing and Fossil Fuel Subsidies

Carbon pricing is an essential element to foster a global level playing field for the transition—these can take form of emission trading, as in the case of the EU, taxation, or other economic and/or fiscal instruments. The EU should increase its efforts in sharing its own experiences in this area with all countries that need to start putting a price on carbon. This will continue to include countries like China and South Korea that are setting up emissions trading systems, as well as a broader range of countries, including all major economies that are deploying renewable energy technologies and improving their energy efficiency policies. While the Paris Agreement is a game changer in the sense of being global, the nationally determined level of effort by countries differs, with a risk of competitive disadvantage for industries if an uneven playing field will remain. The strategic decision by the European Council to preserve the free allocation regime beyond 2020 and the proposed carbon leakage provisions for the EU emission trading system strike the right balance at this point of time, but should be kept under review in the coming decade.

The outlook of carbon and energy pricing is further complicated by the current low oil price situation. This can provide a good opportunity not only to introduce carbon pricing but also to remove fossil fuel subsidies, which, according to the International Energy Agency, amounted worldwide to USD 548 billion in 2013. Such subsidies are the biggest obstacle to innovation in clean technologies, as is recognised in the G20 and G7 calls for the elimination of fossil fuel subsidies. The forthcoming EU energy prices and cost report will look at the latest situation in this respect.

The Role of Cities, Civil Society and Social Partners

Catalysing multi-stakeholder action from civil society—citizens, consumers, social partners, SMEs, innovative start-ups and globally competitive industries is another prerequisite for the transition. The Paris conference and the Lima-Paris Action Agenda, an initiative of the Peruvian and French COP Presidencies aimed at, bringing together an unprecedented number of non-State actors together on a global stage to accelerate cooperative climate action in support of the new agreement. The EU is uniquely placed to mainstream the low carbon transition through all sectors and levels of governance.

Smart cities and urban communities are the place where a big part of the future transformation will actually happen. Work at city level and urban policies will therefore be intensified in 2016, including on supporting actions developed by the integrated and global Covenant of Mayors and the setting up of a "one stop shop" for local authorities. This should allow local authorities to better contribute to the EU's low carbon transition and will provide European companies with worldwide opportunities to use their competitive edge in innovative technologies for smart cities.

Climate Diplomacy and Global Action

Climate action is a major strategic foreign policy challenge with implications for EU external policy making in, for example, development aid and cooperation, neighbourhood and enlargement policies, international scientific and technological co-operation, trade, economic diplomacy and security. Maintaining the positive momentum from Paris will require sustained political and diplomatic mobilisation at global level.

As agreed by the Council,[10] in 2016 climate diplomacy will need to focus on (i) maintaining climate change advocacy as a strategic priority, (ii) supporting implementation of the Paris Agreement and the climate plans, and (iii) increasing efforts to address the nexus of climate change, natural resources, including water, prosperity, stability and migration.

In terms of climate finance, the EU and its Member States are committed to scaling up the mobilisation of climate finance in the context of meaningful mitigation actions and transparency of implementation, in order to contribute their share of the developed countries' goal to jointly mobilise USD 100 billion per year by 2020 from a wide variety of sources, public and private, bilateral and multilateral, including alternative sources of finance. Current trajectories for EU development assistance will substantially contribute to reaching the EU's share of the USD 100 billion goal. In the context of the Multiannual Financial Framework 2014–2020 the EU has undertaken to ensure that 20% of its overall budget is directed to climate-relevant projects and policies. In the context of external expenditure, this more than doubles the amount of climate finance for developing countries and could represent as much as EUR 14 billion in climate spending. An increasing share of these resources will be invested in adaptation and facilitating innovation and in capacity building.

In order to assist developing countries in delivering on their climate plans as of 2020, support programmes (such as the Global Climate Change Alliance+) will be strengthened. In this context the synergies between climate action, the Addis Ababa Action Agenda and the 2030 Agenda with its Sustainable Development Goals need to be exploited fully. This also includes EU participation in the Africa Renewable Energy Initiative. In the framework of enlargement and neighbourhood policies, the EU will continue its political dialogue and support to partner countries. Particular emphasis would be placed on capacity building.

Ongoing bilateral and multilateral negotiations on liberalising trade in green goods and services should be accelerated to facilitate the global action to mitigate climate change and to create business opportunities for European companies. The EU should also continue its leadership in promoting ambitious outcomes in the context of the negotiations in the International Civil Aviation Organisation (ICAO) and the International Maritime Organisation (IMO) to address greenhouse gas emissions, as well as under the Montreal Protocol negotiations.

3.2 The 2030 Energy and Climate Regulatory Framework

After the Paris climate conference, all countries need to turn their commitments into concrete policy actions. In October 2014 the European Council set the 2030 climate and energy policy framework for the EU setting an ambitious economy-wide domestic target of at least 40% greenhouse gas emission reduction for 2030, as well as renewable energy and energy efficiency targets of at least 27%.[11] The Paris Agreement vindicates the EU's approach. Implementing the 2030 energy and climate framework as agreed by the European Council is a priority in follow up to the Paris Agreement.

The Commission has already initiated this process by putting forward a proposal to revise the Emissions Trading System (ETS), covering 45% of the EU's greenhouse gas emissions. The Commission will present during the next 12 months the key remaining legislative proposals to implement the agreed 2030 regulatory framework domestically in a fair and cost-efficient manner, providing maximum flexibility for Member States and striking the right balance between national and EU level action. As the next step, the Commission is working on the preparation of proposals for an Effort-Sharing Decision and on land use, land use change and forestry (LULUCF). The Commission will also propose legislation to set up a reliable and transparent governance mechanism and to streamline the planning and reporting requirements related to climate and energy for the post-2020 period.

Moreover, the Commission will present the necessary policy proposals to adapt the EU's regulatory framework in order to put energy efficiency first and to foster EU's role as a world leader in the field of renewable energy in line with the European Council conclusion of October 2014. This includes a new energy market design to place consumers at the centre of the energy system, enabling demand response and enhancing flexibility. In addition, this year the Commission has already launched the Energy Security Package to address without delay the new challenges to the security of supply posed by the developments in the international energy context.

4. Conclusion

On the road to Paris and in Paris the EU has been at the heart of the High Ambition Coalition of developed and developing countries. To effectively secure the transition to the low carbon economy, the EU needs to keep up this ambition, internally and internationally:

- The Paris Agreement should be signed and ratified as soon as possible. The proposal to sign the Agreement is attached to this communication.
- The EU needs to consolidate the enabling environment for the transition to a low carbon economy through a wide range of interacting policies, strategic frameworks and instruments reflected under the

10 priorities of the Juncker Commission—in particular the Resilient Energy Union with a Forward-Looking Climate Change Project.

- The EU's 2030 energy and climate change regulatory framework needs to be swiftly completed in line with the European Council conclusions of October 2014. The forthcoming legislative proposals should be fast-tracked by the Parliament and the Council.
- All Parties will need to be ready to fully participate in the review processes under the Paris Agreement designed to ensure the achievement of the goal of keeping climate change well below 2°C and pursuing efforts towards 1.5°C.

Notes

1. European Council conclusions of 24 October 2014.

2. The Paris Protocol—A blueprint for tackling global climate change beyond 2020, COM(2015) 81 final.

3. European climate diplomacy after COP21—Council conclusions of 15 February 2016.

4. A Framework Strategy for a Resilient Energy Union with a Forward-Looking Climate Change Policy—COM(2015)80 of 25 February 2015.

5. EUR 114 billion have been programmed from the reformed European Structural and Investment Funds (ESIF) for climate-related actions over the 2014–2020 period. The programming has been carried out in a wider partnership with the relevant stakeholders. The resulting amount is 25% of ESIF showing strong commitment to climate-actions and exceeding the targeted 20% for the overall EU budget. The support goes beyond funding opportunities as it includes strong regional cooperation, capacity building and technical assistance components.

6. Overview of Investment Plan projects in low carbon and energy efficiency: http://ec.europa.eu/priorities/sites/beta-political/files/sector-factsheet-energy_en.pdf

7. State of the Energy Union 2015.

8. Action Plan on Building a Capital Markets Union, COM(2015) 468 final.

9. https://www.esrb.europa.eu/pub/pdf/asc/Reports_ASC_6_1602.pdf

10. Council conclusions on European climate diplomacy after COP 21.

11. The energy efficiency target will be reviewed by 2020, having in mind an EU level of 30%.

The European Commission is the EU's politically independent executive arm of the European Parliament and Council. It draws up proposals for new European legislation and it implements decisions of the Parliament and the Council of the European Union.

Rupert Darwall

 **NO**

Paris: The Treaty That Dare Not Speak Its Name

The agreement adopted in Paris at 7:28 P.M. local time Saturday doesn't call itself a treaty, but in every other respect it is one. Four years ago at the Durban climate conference, climate negotiators decided to launch a process "to develop a protocol, another legal instrument, or an agreed outcome with legal force." If the Paris Agreement is to meet the requirements of the Durban Platform, legal scholar and Clinton-era climate-change coordinator at the State Department Daniel Bodansky states that "the Paris Agreement must constitute a treaty within the definition of the Vienna Convention."

Article Two (a) of the 1969 Vienna Convention on the Law of Treaties defines a treaty as an "international agreement concluded between states in written form and governed by international law." Under the principle of *pacta sunt servanda* ("agreements must be kept"), treaties are binding on the parties and must be performed by them in good faith, Bodansky observes in a recent book. Article 14 of the Paris Agreement establishes a compliance mechanism, and Article 20 duly sets out the process for the depositing of "instruments of ratification, acceptance, approval, or accession."

Article Two of the Constitution of the United States circumscribes the power of the executive to make treaties by stating that the president "shall have the power, by and with the advice and consent of the Senate, to make treaties, provided two-thirds of the Senators present concur." The question then arises whether the Paris Agreement imposes new legally binding obligations on the United States. American negotiators were mindful of this when Secretary of State John Kerry reportedly threatened that the U.S. would walk out unless negotiators removed from the draft treaty the specification that developed countries would begin providing $100 billion a year in climate funding, by 2020. The *Business Standard* of India reported that Kerry said: "I would love to have a legally binding agreement. But the situation in the U.S. is such that legally binding with respect to finance is a killer for the agreement."

Here, one needs to be careful not to be taken in by the hoopla surrounding the agreement. "For the first time, we have a truly universal agreement on climate change," U.N. Secretary General Ban Ki-moon said in an interview after the agreement was adopted. Article Four (2) of the Paris Agreement might appear to signal universal accord—it requires all parties to pursue "domestic mitigation measures," that is to say, policies that slow down or reverse the growth of their greenhouse-gas emissions.

This provision, however, is no more than a reformulation of what the United States and all other parties are already obligated to do under the 1992 U.N. framework convention on climate change, which the Senate approved in a 95–0 vote. All parties to the convention shall "formulate, implement, publish, and regularly update national programs to mitigate climate change," Article Four (1) (b) of the 1992 convention states. Despite what the U.N. secretary general claims, the global scope of the Paris Agreement is not what makes the Paris Agreement legally important or novel, as it merely reiterates what was enacted 23 years ago.

Kerry had compelling political reasons to take the $100 billion a year of climate money out of the text of the treaty. Though not in the treaty, the $100 billion a year—now pushed back to 2025—survives as a formal decision of the Paris conference (paragraph 54 of the decision document). That massive fund would have been not only a killer for the deal, it would also have been a killer for Hillary Clinton's presidential candidacy. As secretary of state, Clinton made the $100 billion pledge at the Copenhagen climate conference to keep those talks from cratering.

Nonetheless, the Paris Agreement does impose a new, expansive, and enormously consequential legally binding obligation on the United States. "The efforts of all Parties will represent a progression over time," Article Three states. Thus the treaty creates a ratchet mechanism where none previously existed. The new obligation is repeated in the next article: "Each Party's successive nationally determined contribution will represent a progression beyond the Party's then current nationally determined contribution and reflect its highest possible ambition." This is the strongest language of the whole document.

The spin coming out of Paris and the Obama administration is that the new treaty bridges the divide between developed and developing nation parties. As President Obama put it in his statement welcoming agreement on the treaty, "We showed it was possible to bridge the old divides between developed and developing nations that had stymied global progress for so long." In fact, the divide between developed and developing countries remains, the same divide that led the Senate (including then-senator Kerry) to pass the 1992 Byrd-Hagel resolution by 95 to 0 that then torpedoed U.S. ratification of the Kyoto Protocol.

Article Four of the Paris Agreement states that developing countries should be provided with support for implementing the agreement, and this provision in turn is tied to Article Nine on the obligation of developed country parties to provide financial contributions to developing countries. India's Intended Nationally Determined Contribution makes this conditionality explicit by tying its mitigation efforts "to the availability and level of *international financing and technology transfer*" (emphasis is mine). By contrast, developed countries are on the hook and obligated to ratchet up their emissions cuts over time—whatever developing countries do or don't do—in perpetuity.

Furthermore, "highest possible ambition" precludes the devising of the escape hatches that Clinton-administration officials viewed as critical. "We agreed there needed to be goals, even aggressive goals," a senior economic adviser to Bill Clinton told the *New York Times* in 1997. "But there also needed to be escape hatches, in case the economic effects turned out to be a lot more damaging than we thought."

In comparison with the Clinton administration and the Kyoto Protocol, the Paris Agreement is reckless and irresponsible. The new treaty replicates the structural flaw of Kyoto in not having reciprocal actions and obligations between developed and developing countries, and it adds a deeply problematic procedural one. This is a vehicle with the gas pedal forever being pushed down and no brakes or steering wheel.

U.S. ratification of the Paris Agreement without congressional approval would raise profound constitutional questions. Attempts to push through a quasi-carbon tax (in the form of Al Gore's BTU tax) and cap-and-trade measures repeatedly failed, and the Senate made clear that it would not ratify the Kyoto Protocol, which in key respects was a less bad treaty than the Paris Agreement. The legal effect of the Paris ratchet would be to constrain the discretion of future administrations. The EPA's Clean Power Plan is the most costly component of the Obama administration's plan to cut emissions by 26 to 28 percent below 2005 levels by 2025.

No one, least of all the EPA, knows how much the Clean Power Plan will eventually cost. In 2013, the German environment minister declared that its renewable program could cost one trillion euros ($1.1 trillion). The U.S. generates nearly seven times as much electricity as Germany does, which gives some idea of the scale of its cost. When the Clean Power Plan turns out to be a disaster, under international law, the U.S. could not simply ditch it, but would have to replace it with something else (a carbon tax or the cap-and-trade plan, for example, that Congress already rejected) otherwise it would be in breach of the Paris Agreement.

President Obama has developed a new presidential doctrine to bind his successors by aiming to make them accountable to the court of international opinion. "Everybody else is taking climate change really seriously," the president said at a press conference shortly before he left Paris. "They think it's a really big problem. . . . So whoever is the next president of the United States . . . is going to need to think this is really important." If the president decides to ratify the Paris Agreement without obtaining congressional approval from two-thirds of the Senate, not only will he ensure that the only way of reversing the agreement will be to put a Republican in the White House, but he also will be subverting Article Two of the U.S. Constitution. For America, the Paris Agreement is a very big deal.

RUPERT DUVALL, who studies at Cambridge University, is a writer and the author of *The Age of Global Warming: A History* (2014).

EXPLORING THE ISSUE

Is the Paris Climate Change Agreement a Good Deal?

Critical Thinking and Reflection

1. Is the debate about whether greenhouse gases are at the heart of global warming been finally settled?
2. Does the Paris agreement on climate change mean that international leaders have now officially accepted the view regarding the cause of global warming and climate change?
3. Do the criticisms from the left that the new agreement has no teeth to address the issue of greenhouse emissions have merit?
4. Do you agree with the excitement and optimism expressed by world leaders at the signing of the agreement that it is truly historic in nature?
5. Is the opposition to the agreement from Republican members of Congress that is based on the lack of a role for Congress the best way for the GOP to attack the agreement?

Is There Common Ground?

Assessing the existence of common ground on the issue of global warming or the broader issue of climate change, let alone the parameters of the new Paris agreement, is difficult. While the scientific community and the general public agree on the fact that the global climate is changing, they disagree on the role that humans play in that process. A much higher percentage of scientists compared to the general public place significant blame on humans for these changes.

A vast majority of members within the scientific community also accept the views found in the various reports of the Intergovernmental Panel on Climate Change (IPCC) concerning the current state of climate change, and its causes and consequences. Reading the research findings of most members of the scientific community would lead one to believe that disagreements of an earlier day regarding the basic questions associated with climate change have been settled. But nonetheless, within the scientific community one finds a handful of scientists who, along with popular nonscientific writers, have cornered the market on books with sensational titles that enjoy wide circulation among the general public. Thus, it is not surprising that one finds a larger lack of consensus among the general public, as well as among politicians, regarding the basic questions associated with climate change. This is so despite the reality that month after month, weather reports show each succeeding month to be hotter than ever before. But world leaders and those other political actors who compete with these leaders seem to accept that the international community needs to do something.

So while there is much within the agreement that both sides can accept, differences do remain over the issue of what responsibility the developed world has to help the developing countries meet their goals and over the extent to which the voluntary aspects of the agreement have enough teeth in them to make a difference.

Additional Resources

Alexander, Ralph B., *Global Warming False Alarm: The Bad Science Behind the United Nations' Assertion that Man-Made CO_2 Causes Global Warming*, 2nd edition (Canterbury Publishing, 2012)

The book focuses on the "flawed science" behind the UN Intergovernmental Panel on Climate Change's position that recent climate change is the result of human activity.

Ban Ki-Moon, "What Was Once Unthinkable Has Now Become Unstoppable," *Vital Speeches of the Day* (vol. 83, no. 2, February 2016)

This speech by the Secretary-General of the United Nations was delivered at the Paris Conference.

Citroen, Samantha, Kempinski, Josh, and Cullen, Zoë, "Life After COP21: What Does the Paris Agreement Mean for Forests and Biodiversity Conservation?" *Oryx* (vol. 50, no. 2, April 2016)

The articles addresses the impact of the new Paris agreement on the world's forests and the concept of biodiversity conservation.

Darwall, Rupert, *Age of Global Warming* (Quartet Books, 2013)

The author focuses on the history of global leaders' policy attempts to address global warming with much skepticism.

Horton, Richard, "Climate—'A Common Concern of Humankind,'" *Lancet* (vol. 386, no. 10012, December 19, 2015)

The article describes several different viewpoints regarding the Paris agreement.

Li, Anthony, "Hopes of Limiting Global Warming? China and the Paris Agreement on Climate Change," *China Perspectives* (no. 1, 2016)

The article describes efforts of China to reach climate change agreement in Paris.

Magrini, Marco, "The Trump Card," *Geographical* (vol. 88, February 2016)

This article addresses the question of the Paris agreement's survival, discussing a number of failures.

Manolas, Evangelos, "The Paris Climate Change Agreement," *International Journal of Environmental Studies* (vol. 73, no. 2, April 2016)

The article compares the differences between the Paris agreement and the Kyoto protocol.

McKibben, Bill, *The Global Warming Reader: A Century of Writing about Climate Change* (Penguin books, 2012)

The book contains more than 35 essays over the last 100 years on global warming.

Page, Michael Le, "Will the Paris Deal Save Our Future?," *New Scientist* (December 19, 2015)

The author argues that the Paris agreement sets out more ambitious goals than expected.

"Paris Climate Change Agreement," *Congressional Record* (vol. 95, no. 2, February 2016)

The article discusses the aspects of the climate change agreement.

"Remarks on the Adoption of the United Nations Framework Convention on Climate Change Paris Agreement," *Daily Compilation of Presidential Documents* (December 12, 2015)

This article presents the speech of President Barack Obama about the Paris agreement.

Robinson, G. Dedrick, *Global Warming: Alarmists, Skeptics & Deniers; A Geoscientist Looks at the Science of Climate Change* (Kindle Edition, 2013)

This is an analysis of global warming from the perspective of a geologist.

Simons, Stefan, "Our Hard Work Starts Now," *The Chemical Engineer* (no. 896, February 2016)

The author examines the role of chemical engineers in addressing climate change in accordance with Paris provisions.

Internet References . . .

EPA: Global Warming

www.epa.gov/climatechange/index.html

Global Warming: Focus on the Future

www.enviroweb.org/edf

Intergovernmental Panel on Climate Change

www.ipcc.ch

National Oceanic and Atmospheric Administration (NOAA) Paleoclimatology Program

www.ngdc.noaa.gov/paleo/globalwarming/home.html

The Heartland Institute

www.heartland.org/studies/ieguide.htm

U.S. Environmental Protection Agency, Global Warming Page

www.epa.gov/globalwarming/

U.S. Government Website on Global Warming

www.epa.gov/climatechange/index.html

Unit 2

UNIT

Expanding Global Forces and Movements

*O*ur ability to travel from one part of the globe to another in a short amount of time or to communicate anywhere in the world in a variety of ways instantaneously has expanded dramatically since the Wright brothers first lifted an airplane off the sand dunes of North Carolina's Outer Banks and Alexander Graham Bell first uttered those initial famous words into a machine. This technological explosion has not only increased the speed of travel and of information dissemination but has also expanded its reach and impact, making any individual with entree to an airline ticket or Internet access a global actor in every sense of the term. Finally, modern political and economic structures, fueled by the shrinking of the globe, have also expanded from occupying small, rather discrete locations to an ever-expanding array of larger and more fluid boundaries that are constantly moving out toward the far reaches of the globe. National borders no longer keep individuals and groups in and they also no longer keep individuals and groups out as they once did a century or even half a century ago.

Many consequences result from this realization as individuals and groups of every kind have expanded their playing field until the world has now reached a point where every place is affected in some fashion by every event anywhere on the planet. And every place has the capacity to react to every event in the world. The flow of money, information, goods and ideas that connect people around the world also creates fissures or conflict that heighten anxieties and cause increased tensions between rich and poor, connected and disconnected, and cultures and regimes. They manifest themselves negatively in a host of specific problems—health pandemics such as the Zika virus, human trafficking, human rights violations by both legitimate governments and armed thugs, and the like. Add to these issues the growing concentration of wealth into fewer and fewer hands, which in turn exacerbates the ability of the global community to address the negative consequences of these global forces.

In the past such problems were localized to a specific locale or at least within the boundaries of a single country. In these instances the local or national government had the capacity (or was assumed to have the capacity) to address these issues as the sole decision-making body. But just as people cross national boundaries easily, so do these problems. Solving them requires the cooperation of national sovereign governments either through direct diplomacy with one another or through international organizations such as the United Nations and its array of functional and geographical bodies.

At the same time, new forms of social media have expanded our ability to communicate from one end of the globe to the other instantaneously, leading to both positive and negative consequences of an ever expanding scope. The nature of influence has expanded as individuals and groups intent on doing good or evil somewhere throughout the world no longer have to rely on traditional armed force or even any kind of military warfare at all. The impact of these new and emerging patterns of access is yet to be fully calculated or realized, but we do know that billions are feeling their impact, and the result is both exhilarating and frightening.

Selected, Edited, and with Issue Framing Material by:
James E. Harf, *Maryville University*
and
Mark Owen Lombardi, *Maryville University*

ISSUE

Will the International Community be Able to Successfully Address the ZIKA Virus Pandemic?

YES: World Health Organization, from "ZIKA: Strategic Response Framework & Joint Operations Plan," United Nations Publications (2016)

NO: Matthew Weaver and Sally Desmond, from "Zika Virus Spreading Explosively, Says World Health Organization," *The Guardian* (2016)

Learning Outcomes
After reading this issue, you will be able to:
• Describe the evolution of the international community's response to global pandemics in general.
• Describe how the World Health Organization (WHO) has been able to develop successful strategies for addressing global pandemic threats.
• Gain an understanding of management system instituted by WHO to combat the Zika virus.
• Grasp the seriousness of the threat posed by the Zika virus.

ISSUE SUMMARY

YES: The WHO report lays out a comprehensive global plan to address the pandemic threat posed by the Zika virus. The WHO-led effort is designed to provide help to affected countries, to build the capacity to stop new outbreaks and to address them successfully when they do occur, and to engage in research to address all aspects of the pandemic.

NO: Matthew Weaver and Sally Desmond report that the Director General of the WHO, in calling an emergency meeting, stated that the "level of alarm (about Zika) is extremely high."

Zika virus—these two words in the summer of 2016 brought fear to individuals in the developed world as the disease made its way across countries of the South toward the richer areas of the world. Of particular concern were pregnant women who were at particularly high risk. This virus is the latest in a long list of diseases that have had the potential to unleash havoc on large segments of a region's or even the entire global population. Hear the words "global pandemics" and one also thinks in far earlier times of the bubonic plague or Black Death of the Middle Ages where an estimated 30 percent of Europe's population died, or the influenza epidemic of 1918 that killed upward of 40 million or one in 20 people worldwide and fully a third of the human race were afflicted. Both incidents seem like stories from a bygone era when modern medicine was unknown, where people were simply at the mercy of the spreading tendencies of the virulent diseases, and where the international community had yet to begin to cooperate to address such outbreaks. The latter did not begin to happen until the 1830s when a board was established in Egypt to track diseases throughout the Mediterranean region. In 1851, European governments gathered formally for the first time in Paris to discuss sanitary matters in light of persistent cholera epidemics plaguing Europe. And the first permanent health organization was founded in 1902 in the Western Hemisphere.

The world of medicine is different today, as the public attention to global health has dramatically grown over the last 50 years, what the Council on Foreign Relations has called a public health revolution. Increases in funding have resulted in a dramatic growth in the number of international organizations devoted to public health worldwide. The WHO has been joined by a "panoply of new multilateral initiatives, public-private partnerships, foundations, faith-based organization, and nongovernmental organizations." As a result, the average global life expectancy has risen from 40 to over 71 years. This leads many to assume that somewhere on the shelves of the local pharmacy or the Centers for Disease Control and Prevention in Atlanta lies a counteragent to whatever killer lurks out there. In 2009, however, the world watched in much the same way as it did 750 years ago or 93 years ago. The reason was the culprit H1N1 swine flu. In April 2009, it was reported that a Mexican boy had flu caused by a mosaic of swine/bird/human flu known as H1N1. On the other side of the ocean in Cairo, the Egyptian government ordered the killing of 300,000 animals as a precaution. Soon in every corner of the planet, officials began to take precautions and deaths began to mount. This 2009 scare followed on the heels of a global scare two decades earlier as a virulent disease of another type, AIDS, began to sweep across Africa to all other sectors of the globe.

The world of travel for both humans and things nonhuman is far different from that of the fourteenth century or even 1918. Globalization is with us. The world has shrunk, literally and figuratively, as the human race's ability to move people, money, goods, information, and also unwanted agents across national boundaries and to the far corners of the globe has increased exponentially. Viruses, germs, parasites, and other virulent disease agents can and do move much more easily than at any time in recorded history. Today's airplane is much faster than yesterday's ship.

The result—fast forward to 2016 where a pandemic was on the rise, moving quickly throughout the globe from the developing world into the most developed countries on the globe. It is the Zika virus; a virus discovered in 1947 in Uganda and long thought to pose little threat to humans. In 1952, the first case was reported in humans. And in May 2015 the Pan American Health Organization issued the first alert of a Zika virus infection in Brazil. Within a year it swept through Latin America and made its way into the United States, bringing with it the threat of neurological complications and birth malformations. Among the groups most feeling this quickly emerging

threat were potential participants at the 2016 Olympic Games in Rio de Janeiro who fear the alarming effects of pregnancies while infected with the virus.

Let us go back and look at the evolution of the international community's attempts to address global pandemics. The word "pandemic" is derived from two Greek words pan meaning "all" and demos meaning "people." Thus, a global pandemic is an epidemic of some infectious disease that can and is spreading at a rapid rate throughout the world. Officially, the WHO labels a disease outbreak a pandemic if community-level outbreaks of a disease are occurring in more than one country in a WHO region of the globe and one additional country in a different WHO region. Throughout history, humankind has fallen victim to many such killers. As early as the Peloponnesian War in fifth-century B.C. Greece, typhoid fever was responsible for the deaths of upward of 25 percent of combatants and civilians alike, necessitating major changes in military tactics. Imperial Rome felt the wrath of a plague thought to be smallpox, as did the eastern Mediterranean during its political height several centuries later.

In the last 100 years, influenza (1918, 1957, and 1968), typhoid, and cholera were major killers. In recent years, other infectious diseases have made front page news: HIV, Ebola virus, SARS, and more recently, avian or bird flu. For a while, the latter flu struck tremendous fear in the hearts of global travelers and governmental policymakers everywhere. WHO Europe predicted that as many as 175–360 million people could fall victim if the 2009 outbreak was severe enough. The bird flu was front page news because more than 150 million birds had died worldwide from one of its earlier strains, H5N1. This strain was first found in humans in 1997, and WHO estimated that the human fatality rate has been 50 percent, with 69 deaths occurring by December 2005. One might be prompted to ask: what was the "big deal, only 134 confirmed cases?" It is not quite so simple.

Unlike previous pandemics that hit suddenly and without little or any warning, the avian flu gave us a clear warning. The loss in poultry had been enormous. And with the jump to humans, with an initial high mortality rate, our senses had been awakened to the potential for global human disaster. But there was good news as well. There was time to prepare for the worst-case scenario and diminish its likelihood. The flu had the attention of all relevant world health agencies and most national agencies, and steps were undertaken to find a way to combat this contagious disease. Whether global preparedness or simply the natural evolution of this particular strain of influenza, the 2009 global scare was not matched by reality, as the resultant mortality rates were not much different from those of the annual flu outbreaks.

WHO, created by the United Nations following World War II, became the first modern international organization in the fight against widespread diseases and it continues to play a leading role against both epidemics and pandemics. It is now joined by a complex network of international governmental and nongovernmental organizations and private foundations. This network has enjoyed great successes in several areas: smallpox, polio, and measles. And it has also proven to becoming effective in more recent challenges of SARS and AIDS.

One of the reasons for success has been the recognition by WHO that successful response to an emerging pandemic not only depends on the cooperation among health professionals and organizations throughout the world but also what WHO calls "the whole of society." The latter include all governments, businesses, and civil society who work to "sustain essential infrastructure and mitigate impacts on the economy and the functioning of society." WHO has spelled out its master plan for a total societal response in Whole-of-Society Pandemic Readiness: WHO Guidelines for Pandemic Preparedness and Response in the Non-Health Sector (July 2009). This plan encompasses five basic principles: a whole-of-society approach, preparedness at all levels, attention to critical interdependencies, a scenario-based response, and respect for ethical norms. Each sector of society must have a flexible response plan in place. National and local governments must provide leadership while local governments stand ready to undertake specific actions. Standard operating principles and detailed communication strategies must be developed by governments. All relevant executive branches, from defense to finance, must be involved. Each provider of essential services, such as water and energy, must know the critical linkages and interdependencies among all providers. Plans for different scenarios, from mild to severe, must be developed. And finally, preparedness and response must be consistent with ethical norms and human rights considerations, with special attention to vulnerable peoples throughout the globe. The latter include people who have no access to health systems, estimated at one billion throughout the globe.

The issue of especially vulnerable people suggests an alternative way to view pandemics. The pandemics in our lifetimes have originated in the developing world. As the developing world lacked the appropriate health infrastructure, these diseases took hold and spread, eventually crossing national boundaries and making their way to the far corners of the earth. Had the diseases been confronted with an adequate health infrastructure with a detailed plan of action and the resources to implement it, the diseases might have been contained within one country. Toba Bryant and Dennis Raphael so suggest in the title of an article on the subject, "The Real Epidemic is Inequality" (2010). The essential thesis of their work is that (1) health inequalities growing out of social inequalities represent the primary health issue in the world, and (2) the pace and scope of epidemics and pandemics are influenced by a function, in part, of these inequalities, particularly in the developing world.

So the world has awakened to the Zika virus. Can the experiences of the international community in marshaling its resources to fight recent global pandemics like HIV/AIDS, avian flu, and Ebola allow it to once again successfully address another global health threat? This global issue addresses pandemics in general and this is reflected in our YES selection. Here the WHO spells out its Emergency Operations Plan whose purpose is to coordinate the international response to Zika. In the NO selection, a variety of leaders in the fight against Zika sound the alarm posed by the Zika virus.

YES ⬅

World Health Organization

Zika: Strategic Response Framework & Joint Operations Plan

Overview of the Situation

This strategy has been developed to guide the international response to the current cluster of congenital malformations (microcephaly) and other neurological complications (Guillain-Barré Syndrome) that could be linked to Zika virus infection.

Background

Zika virus is an emerging viral disease that is transmitted through the bite of an infected mosquito, primarily Aedes aegypti, the same vector that transmits chikungunya, dengue and yellow fever. Zika has a similar epidemiology, clinical presentation and transmission cycle in urban environments as chikungunya and dengue, although it generally causes milder illness.

Symptoms of Zika virus disease include fever, skin rash, conjunctivitis, muscle and joint pain, malaise and headache, which normally last for 2 to 7 days. There is no specific treatment but symptoms are normally mild and can be treated with common pain and fever medicines, rest and drinking plenty of fluids.

Zika virus was first identified in 1947 in a monkey in the Zika forest of Uganda, and was first isolated in humans in 1952 in Uganda and the United Republic of Tanzania. Zika virus has been causing sporadic disease in Africa and Asia. Outbreaks were reported for the first time from the Pacific in 2007 and 2013 in Yap Island (Federated States of Micronesia) and French Polynesia, respectively. There was subsequent spread of the virus to other Pacific islands, including New Caledonia, Cook Islands, Easter Island (Chile), Fiji, Samoa, Solomon Islands, and Vanuatu. The geographical range of Zika virus has been steadily increasing ever since.

Current Situation

In February 2015, Brazil detected cases of fever and rash that were confirmed to be Zika virus in May 2015. The last official report received dated 1 December 2015, indicated 56,318 suspected cases of Zika virus disease in 29 States, with localized transmission occurring since April 2015. Due to the magnitude of the outbreak, Brazil has stopped counting cases of Zika virus. Today the Brazilian national authorities estimate 500,000 to 1,500,000 cases of Zika virus disease. In October 2015, both Colombia and Cape Verde, off the coast of Africa, reported their first outbreaks of the virus. As of 22 January 2016 Colombia had reported 16,419 cases, El Salvador 3,836 cases and Panama 99 cases of Zika virus disease.

As of 12 February, a total of 39 countries in multiple regions have reported autochthonous (local) circulation of Zika virus, and there is evidence of local transmission in six additional countries (Figure 1). Imported cases have been reported in the United States of America, Europe and non-endemic countries of Asia and the Pacific.

Increase in Neurological Syndromes

National health authorities have reported an observed increase of Guillain-Barré syndrome (GBS)[1] in Brazil and El Salvador which coincided with the Zika virus outbreaks.

During the French Polynesia outbreak in 2013/2014, national authorities also reported an observed increase in neurological syndromes in the context of co-circulating dengue virus and Zika virus. Seventy-four patients presented with neurological or auto-immune syndromes after the manifestation of symptoms consistent with Zika virus infection. Of these, 42 were classified as GBS.

On January 22 Brazil reported an increase of GBS at the national level. A total of 1708 GBS cases were registered between January and November 2015. Most of Brazil's states have Zika, chikungunya and dengue virus circulation.

Increase in Congenital Malformations

On 27 January 2016, Brazil reported that of 4180 suspected cases of microcephaly, 270 were confirmed, 462 were discarded and 3448 are still under investigation. This compares to an average of 163 microcephaly cases recorded nationwide per year. Only six of the 270 confirmed cases of microcephaly had evidence of Zika infection. According to the US Centers for Disease Control and Prevention (US CDC) and Ministry of Health Brazil, the results of two specimens taken during autopsy from the brain tissues of microcephalic patients, indicated infection with Zika virus. A placenta was also evaluated and found to be PCR positive for Zika.

Although the microcephaly cases in Brazil are spatio-temporally associated with the Zika virus outbreak, health authorities and agencies are investigating and conducting comprehensive research to confirm a causal link.

Following the Zika outbreak in French Polynesian, health authorities reported an unusual increase in the number of congenital malformations in babies born between March 2014 and May 2015. Eighteen cases were reported, nine of which were diagnosed as microcephaly.

Other countries with current outbreaks (Cape Verde, Colombia, El Salvador, and Panama) have not reported an increase in microcephaly.

Status of Response

The current Zika virus outbreaks and their possible association with an increase in microcephaly, other congenital malformations, and GBS have caused increasing alarm in countries across the world, particularly in the Americas. Brazil announced a national public health emergency in November 2015.

An International Health Regulations (IHR 2005) Emergency Committee met on 01 February 2016, and WHO declared the recent clusters of microcephaly and other neurological disorders in Brazil a Public Health Emergency of International Concern (PHEIC). In the absence of another explanation for the clusters of microcephaly and other neurological disorders, the IHR Emergency Committee recommended enhanced surveillance and research, and aggressive measures to reduce infection with Zika virus, particularly amongst pregnant women and women of childbearing age.

Colombia, Dominican Republic, Ecuador, El Salvador and Jamaica have all advised women to postpone getting pregnant until more is known about the virus and its rare but potentially serious complications. The US CDC has also issued a level 2 travel warning, which includes recommendations that pregnant women consider postponing travel to any area with ongoing Zika virus transmission.

WHO's Regional Office for the Americas (AMRO/PAHO) has been working closely with affected countries in the Americas on the investigation of and response to the outbreak since mid-2015. AMRO/PAHO has mobilized staff and members of the Global Outbreak Alert and Response Network (GOARN) to assist Ministries of Health in strengthening detection of Zika virus through rapid reporting and laboratory testing. A GOARN international team visited health authorities in Brazil to help assess the unprecedented increase in microcephaly cases and their possible association with Zika virus infection, as well as to provide recommendations to the Ministry of Health for surveillance, disease control measures and epidemiological research.

Need for Response

Major, epidemics of Zika virus disease may occur globally since environments where mosquitoes can live and breed are increasing due to recent trends including climate change, rapid urbanization and globalization. For the Americas, it is anticipated that Zika virus will continue to spread and will likely reach all countries and territories where *Aedes aegypti* mosquitoes are found. Other *Aedes* species are believed to be competent vectors for Zika virus and have a much farther geographical reach. For example, *Aedes albopictus* is found in temperate climates.

This strategy provides the basis for close partner coordination and collaboration in addressing this crisis to ensure that national response activities are supported to the fullest extent possible.

Strategic Objectives

The over-arching goal of this strategy is to investigate and respond to the cluster of microcephaly and other neurological complications that could be linked to Zika virus infection, while increasing preventive measures, communicating risks and providing care to those affected.

1. Surveillance

- Provide up to date and accurate epidemiological information on Zika virus disease, neurological syndromes and congenital malformations.

2. Response

- Engage communities to communicate the risks associated with Zika virus disease and promote protective behaviors, reduce anxiety, address stigma, dispel rumors and cultural misperceptions.
- Increase efforts to Control the spread of the Aedes and potentially other mosquito species as well as provide access to personal protection measures equipment and supplies.
- Provide guidance and mitigate the potential impact on women of childbearing age and those who are pregnant, as well as families with children affected by Zika virus.

3. Research

- Investigate the reported increase in incidence of microcephaly and neurological syndromes including their possible association with Zika virus infection.
- Fast-track the research and development (R&D) of new products (e.g. diagnostics, vaccines, therapeutics).

Response Strategy

Surveillance

Provide up-to-date and accurate epidemiological information on Zika virus disease, neurological syndromes and congenital malformations

Surveillance: Primary focus will be on improved understanding of the distribution, spread and nature of Zika virus infection, and trends in microcephaly and GBS. Uniform case definitions, clinical and data collection protocols will be established to improve monitoring of Zika virus infections and its potential complications. Existing vector disease surveillance systems will be adapted and enhanced to track, detect and monitor the Zika virus. Existing facility-based surveillance for detecting suspected complications will be strengthened and expanded in areas of known Zika virus infection circulation and those at highest risk. An integrated global system approach will be established that utilizes and strengthens existing surveillance systems.

Laboratories and diagnostics: Laboratory capacity to test for Zika virus infection will be expanded and other diseases relevant to their national context will be ensured. This includes upgrading existing laboratory capacities, and enabling countries to access and use Real-Time Polymerase Chain Reaction (RT PCR) tests in particular, and other diagnostics tools. Virus sharing between countries will be

encouraged. Serological diagnostics to detect evidence of past infection will be improved and/or expanded, developed, and distributed (see RESEARCH). A diagnostic algorithm will be developed for Zika virus to differentiate between other relevant diseases present in the context of the country (e.g. dengue, chikungunya, dengue, yellow fever). Timely sharing of data using existing networks (e.g. dengue) will also be ensured.

Rapid response: International alert, risk assessment and laboratory capacities (e.g. GOARN, and the French National Research Agency) will be made available to support national efforts for readiness, rapid outbreak response and field investigations.

Response

Engage communities to communicate the risks associated with Zika virus disease and promote healthy behaviours, reduce anxiety, address stigma, dispel rumours and resolve cultural misperceptions and engage in response activities

Public health risk communication: Information will be provided to key stakeholders in affected and non-affected populations, government, media, travellers and partners through systematically updated information related to the Zika virus and its complications in a format they can use and trust. News and social media channels will be monitored and analysed to identify audience concerns, knowledge gaps, rumours and misinformation. Messages will be tailored to specific audiences to ensure comprehensive guidance with special efforts made to reach excluded and the most at risk populations. Rumours and misinformation will be proactively identified and addressed.

Community engagement: Communities will be engaged for vector control and to promote personal protection measures building on existing community mobilization programmes. Potentially high-risk populations (especially pregnant women and those considering pregnancy of childbearing age) will be empowered to access medical care and given real-time information on evolving risks Rapid community assessments on social and behavioural drivers that may increase risk or facilitate protective behaviours will be conducted as needed, especially in high risk and marginalized areas. Community engagement strategies will be developed and modified based on rapid assessments, news and social media monitoring, analysis of public concerns and knowledge gaps. Communication and community engagement activities targeting health workers, teachers and other education personnel, leaders and the general public should emphasize the difficult situation of children living disabilities and other outcomes such as GBS, to minimize the risk of stigma and discrimination faced by families and children with microcephaly.

Health care personnel: Health workers will be trained, empowered and enabled to communicate risk, provide advice and specialized counselling to those affected by Zika virus disease. Family planning and antenatal care units, as well as social services for families will be strengthened and expanded to respond to increased demand for information, counselling and sexual and reproductive health commodities.

Increase efforts to Control the spread of the Aedes mosquito as well as provide access to personal protection measures

Vector control: Existing vector surveillance will be intensified in the context of Integrated Vector Management (IVM), including environmental control activities. Enhanced surveillance and control measures will be implemented in places where Aedes mosquitoes might expand, including intensification of existing control measures at breeding sites, source reduction and adult control measures. In countries where Zika virus was recently detected or has yet to be detected, vector surveillance and control will be strengthened in all border areas and at points of entry. Insecticide resistance will be assessed and advice will be provided on the use of insecticides.

Personal protection: In affected countries, there will be intensified measures to enhance personal protection measures including reducing exposed areas with long pants and shirts, use of insecticidal mosquito nets, and insect repellent. Risk communication will be targeted towards, pregnant women and those of childbearing age, taking into consideration their sexual and reproductive health and rights.

Provide guidance and mitigate the impact on pregnant women and girls and those considering pregnancy, as well as families with children affected by Zika virus

Clinical guidance and protocols: Standard guidelines, case definitions and clinical care and case management algorithms will be updated or developed as needed to help clinicians manage, monitor and understand the natural history of the disease (including the risk period for virus exposure) in pregnant women, patients with neurological syndromes, and congenital anomalies in neonates in Zika infected areas. A multi-country, multi-centre platform and centralized database for rapid

knowledge synthesis will be set up where Zika virus is circulating.

Care for those affected: Health systems will be enabled to contribute to event-based and sentinel surveillance in selected priority areas. Health care professionals involved in pre- and neonatal care should be trained in case reporting, psychosocial support, and communication skills. Surge capacity will be established to manage an increasing number of patients and potential complications, and to increase access to health care in most vulnerable areas to traditionally excluded populations. There will be enhanced access to laboratory equipment, reagents, and intensive, appropriate care for some potential complications and establishment of referral systems for specialized care. Specific focus on health center waste water management will be targeted to ensure elimination of breeding sites.

Pre-pregnancy, maternal and post-natal care: Prenatal care for pregnant women and adolescent girls will be strengthened, including performing basic investigations based on established national protocols, support for pre-natal diagnosis, expanded access to ultrasounds, especially in the third trimester. The capacity to detect and monitor congenital anomalies will be enhanced, when possible, to determine any neurodevelopmental outcomes and to provide enhance care and follow up. Pregnant women who are affected by the Zika virus or those families with babies born with microcephaly, will require specialized counselling and communication skills by trained healthcare workers to disclose the diagnosis and provide psychological support needed to care for affected infants. Additionally, increased access to a range of appropriate social protection services will be provided to mitigate the potential socio-economic impact on those families affected.

Research

Investigate the reported increase in incidence of microcephaly and neurological syndromes, including their possible association with Zika virus infection

Public health research: Partner organizations and other relevant experts will be convened to further define and expand the global research agenda for Zika virus and its potential complications. Research will focus on enhancing current knowledge, pathogenesis and etiology of infection, as well as risk factors in the transmission of Zika virus. In particular, the possible link between Zika virus and potential complications, such as GBS and microcephaly will be examined. In addition, the dynamics of Zika virus transmission—co-circulation with other pathogens, potential modes of transmission immune response, and potential complications—will be further investigated.

Fast-track the research and development of new products (e.g. rapid diagnostics, vaccines, therapeutics)

Research and product development agenda: A prioritized Zika virus research agenda will be developed for potential new approaches, tools, and product development. A landscape analysis will be conducted rapidly and the process for moving candidate vaccines and diagnostics through the R&D pipeline will be accelerated, as well as a process for fast tracking candidate, diagnostic testing.

Country Context

The current cluster of microcephaly and other neurological complications that could be linked to Zika virus infection affect countries differently—the response strategy will be tailored to meet specific needs.

The response strategies outlined above will be implemented through intervention packages tailored for each country context.

In countries where there is spread of Zika virus and increased congenital malformations/neurological syndromes, a full range of response activities will be applied. These include enhanced surveillance and outbreak response, community engagement, vector control and personal protective measures, care for people and families with potential complications, field investigations and public health research towards better understanding risk and mitigation measures.

For countries that are already experiencing the spread of Zika virus or have a documented presence of the Aedes mosquito, the first priority will be to enhance surveillance (for both Zika virus infection and potential complications to establish a baseline) as well as increasing community awareness and engagement in vector control and personal protective measures and understanding the risks associated with the Zika virus. Risk assessment will be conducted to identify areas and populations at high risk of infection and assess the systems and service capacity to respond.

For all other countries, risk communications for the public regarding trade and travel will be the main line of engagement, as well as reducing fear and misconceptions of the virus for those that are imported.

Response Coordination

A coordinated response of partners across sectors and services at the global, regional and national levels is required.

As the scale of the epidemic grows to include new countries and regions and the range of response activities increases—additional coordination mechanisms will be required. These mechanisms will need to cover a range of international and national response activities, including partners and stakeholders such as the GOARN network, UN agencies, public health research partners, national and international NGOs, regional networks and R&D partners. WHO will work closely with the Inter-Agency Standing Committee (IASC) and the UN Office for the Coordination of Humanitarian Affairs (OCHA) to ensure coordination mechanisms are interoperable with existing humanitarian response systems.

To ensure effective coordination of international partners and stakeholders at global level WHO will establish incident management teams at the global, regional and country level, as required. These teams will ensure regular communication between incident managers at different levels and close operational coordination with partners across all sectors and services at all levels.

In countries where there is spread of Zika virus and increased congenital malformations/neurological syndromes, or have a documented presence of the Aedes mosquito, WHO will support the national response efforts to increase surveillance and public health research through the GOARN partner networks and implement community engagement/risk communications, vector control/personal protective measures and health and social protections systems strengthening activities in coordination with the UN country teams as required.

For all other countries, WHO will work through the regional incident management teams to provide guidance and assist with the implementation of enhanced surveillance recommendations and risk communications for the public regarding trade and travel and management of imported cases.

In addition to these coordination mechanisms, the knowledge and lessons learned from the response in countries affected by the Zika virus and the possible links to cases of neurological syndromes and congenital malformations will need to be leveraged. The response will also need to learn from, and integrate with, existing prevention and control programmes for similar vector borne viruses such as dengue and chikungunya and dengue.

Response Monitoring

Effective response operations depend on continuous, regular and detailed surveillance and response monitoring, analysis and reporting.

Response monitoring will enable all partners across the response to have a common understanding of the situation, examine whether sufficient progress has been made against plan to reach the strategic objectives, and make evidence-based decisions for the direction of the response. The proposed response monitoring indicators for the three strategic objectives (Surveillance, Response, Research) are outlined on the following page.

Surveillance and response monitoring data and analysis will provide an overview of trends and will be used to adjust needs, targets and funding required. Response monitoring data and analysis will also allow leadership to review the general direction of the overall response and make adjustments, as necessary.

WHO is working to provide Member States with recommendations on strengthening surveillance and reporting systems in the context of the Zika virus outbreak. As part of comprehensive response monitoring, WHO also encourages and requests partners to regularly report on their response activities taking place at the global, regional and country levels.

As part of the regular strategic response plan and monitoring cycle, this Strategic Response Framework will initially be updated every six weeks, or as the need arises based on a change in circumstances or the discovery of new evidence. An overview of needs and requirements will follow the publication of this Strategic Response Framework. WHO will also publish and distribute a global situation report on a weekly basis, both through the WHO website and through email to partners. To be added to the situation report distribution list or provide response monitoring information, please email: zikainfo@who.int

Surveillance Indicators[2]

- Number of Zika virus cases
- Number of Zika virus deaths
- Number of GBS cases
- Number of GBS deaths
- Number of microcephaly cases
- Number of other malformation cases
- Number of countries with autochthonous transmission
- Number of newly affected countries

- Number and % countries with lab capacity (RT-PCR/PRNT)
- Number and % of countries with surveillance system in place for neurological complications or birth defects

Response Indicators

- Number and % countries with guidelines on case management
- Number and % countries with guidelines on management of neurological complications
- Number and % countries with guidelines on management of congenital complications
- Number and % countries with recs to public especially pregnant women for risk reduction
- Number and % affected communities with outreach and surveillance activities communicating risks and prevention measures
- Number and % of healthcare facilities with Zika clinical guidelines counselling services
- Number or and % of health care facilities with Zika laboratory and surveillance services
- Number and % of districts with active vector control programmes
- Number of call in centers with guidance on the Zika virus
- Number of countries or regions with vector surveillance
- House index (% of houses positive for mosquito breeding)
- Container index (% of containers positive for mosquito breeding)
- Pupal indice

Research Indicators

- Number of large, prospective clinical trials (>300 participants)
- Research network set up
- Research agenda defined

Summary of Requirements

WHO is currenty working with all partners to consolidate the needs and requirements across the response based on the strategic response framework. The tables below summarize the needs and requirements identified to date by response strategy and organazation. Part III of this document provides further details. The strategic response framework and needs and requirements will be reveiwed and updated as the response evolves.

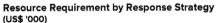

Resource Requirement by Response Strategy (US$ '000)

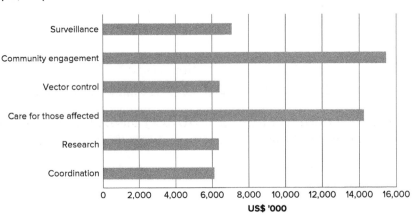

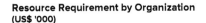

Resource Requirement by Organization (US$ '000)

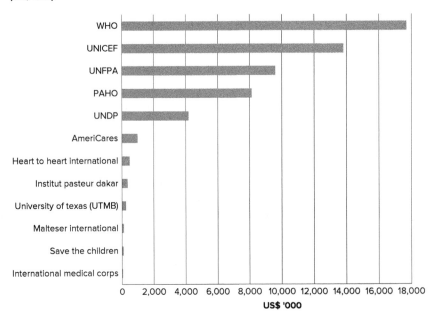

Notes

1. Guillain-Barré syndrome is a rare condition in which a person's immune system attacks their peripheral nervous system. The syndrome can affect the nerves that control muscle movement as well as those that transmit feelings of pain, temperature and touch. This can result in muscle weakness and loss of sensation in the legs and/or arms. The cause of Guillain-Barré cannot always be determined, but it is often triggered by an infection (such as HIV, dengue, or influenza) and less commonly by immunization, surgery, or trauma.

2. Indicators to be disaggregated by age and sex where appropriate/available.

WHO, THE WORLD HEALTH ORGANIZATION, is a specialized agency of the United Nations and is headquartered in Geneva, Switzerland. It focuses on world health issues.

**Matthew Weaver
and Sally Desmond**

 **NO**

Zika Virus Spreading Explosively, Says World Health Organization

Director General Convenes Emergency Committee, Saying 'Level of Alarm Is Extremely High' as Virus Has Now Been Detected in 23 Countries

The World Health Organisation has convened an emergency committee to discuss the "explosive" spread of the Zika virus, which has been linked to thousands of birth defects in Latin America.

"Last year the disease was detected in the Americas, where it is spreading explosively," Margaret Chan, the WHO director general, said at a special briefing in Geneva. It was "deeply concerning" that the virus had now been detected in 23 countries in the Americas, she added.

One WHO scientist estimated there could be 3-4m Zika infections in the Americas over the next year.

The spread of the virus has prompted governments across the world to advise pregnant women against going to the areas where it has been detected. There is no vaccine or cure for Zika, which has been linked to microcephaly, a serious condition that can cause lifelong developmental problems.

Chan said: "The level of alarm is extremely high. Arrival of the virus in some cases has been associated with a steep increase in the birth of babies with abnormally small heads."

She added: "A causal relationship between Zika virus and birth malformations and neurological syndromes has not yet been established—this is an important point—but it is strongly suspected.

"The possible links have rapidly changed the risk profile of Zika from a mild threat to one of alarming proportions. The increased incidence of microcephaly is particularly alarming as it places a heartbreaking burden on families and communities."

Chan outlined four reasons for alarm: "First, the possible association of infection with birth malformations and neurological syndromes. Second, the potential for further international spread given the wide geographical distribution of the mosquito vector. Third, the lack of population immunity in newly affected areas. Fourth, the absence of vaccines."

This year's El Niño weather patterns meant mosquito populations were expected to spread, Chan added. "For all these reasons, I have decided to convene an emergency committee under the international health regulations," she said.

The committee will meet on Monday and will advise on the international responses and specific measures in affected countries and elsewhere.

Brazilian authorities estimate the country could have up to 1m Zika infections by now, and since September, the country has registered nearly 4000 cases of babies with microcephaly.

The Zika outbreak and spike in microcephaly cases have been concentrated in the poor and underdeveloped north-east. But the south-east, where São Paulo and Rio de Janeiro are located, is the nation's second hardest-hit region. Rio de Janeiro is of particular concern, since it will host the Olympic games this summer.

The president of the International Olympic Committee, Thomas Bach, said the IOC was in "close contact" with Brazilian authorities and the WHO, and that all national Olympic bodies would be advised on how to deal with the virus before the Games started.

The Brazilian president, Dilma Rousseff, has pledged to wage war against the *Aedes aegypti* mosquito that spreads the virus, focusing on getting rid of the insect's breeding grounds.

The US Centers for Disease Control and Prevention said there had been 31 cases of Zika infection among US citizens who travelled to areas affected by the virus, but so far there had been no cases of transmission of the virus through mosquitoes in the US itself. The White House said its experts were most concerned about its potential impact on women who are pregnant or could become pregnant.

US officials said the country had two potential candidates for a vaccine, and might begin clinical trials in people by the end of this year. But experts in disease control have warned they do not expect to have a vaccine available in 2016.

Dr Anthony Fauci, director of the National Institute of Allergy and Infectious Diseases, said on Thursday that previous research into dengue fever, the West Nile virus and the chikungunya virus would give scientists an "existing vaccine platform" which could be used as "a jumping-off point" for finding a cure to the Zika virus.

"It is important to note that we will not have a widely available safe and effective Zika vaccine this year and probably not in the next few years," Fauci said, before adding that scientists might be able to begin "a phased clinical trial in this calendar year."

Addressing the global threat, Lawrence Gostin, a public health law expert from Georgetown University, warned that Zika had an "explosive pandemic potential."

Speaking to the BBC's World Service, Gostin, a member of a commission that criticised the WHO for its response to Ebola, said: "With the Rio Olympics on our doorstep I can certainly see this having a pandemic potential."

He said every review of the WHO's response to Ebola found that it was "too little, too late."

Interviewed minutes before Chan's announcement, he said: "I'm disappointed that the WHO has not been acting proactively. They have not issued any advice about travel, about surveillance, about mosquito control.

"The very first thing I would propose is a global mosquito eradication effort, particularly in areas with ongoing Zika transmission. We really need to declare war on this species of mosquito."

The WHO's leadership admitted last April to serious missteps in its handling of the Ebola crisis, which was focused mostly on three west African countries and killed more than 10,000 people.

Some critics have said the WHO's slow response played a major role in allowing the epidemic to balloon.

Zika is related to yellow fever and dengue. An estimated 80% of people that have it have no symptoms, making it difficult for pregnant women to know whether they have been infected.

Anne Schuchat, principal deputy director of the Centers for Disease Control and Prevention, said: "We really do expect there to be a lot more travel-associated cases. For the average American who is not travelling this is not something they need to worry about. For people who are pregnant and considering travel to the affected areas, please take this seriously."

She said living conditions in the US, including lower human density in urban areas and access to air conditioning, meant people were at less of a risk of contracting the virus than those living in South and Central American cities.

"We don't have local transmission of the virus in the US right now," added Fauci. "There's essentially no risk at all because we don't have locally transmitted Zika virus in the US."

"We believe this is a time-limited infection in women, men and children," said Schuchat. "People have symptoms for up to about a week, not months and years of chronic viral infection. We know four out of five people with this infection have no symptoms."

She added the Zika virus passed very quickly through the bloodstream and in most cases the virus would clear from the bloodstream within about a week.

Asked when the babies of pregnant mothers could become infected with the virus, Schuchat said the foetus was most at risk of contracting microcephaly through Zika in the first trimester of the pregnancy. Scientists did not yet have "sufficient knowledge to know what effects in the second and third trimester," she cautioned.

There has been one reported case of the Zika virus through "possible sexual transmission," while a second case was found in a man's semen. However, Schuchat highlighted that scientific research clearly showed Zika was "primarily transmitted through the bite of an infected mosquito."

MATTHEW WEAVER AND SALLY DESMOND are reporters who write for *The Guardian*'s website. *The Guardian* is a British national daily newspaper founded in 1821.

EXPLORING THE ISSUE

Will the International Community be Able to Successfully Address the ZIKA Virus Pandemic?

Critical Thinking and Reflection

1. Is the challenge of successfully combating global pandemics simply too complex to succeed?
2. Does the Emergency Operations Plan of WHO appear to provide a comprehensive strategy against the Zika virus that has a strong potential to work?
3. Is the developing world at greater risk against global pandemics such as the Zika virus than the developed world?
4. Do you believe that the developed world's governments have adequately taken notice of the nature and scope of the Zika virus and have begun to act to address the issue?

Is There Common Ground?

There is much common ground on the issue of global pandemics. There is much consensus regarding the threat posed by the Zika virus. The global community is also in agreement about the need for comprehensive planning against any pandemic and more specifically against the Zika virus among all governments if not all segments of society. And increasingly, the developed world now understands the inequalities existing in the developing world and that these lead to greater probability of the rise and spread of pandemics originating in these poorer countries.

Additional Resources

Al-Qahtani, Ahmed A., et al., "Zika Virus: A New Pandemic Threat," *Journal of Infection in Developing Countries* (vol. 10, no. 3, March 2016)

This article provides basic information about the virus, including its background.

Brown, Jennifer J., "10 Essential Facts about Zika Virus," *EverydayHealth* (www.everydayhealth.com/news/10-essential-facts-about-Zika-virus/, May 12, 2016)

This article highlights 10 important facts about the Zika virus.

Gallagher, James, "Zika Outbreak: What You Need to Know," *BBC News* (www.bbc.com/news/health-3570848, April 13, 2016)

This article describes the most important aspects of the Zika virus and its effects.

Harris, Lisa H., et al., "The Paradigm of the Paradox: Women, Pregnant Women, and the Unequal Burdens of the Zika Virus Pandemic," *American Journal of Bioethics* (vol. 16, no. 5, May 2016)

The article critiques paradoxes of the Zika virus and describes its burdens on women.

Idris, Fauziah Mohamad, "Zika: A Pandemic in Progress?" *Malaysian Journal of Medical Sciences* (vol. 23, no. 2, March/April 2016)

This article focuses on the ability of the Zika virus to become a pandemic.

Lucey, Daniel R. and Gostin, Lawrence O., "The Emerging Zika Pandemic: Enhanced Preparedness," *JAMA: Journal of the American Medical Association* (vol. 315, no. 9, March 1, 2016)

This article describes the origin of the virus and how to prepare for it.

Mackenzie, Debora, "How Zika became a Threat," *New Scientist* (vol. 229, no. 3058, January 30, 2016)

The article focuses on the pandemic in 2016, describing its current level of threat.

Russell, Philip K., "The Zika Pandemic: A Perfect Storm," *PloS Neglected Tropical Diseases* (vol. 10, no. 3, March 18, 2016)

This article focuses on the epidemiology of the Zika virus.

Internet References . . .

Avian and Pandemic Influenza Research Link

www.avianflu.aed.org/globalpreparedness.htm

Centers for Disease Control and Prevention

www.cdc.gov

European Centre for Disease Prevention and Control

Eclc.europa.eu/en/healthtopics/zika-virus-infection/

Indiana University Center for Bioethics

www.bioethics.iu.edu/reference-center
/pandemics-influenza/

United Nations Food and Agricultural Organization

www.fao.org/zika-virus/en/

WebMD

www.webmd.com

World Health Organization

http://www.who.int/en/

Selected, Edited, and with Issue Framing Material by:
James E. Harf, *Maryville University*
and
Mark Owen Lombardi, *Maryville University*

ISSUE

Is the International Community Making Effective Progress in Securing Global Human Rights?

YES: **The Council on Foreign Relations**, from "The Global Human Rights Regime," *The Council on Foreign Relations* (2012)

NO: **Amnesty International**, from "Amnesty International Report 2014/15: The State of the World's Human Rights," Amnesty International, London (2015)

Learning Outcomes

After reading this issue, you will be able to:

- Develop a basic understating of the evolution of the international community's concern for human rights.
- Distinguish between how the international community approached the human rights issue prior to and after World War II.
- Briefly describe the basic role played by the United Nations in the advancement of human rights between 1945 and 2006.
- Briefly discuss the tension between universal human rights and national sovereignty.
- Briefly discuss the different approaches to human rights taken by the Western and communist blocs during the Cold War.
- Describe recent progress made by the international community in advancing universal human rights.
- Describe current challenges faced by the international community in advancing human rights.
- Describe Amnesty International's analysis of recent failures in the international advancement of human rights, with special emphasis on the impact of war on human rights.

ISSUE SUMMARY

YES: The Council on Foreign Relations, an independent nonpartisan and essentially American think tank, in an *Issue Brief* summarizes the development of an elaborate global system of governmental and nongovernmental organizations developed primarily over the past few decades to promote human rights throughout the world, while recognizing that the task is still far from complete.

NO: Amnesty International's annual report on the state of human rights around the world suggests major failures in all regions, with suffering by many from conflict, displacement, discrimination, and/or repression.

Human rights has become such a common phrase in our everyday vocabulary that one might think that the seeking and practice of human rights have been part of human behavior for well over a millennium. This view is probably reinforced by the fact that one of the enduring events of a high school world history class is the signing of the Magna Carta in 1215, acknowledged by most as the first successful effort by humans to limit the power of a sovereign and thus protect some of its subjects.

But indeed, we can go back to the ancient Greeks for the first references to the idea that humans have fundamental rights and freedoms simply by virtue of the fact that they are members of the human race. The Greek notion of universal rights, championed by Greek philosophers or Stoics, influenced Roman thinking, which in turn ultimately spread throughout Europe. Those who read political philosophy are familiar with men like John Locke, Immanuel Kant, Hugo Grotius, Rousseau, and Voltaire.

These men's writings influenced the early attempts of citizens to rise up against their national rulers demanding protection of rights that they believed originated not with these human rulers but with a higher order. English history is replete with references to citizen demands for sovereign recognition of their rights, culminating with a revolution against the king in 1688. But it is the American revolution of 1776 and the French Revolution of 1789 that have captured the headlines in the early struggle for human rights. The former was a rather modest revolution, conducted on the basic principle that individuals have a right to determine their own form of government and also their ruler. The French episode was much broader, representing a struggle for equality throughout society.

These revolutions and others that followed were characterized by one major fact. They were staged within the confines of a single country. That is, they were domestic battles launched against a national ruler and fought within the context of the nation-state system. The latter international arrangement, emerging in the early sixteenth century, laid out the norms of national governments and the behavior among them. Chief among these rules was the idea that national leaders were sovereign. That is, they had complete authority over how they ruled, including how they treated their citizens. And the rest of the world was not to interfere in the internal affairs of the country, no matter how egregious the ruler's behavior.

By the nineteenth century, however, cracks began to appear in the extreme idea of total national sovereignty, first on the issue of slavery and later on the issue of soldiers, prisoners, and other victims of war. But it was not until World War II and the initial evidence of Nazi atrocities that national leaders began to think and talk seriously about the need for global human rights. President Franklin Roosevelt in early 1941 announced to Congress America's commitment to four freedoms—of speech and expression, of religion, from economic hardship, and from fear. These freedoms were formalized later that year in the Atlantic Charter by Roosevelt and his British counterpart, Winston Churchill.

As World War II was winding down in spring 1945, world leaders met in San Francisco and created a new international organization, the United Nations, giving it responsibility for not only maintaining and restoring peace throughout the world but also entrusting it with other important responsibilities, among them identifying and securing human rights for citizens throughout the globe. Indeed, the UN Charter mentions human rights many times through the document.

The first step in identifying human rights was the UN's adoption of the Universal Declaration of Human Rights in 1948, which lays out 29 separate categories of human rights and claims them to be universal, that is, relevant for all citizens of the globe no matter where they live and no matter the form of government under which they live. How a government treats its own citizens was now assumed to be a legitimate international concern that ought be addressed by appropriate international bodies such as the United Nations. It soon became evident that the two major players in the international arena, the Western bloc and the communist or Soviet bloc of nations, had different ideas over precisely which kinds of human rights were important, however. In short, the West focused on political and civil rights, while the Eastern bloc favored economic and social rights, thus creating tensions between the two blocs.

The other stumbling block in global efforts to ensure human rights for everyone was the clash between the concept of the universality of human rights and the basic principle of the nation-state system, national sovereignty. This was not a fight between East and West. Rather it was a struggle between democratic governments and nongovernments everywhere. The former suggested that government was limited and its powers flowed from its own citizens, while the latter argued that national rulers had few if any limitations on their capacity to rule their subjects, including freedom from outside interference. More recently, the concept of universality itself has been once again called into question, as some believe that earlier documents were biased in favor of Western thinking and also ignored other religious and national cultures.

Despite the challenges caused by both the Cold War and the national sovereignty versus universal human rights argument, the international community became increasingly interested in how governments treated their own citizens as well as how others in positions of authority treated those individuals under them. In fact, the United Nations was immediately busy with the task of codifying national behavior in many different areas of human rights. Its initial action was a Convention on Genocide, passed in 1948 as the Universal Declaration was in its last day of debate. This was the first of over 60 binding human rights treaties in effect today. It also immediately created a permanent body, the United Nations Commission on Human Rights (UNCHR), which focused on the promotion and protection of human rights. In 2006, the UN replaced UNCHR with a new body, the UN Human Rights Council. The work of UNCHR can be divided into two distinct phases. From 1947 until 1967, the promotion rather than the protection of human rights was emphasized, but after 1967 it focused on an interventionist strategy of protection. UNCHR was not without its critics during its 60-year existence, as the organization was often attacked for including member states whose record of human rights compliance within their own national boundaries was horrible, and for its selective "finger-pointing," with first South Africa and later Israel being all too frequent targets of attention. The move to disband UNCHR and replace it with another UN agency gained steam and was finally accomplished in 2006, but critics have not been silenced.

The United Nations has not been the only global body to concern itself with human rights. Increasingly, other international regional governmental agencies as well as international nongovernmental organizations, many with a secular focus on one area such as child abuse or abuse of women, have joined the fight against human rights violators. It is not surprising that nongovernmental organizations are playing increasingly important roles, as its citizen members strive to fill the void caused by inaction of reluctant leaders of national governments.

One alternative approach to analyzing progress on human rights is to focus on very specific areas within the broad range of political, economic, and social rights, for progress has been significant in selected categories of rights. Two such successful examples relate to the international community's handling of perpetrators of war crimes, as the example of Slobodan Milošević of Serbia shows, or a

revolutionary government's handling of deposed dictators such as revealed in the case of Hosni Mubarak of Egypt or Saddam Hussein of Iraq, not to mention half a dozen other ousted or soon-to-be-ousted dictators. It was not too long ago when those national leaders who engaged in unjust wars or abused their own citizens were never made to account for their behavior.

Another alternative approach to the question of international progress on human rights is to focus on the extent to which the global community has made significant progress on meeting the eight Millennium Development Goals spelled out at the turn of the century by the United Nations. The latter organization, in fact, used a human rights framework to report in 2010 on the extent to which nations of the globe have met these goals.

In the YES selection, the Council on Foreign Relations suggests an on-balance "glass is half-full" viewpoint, namely that an "elaborate global system is being developed" for the protection of human rights. While alluding to the fact that more national governments are promoting human rights, it suggests that an increasing "dynamic and decentralized network of civil society actors" is beginning to make a significant difference. In the NO selection, Amnesty International's annual report (2015) sums up the current state of progress in human rights advocacy by asserting that it "has been a devastating year for those seeking to stand up for human rights and for those caught up in the suffering of war zones." In war-torn zones, non-combatants bear a much high burden than in decades past. A case in point is Syria where the numbers for civilians killed or displaced are shockingly high. Amnesty International also points out the dire situations in Iraq, Gaza, Nigeria, the Central Africa Republic, and South Sudan as examples of the devastation of civilians in war zones.

YES ⬑

The Council on Foreign Relations

The Global Human Rights Regime

Scope of the Challenge

Although the concept of human rights is abstract, how it is applied has a direct and enormous impact on daily life worldwide. Millions have suffered crimes against humanity. Millions more toil in bonded labor. In the last decade alone, authoritarian rule has denied civil and political liberties to billions. The idea of human rights has a long history, but only in the past century has the international community sought to galvanize a regime to promote and guard them. Particularly, since the United Nations (UN) was established in 1945, world leaders have cooperated to codify human rights in a universally recognized regime of treaties, institutions, and norms.

An elaborate global system is being developed. Governments are striving to promote human rights domestically and abroad, and are partnering with multilateral institutions to do so. A particularly dynamic and decentralized network of civil society actors is also involved in the effort.

Together, these players have achieved marked success, though the institutionalization and implementation of different rights is progressing at varying rates. Response to mass atrocities has seen the greatest progress, even if enforcement remains inconsistent. The imperative to provide people with adequate public health care is strongly embedded across the globe, and substantial resources have been devoted to the challenge. The right to freedom from slavery and forced labor has also been integrated into international and national institutions, and has benefited from high-profile pressure to combat forced labor. Finally, the steady accumulation of human rights-related conventions has encouraged most states to do more to implement binding legislation in their constitutions and statutes.

Significant challenges to promoting human rights norms remain, however. To begin with, the umbrella of human rights is massive. Freedom from slavery and torture, the imperative to prevent gender and racial persecution, and the right to education and health care are only some of the issues asserted as human rights. Furthermore, nations continue to dispute the importance of civil and political versus economic, social, and cultural rights. National governments sometimes resist adhering to international norms they perceive as contradicting local cultural or social values. Western countries—especially the United States—resist international rights cooperation from a concern that it might harm business, infringe on autonomy, or limit freedom of speech. The world struggles to balance democracy's promise of human rights protection against its historically Western identification.

Moreover, implementing respect for established human rights is problematic. Some of the worst violators have not joined central rights treaties or institutions, undermining the initiatives' perceived effectiveness. Negligence of international obligations is difficult to penalize. The UN Charter promotes "fundamental freedoms," for example, but also affirms that nations cannot interfere with domestic matters. The utility of accountability measures, such as sanctions or force, and under what conditions, is also debatable. At times, to secure an end to violent conflict, negotiators choose not to hold human rights violators accountable. Furthermore, developing nations are often incapable of protecting rights within their borders, and the international community needs to bolster their capacity to do so—especially in the wake of the Arab Spring. Finally, questions remain over whether the UN, regional bodies, or other global actors should be the primary forums to advance human rights.

In the long term, strengthening the human rights regime will require a broadened and elevated UN human rights architecture. A steady coalition between the global North and South to harmonize political and economic rights within democratic institutions will also be necessary. In the meantime, regional organizations and non-governmental organizations must play a larger role from the bottom up, and rising powers must do more to lead. Together, these changes are the world's best hope for durable and universal enjoyment of human rights.

Human Rights: Strengths and Weaknesses

Overall Assessment: *Heightened Attention, Uneven Regional Efforts, Weak Global Compliance*

The international human rights regime has made several welcome advances—including increased responsiveness in the Muslim world, attention to prevention and accountability for atrocities, and great powers less frequently standing in the way of action, notably at the UN Security Council (UNSC). Yet, despite responses to emerging cases demanding action, such as Sudan and Libya, global governance in ensuring human rights has faltered.

Many experts credit intergovernmental organizations (IGOs) for advances—particularly in civil and political rights. These scholars cite the creation of an assortment of secretariats, administrative support, and expert personnel to institutionalize and implement human rights norms. Overall, the United Nations (UN) remains the central global institution for developing international norms and legitimizing efforts to implement them, but the number of actors involved has grown exponentially.

The primary mechanisms include UNSC action, the UN Human Rights Council (UNHRC), committees of elected experts, various rapporteurs, special representatives, and working groups. War crimes tribunals—the International Criminal Court (ICC), tribunals for the former Yugoslavia and Rwanda, and hybrid courts in Sierra Leone and Cambodia—also contribute to the development and enforcement of standards. All seek to raise political will and public consciousness, assess human rights–related conduct of states and warring parties, and offer technical advice to states on improving human rights. . . .

Of all UN bodies with a similar focus, the UNHRC receives the most attention. In its former incarnation as the Commission on Human Rights, it developed a reputation for allowing the participation—and even leadership—of notorious human rights abusers, which undermined its legitimacy. Reconstituted as the UNHRC in 2006, the new forty-seven member body has a higher threshold for membership as well as a universal periodic review process, which evaluates the human rights records of states, including those on the council. Nevertheless, the UNHRC's effectiveness has been uneven. On the one hand, it took the unprecedented decision to vote to suspend Libya in 2011 and passed a pioneering resolution on sexual orientation. On the other, it maintains a disproportionate focus on Israel, ignores major abuses in other nations, avoids condemning specific rights issues, and still includes serial rights abusers in its ranks. . . .

Increasingly, the locus of activity on human rights is moving to the regional level, but at markedly different paces from place to place. Regional organizations and powers contribute to advancing human rights protections in their neighborhoods by bolstering norms, providing mechanisms for peer review, and helping countries codify human rights stipulations within domestic institutions. Regional organizations are often considered the first lines of defense, and better able to address rights issues unique to a given area. This principle is explicitly mentioned in the UN Charter, which calls on member states to "make every effort to achieve pacific settlement of local disputes through such regional arrangements or by such regional agencies" before approaching the UNSC.

Major regional organizations in the Western Hemisphere, Europe, and Africa—such as the Organization of American States (OAS), the European Union (EU), and the African Union (AU)—have integrated human rights into their mandate and established courts to which citizens can appeal if a nation violates their rights. This has led to important rulings on slavery in Niger and spousal abuse in Brazil, for example, but corruption continues to hamper implementation throughout Latin America and Africa, and a dearth of leadership in African nations has slowed institutionalization.

Meanwhile, organizations in the Middle East and Asia, such as the Association of Southeast Asian Nations (ASEAN) and the South Asian Association for Regional Cooperation, focus primarily on economic cooperation and have historically made scant progress on human rights. The Arab League, however, departed from its indifference to human rights in 2011, backing UN action against Libya and sanctioning Syria, and may prove more committed to protecting human rights in the wake of the Arab Spring.

Civil society efforts have achieved the most striking success in human rights, though they often interact with international institutions and many national governments. Nongovernmental organizations [NGOs] provide valuable data and supervision, which can assist both states and international organizations. NGOs also largely rely on international organizations for funding, administrative support, and expert assistance. Indeed, more than 3,000 NGOs have been named as official consultants to the UN Economic and Social Council alone, and many more contribute in more abstract ways. Domestic NGOs understand needs on the ground far better than their international counterparts. That international NGOs are beginning to recognize this is clear in two recent developments: the first is financier-philanthropist George Soros's $100 million donation to Human Rights Watch to develop field offices staffed by locals, which enabled the organization to increase its annual operating budget to $80 million. Second, the number of capacity-building partnerships between Western-based NGOs and NGOs indigenous to a country is increasing. That said, NGOs have to date been more successful in advocacy—from achieving passage of the Anti-Personnel Mine Ban Convention to calling attention to governments' atrocities against their own citizens. Yet NGOs devoted to implementing human rights compliance have been catching up—on issues from democratic transitions to gender empowerment to protecting migrants.

Norm and Treaty Creation: *Prodigious but Overemphasized*

The greatest strength of the global governance architecture has been creating norms. Myriad treaties, agreements, and statements have enshrined human rights on the international community's agenda, and some regional organizations have followed suit. These agreements lack binding clauses to ensure that action matches rhetoric, however, and many important violators have not signed on. In addition, states often attach qualifiers to their signatures that dilute their commitments.

The array of treaties establishing standards for human rights commitments is broad—from political and civil liberties to economic, social, and cultural rights to racial discrimination to the rights of women, children, migrant workers, and more recently the disabled. Other global efforts have focused on areas such as labor rights and human trafficking. Regional organizations, most notably the Council of Europe and the Organization of American States, have also promulgated related instruments, although less uniformly. In addition, member states have articulated declarations and resolutions establishing human rights standards, and increasingly so in economic affairs. The United Nations Human Rights Council, in a departure from the premise that states are to be held accountable for human rights conduct, in 2011 even passed formal guidelines for related business responsibilities. . . .

Rights Monitoring: *Proliferating Experts, Increasing Peer-Based Scrutiny*

Monitoring is imperative to matching rhetoric with action. Over the years, human rights monitoring has matured and developed considerably, though serious challenges remain, such as ensuring freedom from torture for suspected terrorists, and uniformly protecting and promoting human rights despite the biases of rights organizations or officials entrusted with doing so.

The original United Nations Commission on Human Rights and its successor Human Rights Council (UNHRC) both authorized a wide array of special procedures to monitor human rights protection in functional areas and particular countries. Since the UNHRC was established in 2006, country-specific mandates have decreased, and functional monitors addressing economic and social rather than political and civil liberties have increased.

In addition, each UN human rights treaty has an elected body of experts to which state parties must report at regular intervals on implementation. For instance, the Human Rights Committee (not to be confused with the Council) is charged with receiving reports about the implementation of the International Covenant on Civil and Political Rights (ICCPR) and making nonbinding "concluding observations" about states' overall compliance. The Human Rights Committee can also receive complaints from individuals regarding state compliance with the accord. The UN Convention against Torture's monitoring mechanism, the Committee against Torture, is similar but can also send representatives to inspect areas where torture is suspected. However, the applicability of this unique monitoring tool is limited because investigations can be undertaken only in states that have ratified the First Optional Protocol to the UN Convention against Torture, which only sixty-one of 193 UN member states have done. . . .

Capacity Building: *Vital but Underemphasized*

Capacity building—especially for human rights—is often expensive and daunting, viewed with suspicion, and the success of assistance is notoriously hard to measure. In many cases, national governments have signed international commitments to promote and protect human rights, and earnestly wish to implement them, but are incapable of doing so. For example, many experts have noted that Libya may require an entirely new judicial system, following the collapse of Muammar al-Qaddafi's regime. On the other hand, some states refuse assistance from nongovernmental organizations (NGOs) and international organizations (IGOs), suspecting that it might interfere with domestic affairs. On balance, it also remains far easier, and less costly, for the international community to condemn, expose, or shame human rights abusers rather than provide material aid for human rights capacity building.

The international community has developed various ways to offer technical assistance. Most notable is the Office of the High Commissioner for Human Rights (OHCHR), established in 1993. In addition to providing an institutionalized moral voice, OHCHR offers technical assistance to states through an array of field offices—for example, by providing training to civilian law enforcement and judicial officials through its country office in Uganda, strengthening the Cambodian legal and institutional framework for human rights, and assisting Mexico with development of a National Program on Human Rights. Such work is undercut, however, by member states' propensity to prefer unilateral support for capacity building, to favor naming and shaming over capacity building, or to oppose human rights capacity building as either a threat to sovereignty or tantamount to neocolonialism.

Regional organizations such as the Organization for Security Cooperation in Europe (OSCE), Council of Europe, Organization of American States (OAS), European Union, and to some extent the African Union, may be more effective than the United Nations in sharing best practices and providing capacity-building advice to states. Often capacity building entails training human rights protectors and defenders, but it may also include legal framework building or addressing countries' specific capacity deficits. The OSCE, for instance, collaborates with member states on election monitoring and offers training and education to human rights defenders through its Office for Democratic Institutions and Human Rights. In another example, the OAS collaborates with European partners in its judicial facilitators program, which trains judicial officials in rural areas with limited access to justice, and assisted Haiti in establishing a civil registry. Still, sharing resources and coordinating between IGOs and NGOs in capacity building are limited. As mentioned, norm creation has outstripped both monitoring and implementing norms.

Human rights capacity building also occurs on a bilateral basis. Indeed, some developed states prefer providing bilateral assistance to working with IGOs and multilateral institutions because resources can be better monitored and projects more carefully tailored to support donor state interests. For instance, the U.S. Foreign Assistance Act of 1961, which laid the basis for the creation of the U.S. Agency for International Development (USAID), calls for the use of development assistance to promote economic and civil rights. Since its inception, USAID has provided billions of dollars to support good governance, transparency building, and civil society projects worldwide. It recently gave hundreds of millions of dollars to Liberia to train judges, promote the rule of the law, and increase government transparency.

Meanwhile, other multilateral institutions like the World Bank, International Monetary Fund, and World Trade Organization also support human rights promotion, but tend to do so more indirectly, through poverty alleviation and community enhancement schemes. Together, though, these institutions face new constraints as the international community continues to grapple with the global financial crisis and unprecedented budget deficits. . . .

As a whole, successful capacity building forms the core of long-term efforts to improve human rights in countries. Regardless, human rights capacity building is often underemphasized both in states with the poorest of human rights as well as among countries or intergovernmental organizations that are most in a position to help. While NGOs are crucial contributors to capacity-building efforts, they cannot—and should not—shoulder the entire burden. Broad, crosscutting partnerships are essential for such efforts to enjoy success and produce sustainable human rights reform.

Response to Atrocities: *Significant Institutionalization, Selective Action*

Atrocities of all sorts—whether war crimes, genocide, crimes against humanity, or ethnic cleansing—have been a major focus in the international community over the last two decades. A number of regional and country-specific courts, as well as the International Criminal Court (ICC), provide potential models for ending impunity. However, these courts have unevenly prosecuted violators of human rights, and have been criticized for focusing on some abuses or regions while ignoring others.

In the aftermath of the Balkans and Rwanda in the 1990s, where UN peacekeepers on the ground failed to prevent mass killing and sexual violence, efforts to establish preventive and responsive norms to atrocities accelerated. To hold perpetrators accountable, the Rome Statute established the ICC as the standing tribunal for atrocities. The ICC was largely considered an alternative to ad hoc tribunals like those for the former Yugoslavia and Rwanda, which were criticized for proceeding too slowly and for requiring redundant and complex institution building. The ICC is the result of UN efforts to evaluate the prospects for an international court to address crimes like genocide as early as 1948. . . .

As for preventive action, former UN secretary-general Kofi Annan championed stronger norms for intervention against ongoing atrocities. In the wake of the Kosovo crisis, Annan cited the need for clarifying when international intervention should legally be used to prevent atrocities in states. In response, the Canadian-sponsored

International Commission on Intervention and State Sovereignty promoted the concept of the "responsibility to protect" (R2P) in 2000 and 2001. This principle sought to reframe the debate over humanitarian intervention in terms of state sovereignty. Specifically, it placed the primary responsibility on states to protect their own citizens. When states failed, responsibility would fall to the international community. Annan's *In Larger Freedom* report picked up on this concept, and R2P informed two paragraphs in the Outcome Document of 2005 UN World Summit. The latter also included an emphasis on the importance of capacity-building assistance to help states meet their R2P obligations. In the UN Security Council (UNSC), the R2P doctrine has been invoked repeatedly— first generically affirmed, then raised in semi-germane cases in 2008 (in Myanmar after a cyclone and in Kenya during post-election violence), and then more conclusively in 2011 (UNSC Resolution 1973 on Libya). . . .

In short, the international community has taken its greatest step by redefining sovereignty as answerable to legal international intervention should a state fail to shield its citizens from atrocities, or worse yet, sponsor them. However, state practice has not matched these norms, and it remains to be seen whether consensus about Libya was sui generis.

Political and Civil Rights: *Disproportionately Institutionalized, Backlash on Free Expression and Association*

Treaties that define political and civil liberties are widely ratified, but many countries have not signed on to enforcement protocols, and many continue to violate the rights of their citizens regardless of treaties. In addition, the right of people to choose their leaders and freedom of the press, religion, and association has backslid in recent years. At the same time, however, people are increasingly demanding rights and attempting to bypass repression of illiberal regimes. New technology (such as cell phones, social media, and satellite television) is also providing unprecedented opportunities to publicize abuse and organize protests, though repressive regimes are closely following with practices to censor new technology.

States resisting the spread of political and civil liberties have been challenged more by civil society than by other states or by intergovernmental organizations (IGOs). Using information and communications technology, and with the support of global nongovernmental organizations (NGOs) and occasionally the private sector, civil society have taken their demands to a new level.

China's effort to control dissent, for example, has been greatly challenged by Uighur dissenters in Xinjiang, Falun Gong groups, and the decision by Google to refuse to implement comprehensive censorship in China. However, international pressure remains relevant. For example, the Obama administration's recent statement that censorship practices in China may violate World Trade Organization rules has increased pressure on China to reform.

In the United Nations, the number of member states, organs, and generic mandates related to freedom of expression and association have increased. For instance, the UN General Assembly adopted a resolution in 2007 calling for the end of capital punishment. In September 2010, the UN Human Rights Council (UNHRC) adopted another resolution, creating a special rapporteur on rights to freedom of peaceful assembly and of association. This occurred in the wake of a multiyear backlash against domestic NGOs and their international philanthropic and civil society backers in a series of autocracies. . . .

Economic Rights and Business Responsibilities: *Increased Corporate Focus and Engagement*

A long-standing debate between the global North and global South has been over whether to prioritize negative obligations of states to avoid restricting political and civil liberties or positive obligations to deliver economic and social benefits. Indicators, however, show a subtle yet important shift in the last ten of the forty-year debate.

Until the end of the twentieth century, international law frameworks placed human rights obligations on the shoulders of states. Not least through former UN secretary-general Kofi Annan's role as an ideas entrepreneur, notions of the obligation of businesses on human rights have blossomed. First, in 2000, Annan and his Harvard-based scholar-adviser John Ruggie crafted the UN Global Compact, which enumerates voluntary principles for business related to human rights and environmental stewardship. The UN then created a mandate for a special representative of the secretary-general to assess state, business, and civil society stakeholders on business conduct and human rights. In July 2011, the UN Human Rights Council (UNHRC) adopted guidelines that delineate state obligations to protect human rights, business obligations to respect them, and a joint role to provide remedies to people robbed of them. These successes do not come without challenges, however. Ruggie, who has been at the forefront of business and human rights, completed his term as special representative in mid-2011, raising the prospect that UN efforts may stall in his absence. Further, although the UN Security Council's adoption of the Global Compact guidelines is significant, implementation will be a difficult next step. Additionally, the International Labor Organization (ILO) and its counterpart, the International Organization of Employers, have jointly engaged businesses on best practices on human rights.

Nevertheless, businesses' decisions to uphold human rights standards remain largely voluntary and thus subject to market—rather than moral—forces. Even when businesses make commitments to corporate responsibility programs, no actor exists to enforce such commitments. Civil society can play a critical role in mitigating these challenges, however, by publicizing corporate human rights abuses and working directly with businesses on corporate responsibility. NGOs such as Human Rights Watch, the Institute for Human Rights and Business, the International Federation for Human Rights, Global Witness, and the International League for Human Rights exemplify these efforts. Additionally, even where businesses act in violation of domestic laws or international conventions protecting human rights, limited domestic law enforcement capabilities undermine the force of accountability standards.

The international community's efforts to address economic and social rights have advanced. Some measures evidence a redefinition of human rights, such as the mandate from the UNHRC on toxic waste. Some entail ambitious norm setting, such as the UN Convention on the Rights of Persons with Disabilities, negotiated during the George W. Bush administration and signed by the Obama administration. Most important have been efforts to address economic and social rights with tangible programming. The Global Fund to Fight AIDS, Tuberculosis, and Malaria is a landmark achievement for bridging health, economic, and discriminatory ills; for mobilizing significant resources beyond regular assessed budgets of the UN; and for involving an array of UN, private sector, philanthropic, and civil society actors in a concerted partnership. It is worth noting that the global North (and its greatest skeptic on economic and social rights, the United States) have championed this effort, supplementing it heavily through the U.S. President's Emergency Plan for AIDS Relief (PEPFAR).

Child labor, forced labor, human trafficking, and contemporary slavery have also become a focus of global governance efforts since the beginning of the twenty-first century. Such abridgments of freedom and autonomy signal a tragic combination of economic desperation, weak rule of law, and discrimination. The ILO's work to address forced labor and the most acute forms of child labor through conventions and preventive programs has now been supplemented by other efforts. New energy has been directed to mitigating the most coercive of labor practices as a result of the near simultaneous enactment of the Palermo Protocol to the UN Crime Convention on Trafficking in Persons (TIP) and the U.S. Victims of Trafficking and Violence Protection Act in 2000.

The UNHRC has also authorized special rapporteurs on both human trafficking and contemporary slavery. States, intergovernmental organizations, and NGOs have developed partnerships to address child labor, forced labor, and human trafficking. Businesses are also joining global governance efforts, moving from sector-specific partnerships (such as the travel and hospitality sector on child sex trafficking and chocolate companies on child labor in West Africa) to cross-sectoral ones (such as the Athens Ethical Principles and emerging thought-leader coalitions).

Women's and Children's Rights: *Institutional Progress but Holdouts on Implementation*

The rights of women have advanced incrementally. The United Nations (UN) system has moved beyond creating norms, such as the Convention on the Elimination of All Forms of Discrimination against Women and the Convention on the Rights of the Child to more assertive leadership and calls for implementation efforts among national governments. However, despite marked success on various fronts, the UN estimates that women continue to make up less than 10 percent of world leaders and less than one-fifth of parliamentarians. Moreover, it remains to be seen whether the Arab Spring will help or hinder the cause of gender equality. Efforts to enhance the economic and social wellbeing of women and children have also improved, but remain at risk as a result of tightened national and international aid budgets.

Arguably, the decision of the UN Development Program to commission reports by Arab experts to link gender inequality and reduced development in the Arab world, published in 2005, was an important step forward. The formation of the UN Entity for Gender Equality and the Empowerment of Women (UN Women), amalgamating four existing agencies, received an additional boost when Chile's Michelle Bachelet was appointed its first leader. The remaining question is whether the consolidation of women's rights functions will mainstream or silo them. Around the world, more women have become involved in political participation—from the first woman elected head of state in Africa to the franchise in Gulf States.

The essential role of women in peace and consensus building has moved from statements like UN Security

Council Resolution (UNSCR) 1325, which recognized that women are not adequately consulted and integrated into peace processes, to reality. In December 2011, for example, the United States joined thirty-two other countries in publishing a National Action Plan (NAP) on Women, Peace and Security designed to integrate governmental efforts to implement UNSCR 1325. Ellen Johnson Sirleaf's leadership in postconflict Liberia and the July 2010 establishment of UN Women provide further evidence of the international community's improving recognition of the indispensable role of women in postconflict situations.

Moreover, attention to the acute problem of violence against women has advanced, even if it has been significantly curtailed in practice. In 1998, The International Criminal Tribunal for the former Yugoslavia (ICTY), along with the Rome Statute, established the precedent that targeted rape is a crime against humanity, though the practice has continued largely unabated in Darfur, the Democratic Republic of the Congo, Burma, and Zimbabwe. The degree to which prostitution of girls and sex trafficking of women is an act of violence is beginning to be better understood around the world. . . .

Other Group Rights: *Heightened Focus, Selective Bias*

Dedicated efforts to address the rights of particular groups have advanced for some, but stalled for others. Racism and other forms of xenophobia have been a major focus. Organization of American States (OAS) members have been negotiating over an antiracism convention proposed by Brazil since 2005, to follow in the footsteps of the United Nations Convention on the Elimination of All Forms of Racism and monitoring regime. The UN process, despite the 1991 repeal of UN General Assembly Resolution 3379 (classifying Zionism as a form of racism), has been sidetracked by the issue of Israel and its occupation of Palestinian territories. The 2001 UN World Conference against Racism in Durban came close to declaring Israel to be racist, and follow-on efforts, such as at the 2009 Review Conference, had a similarly skewed focus. In practice, however, certain great exemplars of antiracism have transcended, from South Africa's reconciliation under Nelson Mandela to Barack Obama's election in a nation in which segregation was widely institutionalized a half century earlier. Sadly, many varied instances of racism and xenophobia remain, from anti-Semitic violence in Europe to anti-white land seizure policies in Zimbabwe. . . .

In short, an increasing number of groups have been recognized by multilateral bodies, states, and publics as deserving equal access to justice. Implementation efforts are spottier. Second, cultural legacies of prejudice may persist as more and more groups lobby for rights.

The Council on Foreign Relations is an independent nonpartisan, essentially American membership organization, think tank, and publisher, founded in 1921 for the purpose of serving as a resource for its members, government officials, business executives, journalists, educators and students, civic and religious leaders, and other interested citizens to better help them understand the world and the foreign policy choices facing the United States and other countries.

Amnesty International    **NO**

Amnesty International Report 2014/15: The State of the World's Human Rights

The *Amnesty International Report 2014/15* documents the state of the world's human rights during 2014. Some key events from 2013 are also reported.

The foreword, five regional overviews and survey of 160 countries and territories bear witness to the suffering endured by many, whether it be through conflict, displacement, discrimination or repression. The Report also highlights the strength of the human rights movement, and shows that, in some areas, significant progress has been made in the safeguarding and securing of human rights.

While every attempt is made to ensure accuracy of information, information may be subject to change without notice.

This has been a devastating year for those seeking to stand up for human rights and for those caught up in the suffering of war zones.

Governments pay lip service to the importance of protecting civilians. And yet the world's politicians have miserably failed to protect those in greatest need. Amnesty International believes that this can and must finally change.

International humanitarian law—the law that governs the conduct of armed conflict—could not be clearer. Attacks must never be directed against civilians. The principle of distinguishing between civilians and combatants is a fundamental safeguard for people caught up in the horrors of war.

And yet, time and again, civilians bore the brunt in conflict. In the year marking the 20th anniversary of the Rwandan genocide, politicians repeatedly trampled on the rules protecting civilians—or looked away from the deadly violations of these rules committed by others.

The UN Security Council had repeatedly failed to address the crisis in Syria in earlier years, when countless lives could still have been saved. That failure continued in 2014. In the past four years, more than 200,000 people

have died—overwhelmingly civilians—and mostly in attacks by government forces. Around 4 million people from Syria are now refugees in other countries. More than 7.6 million are displaced inside Syria.

The Syria crisis is intertwined with that of its neighbour Iraq. The armed group calling itself Islamic State (IS, formerly ISIS), which has been responsible for war crimes in Syria, has carried out abductions, execution-style killings, and ethnic cleansing on a massive scale in northern Iraq. In parallel, Iraq's Shi'a militias abducted and killed scores of Sunni civilians, with the tacit support of the Iraqi government.

The July assault on Gaza by Israeli forces caused the loss of 2,000 Palestinian lives. Yet again, the great majority of those—at least 1,500—were civilians. The policy was, as Amnesty International argued in a detailed analysis, marked by callous indifference and involved war crimes. Hamas also committed war crimes by firing indiscriminate rockets into Israel causing six deaths.

In Nigeria, the conflict in the north between government forces and the armed group Boko Haram burst onto the world's front pages with the abduction, by Boko Haram, of 276 schoolgirls in the town of Chibok, one of countless crimes committed by the group. Less noticed were horrific crimes committed by Nigerian security forces and those working with them against people believed to be members or supporters of Boko Haram, some of which were recorded on video, revealed by Amnesty International in August; bodies of the murdered victims were tossed into a mass grave.

In the Central African Republic, more than 5,000 died in sectarian violence despite the presence of international forces. The torture, rape and mass murder barely made a showing on the world's front pages. Yet again, the majority of those who died were civilians.

And in South Sudan—the world's newest state—tens of thousands of civilians were killed and 2 million fled their homes in the armed conflict between government and opposition forces. War crimes and crimes against humanity were committed on both sides.

The above list—as this latest annual report on the state of human rights in 160 countries clearly shows—barely begins to scratch the surface. Some might argue that nothing can be done, that war has always been at the expense of the civilian population, and that nothing can ever change.

This is wrong. It is essential to confront violations against civilians, and to bring to justice those responsible. One obvious and practical step is waiting to be taken: Amnesty International has welcomed the proposal, now backed by around 40 governments, for the UN Security Council to adopt a code of conduct agreeing to voluntarily refrain from using the veto in a way which would block Security Council action in situations of genocide, war crimes and crimes against humanity.

That would be an important first step, and could save many lives.

The failures, however, have not just been in terms of preventing mass atrocities. Direct assistance has also been denied to the millions who have fled the violence that has engulfed their villages and towns.

Those governments who have been most eager to speak out loudly on the failures of other governments have shown themselves reluctant to step forward and provide the essential assistance that those refugees require—both in terms of financial assistance, and providing resettlement. Approximately 2% of refugees from Syria had been resettled by the end of 2014—a figure which must at least triple in 2015.

Meanwhile, large numbers of refugees and migrants are losing their lives in the Mediterranean Sea as they try desperately to reach European shores. A lack of support by some EU Member States for search and rescue operations has contributed to the shocking death toll.

One step that could be taken to protect civilians in conflict would be to further restrict the use of explosive weapons in populated areas. This would have saved many lives in Ukraine, where Russian-backed separatists (despite unconvincing denials by Moscow of its involvement) and pro-Kyiv forces both targeted civilian neighbourhoods.

The importance of the rules on protection of civilians means that there must be true accountability and justice when these rules are violated. In that context, Amnesty International welcomes the decision by the UN Human Rights Council in Geneva to initiate an international inquiry into allegations of violations and abuses of human rights during the conflict in Sri Lanka, where in the last few months of the conflict in 2009, tens of

thousands of civilians were killed. Amnesty International has campaigned for such an inquiry for the past five years. Without such accountability, we can never move forward.

Other areas of human rights continued to require improvement. In Mexico, the enforced disappearance of 43 students in September was a recent tragic addition to the more than 22,000 people who have disappeared or gone missing in Mexico since 2006; most are believed to have been abducted by criminal gangs, but many are reported to have been subjected to enforced disappearance by police and military, sometimes acting in collusion with those gangs. The few victims whose remains have been found show signs of torture and other ill-treatment. The federal and state authorities have failed to investigate these crimes to establish the possible involvement of state agents and to ensure effective legal recourse for the victims, including their relatives. In addition to the lack of response, the government has attempted to cover up the human rights crisis and there have been high levels of impunity, corruption and further militarization.

In 2014, governments in many parts of the world continued to crack down on NGOs and civil society—partly a perverse compliment to the importance of civil society's role. Russia increased its stranglehold with the chilling "foreign agents law," language resonant of the Cold War. In Egypt, NGOs saw a severe crackdown, with use of the Mubarak-era Law on Associations to send a strong message that the government will not tolerate any dissent. Leading human rights organizations had to withdraw from the UN Human Rights Council's Universal Periodic Review of Egypt's human rights record because of fears of reprisals against them.

As has happened on many previous occasions, protesters showed courage despite threats and violence directed against them. In Hong Kong, tens of thousands defied official threats and faced down excessive and arbitrary use of force by police, in what became known as the "umbrella movement," exercising their basic rights to freedoms of expression and assembly.

Human rights organizations are sometimes accused of being too ambitious in our dreams of creating change. But we must remember that extraordinary things are achievable. On 24 December, the international Arms Trade Treaty came into force, after the threshold of 50 ratifications was crossed three months earlier.

Amnesty International and others had campaigned for the treaty for 20 years. We were repeatedly told that such a treaty was unachievable. The treaty now exists, and will prohibit the sale of weapons to those who may use them to commit atrocities. It can thus play a crucial role in the years to come—when the question of implementation will be key.

2014 marked 30 years since the adoption of the UN Convention against Torture—another Convention for which Amnesty International campaigned for many years, and one reason why the organization was awarded the Nobel Peace Prize in 1977.

This anniversary was in one respect a moment to celebrate—but also a moment to note that torture remains rife around the world, a reason why Amnesty International launched its global Stop Torture campaign this year.

This anti-torture message gained special resonance following the publication of a US Senate report in December, which demonstrated a readiness to condone torture in the years after the 11 September 2001 attacks on the USA. It was striking that some of those responsible for the criminal acts of torture seemed still to believe that they had nothing to be ashamed of.

From Washington to Damascus, from Abuja to Colombo, government leaders have justified horrific human rights violations by talking of the need to keep the country "safe". In reality, the opposite is the case. Such violations are one important reason why we live in such a dangerous world today. There can be no security without human rights.

We have repeatedly seen that, even at times that seem bleak for human rights—and perhaps especially at such times—it is possible to create remarkable change.

We must hope that, looking backward to 2014 in the years to come, what we lived through in 2014 will be seen as a nadir—an ultimate low point—from which we rose up and created a better future.

Salil Shetty, Secretary General

Africa Regional Overview

As Africa remembers the 20th anniversary of the Rwandan genocide, violent conflicts dogged much of the continent throughout 2014—unfolding or escalating in a particularly bloody way in the Central African Republic (CAR), South Sudan and Nigeria, and continuing unresolved in the Democratic Republic of the Congo (DRC), Sudan and Somalia.

These conflicts were enmeshed with persistent patterns of gross violations of international human rights and humanitarian law. Armed conflicts bred the worst crimes imaginable, injustice and repression. Marginalization, discrimination and persistent denial of other fundamental freedoms and basic socioeconomic rights have

in turn created fertile grounds for further conflict and instability.

In many ways, Africa continued to be viewed as a region on the rise. The development context and landscape in many countries is changing. Throughout 2014, rapid social, environmental and economic change continued to sweep across the continent. A fast growing population, rapid economic growth and urbanization combined to alter people's lives and livelihoods at a remarkable pace. Many African states have made remarkable progress towards achieving the UN Millennium Development Goals (MDGs) despite steep challenges. The Africa MDG Report 2014 reveals that eight of the world's top 10 best performers in accelerating rapidly towards the goals are in Africa.

However, many indicators left bitter reminders that rapid economic growth has failed to improve living conditions for many. While the overall poverty rate in Africa has dropped in the past decade, the total number of Africans living below the poverty line (US$1.25 per day) has increased. Two of the conflict-plagued nations, Nigeria (25.89%) and DRC (13.6%), account for almost 40% of the continent's poor. Africa has one of the highest youth unemployment rates in the world and it remains the second most unequal region in the world, after Latin America. All these point towards the nexus between conflicts and fragility on the one hand, and the denial of basic socioeconomic rights, social exclusion, inequality and deepening poverty on the other.

The impacts of repression and persistent denial of fundamental human rights in contributing to instability and violent conflicts were vivid in 2014, as demonstrated in Burkina Faso, CAR, South Sudan and Sudan. A trend of repression and shrinking of political space continued in many African countries during the year. In several, security forces responded to peaceful demonstrations and protests with excessive force. In far too many places, freedoms of expression, association and peaceful assembly continued to be severely curtailed. The trend was not only visible in countries ruled by authoritarian governments but also in those which are less authoritarian and in the process of or preparing for political transition.

Many African countries, including Kenya, Somalia, Nigeria, Mali, and countries in the Sahel region, faced serious security challenges in 2014, as a direct result of increased violence by radical armed groups, including al-Shabab and Boko Haram. Tens of thousands of civilians have lost their lives, hundreds have been abducted and countless others continue to live in a state of fear and insecurity. But the response of many governments has been equally brutal and indiscriminate, leading to mass arbitrary arrests and detentions, and extrajudicial executions. The year ended with Kenya enacting the Security Laws (Amendment) Act 2014, which amended 22 laws and which has far-reaching human rights implications.

Another common element in conflict situations across the Africa region has been impunity for crimes under international law committed by security forces and armed groups. 2014 not only saw a cycle of impunity continuing unabated, including in CAR, DRC, Nigeria, Somalia, South Sudan and Sudan, but it was also a year marked with a serious political backlash against the International Criminal Court (ICC). There was also an unprecedented political momentum in Africa championing immunity from prosecution for serving heads of state and officials for crimes against humanity and other international crimes. This culminated in a retrogressive amendment to the Protocol on the Statute of the African Court of Justice and Human Rights, granting immunity to serving heads of state or other senior officials before the Court.

2014 marked the 10th anniversary of the establishment of the AU Peace and Security Council (PSC), the AU's "standing decision-making organ for the prevention, management and resolution of conflicts" in Africa. The AU and its PSC have taken some remarkable steps in response to the emerging conflicts in Africa, including the deployment of the International Support Mission to the Central African Republic (MISCA), the establishment of a Commission of Inquiry on South Sudan, the Special Envoy for Women, Peace and Security, and several political statements condemning violence and attacks on civilians. But in many cases, these efforts appeared too little and too late, pointing to capacity challenges of the AU in responding to conflicts. In some instances, complicity by AU peacekeeping missions in serious human rights violations was also alleged, as with MISCA and specifically its Chadian contingent which withdrew from the mission in CAR following such allegations.

Nonetheless, failure to address conflict challenges in Africa goes beyond the level of the AU. In CAR, for example, the UN dragged its heels before eventually sending in a peacekeeping force that, although saving many lives, did not have the full resources needed to stem the continued wave of human rights violations and abuses. At other times there was silence. The UN Human Rights Council failed to respond effectively to the conflicts in Sudan, for example, despite a critical need for independent human rights monitoring, reporting and accountability. In Darfur, a review of investigations into the UN Mission in Darfur (UNAMID) was announced by the UN Secretary-General in July, in response to allegations that UNAMID staff had covered up human rights abuses.

Addressing the mounting challenges of conflicts in Africa calls for an urgent and fundamental shift in political will among African leaders, as well as concerted efforts at national, regional and international levels to end the cycle of impunity and address the underlying causes of insecurity and conflicts. Otherwise the region's vision of "silencing the guns by 2020" will remain a disingenuous and unachievable dream.

. . .

Americas Regional Overview

Across the Americas, deepening inequality, discrimination, environmental degradation, historical impunity, increasing insecurity and conflict continued to deny people the full enjoyment of their human rights. Indeed, those at the forefront of promoting and defending those rights faced intense levels of violence.

2014 saw mass public responses to these human rights violations the length and breadth of the continent, from Brazil to the USA and from Mexico to Venezuela. In country after country, people took to the streets to protest against repressive state practices. The demonstrations were a very public challenge to high levels of impunity and corruption and to economic policies that privilege the few. Hundreds of thousands of people joined these spontaneous mobilizations using new technologies and social media to rapidly bring people together, share information and expose human rights abuses.

These outpourings of dissatisfaction and demands that human rights be respected took place against the backdrop of an erosion of democratic space and continuing criminalization of dissent. Violence by both state and non-state actors against the general population, and in particular against social movements and activists, was on the rise. Attacks on human rights defenders increased significantly in most countries in the region, both in terms of sheer numbers and in the severity of the violence inflicted.

This growing violence was indicative of an increasingly militarized response to social and political challenges in recent years. In many countries in the region, it has become commonplace for the authorities to resort to the use of state force to respond to criminal networks and social tension, even where there is no formal acknowledgement that conflict exists. In some areas, the increasing power of criminal networks and other non-state actors, such as paramilitaries and transnational corporations, posed a sustained challenge to the power of the state, the rule of law and human rights.

Grave human rights violations continued to blight the lives of tens of thousands of people throughout the region. Far from making further advances in the promotion and protection of human rights for all, without discrimination, the region appeared to be going backwards during 2013 and 2014.

The UN High Commissioner for Human Rights recorded 40 killings of human rights defenders in Colombia during the first nine months of 2014.

In October, the Dominican Republic publicly snubbed the Inter-American Court of Human Rights after the Court condemned the authorities for their discriminatory treatment of Dominicans of Haitian descent and Haitian migrants.

In September, 43 students from the Ayotzinapa teacher training college were subjected to enforced disappearance in Mexico. The students were detained in the town of Iguala, Guerrero state, by local police acting in collusion with organized criminal networks. On 7 December, the Federal Attorney General announced that the remains of one of the students had been identified by independent forensic experts. By the end of the year, the whereabouts of the other 42 remained undisclosed.

In August, Michael Brown, an 18-year-old unarmed African American man, was fatally shot by a police officer, Darren Wilson, in Ferguson, Missouri, USA. People took to the streets following the shooting and in November to protest against a grand jury decision not to indict the officer. The protests spread to other major cities in the country, including New York in December, after a grand jury declined to indict a police officer for the death of Eric Garner in July.

Also in August, prominent *campesino* (peasant farmer) leader Margarita Murillo was shot dead in the community of El Planón, northwestern Honduras. She had reported being under surveillance and receiving threats in the days immediately prior to the attack.

In February, 43 people died, including members of the security forces, and scores more were injured in Venezuela during clashes between anti-government protesters, the security forces and pro-government supporters.

In El Salvador in 2013, a young woman known as Beatriz was refused an abortion despite the imminent risk to her life and the fact that the foetus, which lacked part of its brain and skull, could not survive outside the womb. Beatriz' situation provoked a national and international outcry and weeks of sustained pressure on the authorities. She was finally given a caesarean in her 23rd week of pregnancy. The total ban on abortion in El Salvador continues

to criminalize girls' and women's sexual and reproductive choices, putting them at risk of losing their lives or freedom. In 2014, 17 women sentenced to up to 40 years' imprisonment for pregnancy-related issues requested pardons; a decision on their cases was pending at the end of the year.

In May 2013, former Guatemalan President General Efrain Rios Montt was convicted of genocide and crimes against humanity. However, the conviction was quashed just 10 days later on a technicality, a devastating outcome for victims and their relatives who had waited for more than three decades for justice. Rios Montt was the President and Commander-in-Chief of the Army in 1982–1983 when 1,771 Mayan-Ixil Indigenous people were killed, tortured, subjected to sexual violence or displaced, during the internal armed conflict.

This long list of grave human rights abuses shows how, despite the fact that states in the region have ratified and actively promoted most regional and international human rights standards and treaties, respect for human rights remains elusive for many throughout the region.

. . .

Asia-Pacific Regional Overview

The Asia-Pacific region covers half the globe and contains more than half its population, much of it young. For years, the region has grown in political and economic strength and is rapidly changing the orientation of global power and wealth. China and the USA tussle for influence. Dynamics among large powers in the region, such as between India and China and the Association of Southeast Asian Nations (ASEAN), were also significant. Trends in human rights must be read against this background.

Despite some positive developments in 2014, including elections of some governments that have promised improvements in human rights, the overall trend was regressive due to impunity, continuing unequal treatment of and violence against women, ongoing torture and further use of the death penalty, crackdowns on freedom of expression and assembly, pressure on civil society and threats against human rights defenders and media workers. There were worrying signs of rising religious and ethnic intolerance and discrimination with authorities either being complicit or failing to take action to combat it. Armed conflict in parts of the region continued, particularly in Afghanistan, the Federally Administered Tribal Areas (FATA) in Pakistan, and in Myanmar and Thailand.

The UN released a comprehensive report on the human rights situation in the Democratic People's Republic of Korea (North Korea), which gave details on the systematic violation of almost the entire range of human rights. Hundreds of thousands of people continued to be detained in prison camps and other detention facilities, many of them without being charged or tried for any internationally recognizable criminal offence. At the end of the year these concerns were recognized in the UN General Assembly and discussed in the Security Council.

Refugees and asylum-seekers continued to face significant hardship. Several countries, such as Malaysia and Australia, violated the international prohibition of *refoulement* by forcibly returning refugees and asylum-seekers to countries where they faced serious human rights violations.

The death penalty continued to be imposed in several countries in the region. In December, the Pakistani Taliban-led attack on Army Public School in Peshawar resulted in 149 deaths, including 132 children, making it the deadliest terrorist attack in Pakistan's history. In response, the government lifted a moratorium and swiftly executed seven men previously convicted for other terror-related offences. The Prime Minister announced plans for military courts to try terror suspects, adding to concerns over fair trials.

Homosexuality remained criminalized in several countries in the region. In India, the Supreme Court granted legal recognition to transgender people and in Malaysia the Court of Appeal ruled that a law making cross-dressing illegal was inconsistent with the Constitution. However, cases of harassment and violence against transgender people continued to be reported.

An increase in activism by younger populations, connected by more affordable communications technologies, was positive. However, in the face of this group claiming their rights, authorities in many countries resorted to putting restrictions on freedom of expression, association and peaceful assembly and attempted to undermine civil society.

. . .

Europe and Central Asia Regional Overview

November 9, 2014 marked the 25th anniversary of the fall of the Berlin Wall, the end of the Cold War and, according to one commentator, "the end of history". Celebrating the anniversary in Berlin, German Chancellor Angela Merkel declared "the fall of the Wall has shown us that dreams can come true"—and, for many in communist Europe, indeed they did. But a quarter of a century later, the dream of greater freedom remained as distant as ever for millions more in the former Soviet Union, as the opportunity for change has been ripped from people's hands by the new elites that emerged, seamlessly, from the old.

2014 was not another year of stalled progress. It was a year of regression. If the fall of the Berlin War marked the end of history, the conflict in eastern Ukraine and the Russian annexation of Crimea clearly signalled its resumption. Speaking on the same day as Angela Merkel, former leader of the Soviet Union Mikhail Gorbachev put it bluntly: "The world is on the brink of a new Cold War. Some are even saying that it's already begun."

The dramatic events in Ukraine exposed the dangers and difficulty of dreaming. Over 100 people were killed as the EuroMaydan protest reached its bloody conclusion in February. By the end of the year, over 4,000 more had died in the course of the fighting in eastern Ukraine, many of them civilians. Despite the signing of a cease-fire in September, localized fighting continued and there was little prospect of a rapid resolution by the end of the year. Russia continued to deny that it was supporting the rebel forces with both troops and equipment, in the face of mounting evidence to the contrary. Both sides were responsible for a range of international human rights and humanitarian law violations including indiscriminate shelling, which resulted in hundreds of civilian casualties. As law and order progressively broke down along the lines of conflict and in rebel-held areas, abductions, executions and reports of torture and ill-treatment proliferated, both by rebel forces and pro-Kyiv volunteer battalions. Neither side showed much inclination to investigate and rein in such abuses.

The situation in Crimea deteriorated along predictable lines. With its absorption into the Russian Federation, Russian laws and practices were employed to restrict freedoms of expression, assembly and association of those opposed to the change. Pro-Ukrainian activists and Crimean Tatars were harassed, detained and, in some cases, disappeared. In Kyiv, the huge task of introducing the reforms needed to strengthen the rule of law, eliminate abuses in the criminal justice system and combat endemic corruption was delayed by Presidential and Parliamentary elections and the inevitable distractions of the conflict still raging in the east. Little progress had been made in investigating the killings of EuroMaydan protesters by the end of the year.

The rupturing of the geopolitical fault line in Ukraine had numerous consequences in Russia, simultaneously boosting President Putin's popularity and rendering the

Kremlin more wary of dissent. The breakdown in east-west relations was reflected in the aggressive promotion of anti-western and anti-Ukrainian propaganda in the mainstream media. At the same time, the space to express and communicate dissenting views shrunk markedly, as the Kremlin strengthened its grip on the media and the internet, clamped down on protest and harassed and demonized independent NGOs.

Elsewhere in the former Soviet Union, the hopes and ambitions unleashed by the fall of the Berlin Wall receded further. In Central Asia, authoritarian governments remained entrenched in Kazakhstan, and even more so in Turkmenistan. Where they appeared to wobble slightly, as in Uzbekistan, it was more the result of in-fighting among the ruling elite than in response to wider discontent, which continued to be suppressed. Azerbaijan proved particularly aggressive in its repression of dissent; by the end of the year Amnesty International recognized a total of 23 prisoners of conscience in Azerbaijan, including bloggers, political activists, civil society leaders and human rights lawyers. Azerbaijan's presidency of the Council of Europe in the first half of the year failed to induce restraint. Indeed, more broadly in Azerbaijan, but also elsewhere in Central Asia, strategic interests consistently prevailed over principled international criticism and engagement on widespread human rights violations. Even for Russia, international criticism of the growing clampdown on civil and political rights remained strangely muted.

If Russia remained the market leader in popular, "democratic" authoritarianism, the trend was also observable elsewhere in the region. In Turkey, Recep Erdoğan demonstrated his vote-winning powers once again by comfortably winning the Presidential elections in August, despite a series of high-profile corruption scandals implicating him and his family directly. His response to these, as it had been to the Gezi protests the year before, was unflinching: hundreds of prosecutors, police officers and judges suspected of being loyal to one-time ally Fetullah Gülen were transferred to other posts. The blurring of the separation of powers in Hungary continued after the re-election of the ruling Fidesz party in April and, in moves that echoed developments further east, critical NGOs were attacked for supposedly acting in the interests of foreign governments. By the end of the year, a number of NGOs faced the threat of criminal prosecution for alleged financial irregularities.

Across the European Union (EU), entrenched economic difficulties and the dwindling confidence in mainstream political parties prompted a rise in populist parties at both ends of the political spectrum. The influence of nationalist, thinly-veiled xenophobic attitudes was particularly evident in increasingly restrictive migration policies, but it was also reflected in the growing distrust of supra-national authority. The EU itself was a particular target, but so too was the European Convention on Human Rights. The UK and Switzerland led the charge, with ruling parties in both countries openly attacking the European Court of Human Rights and discussing withdrawal from the Convention system.

In short, at no time since the fall of the Berlin Wall had the integrity of, and support for, the international human rights framework in the Europe and Central Asia region appeared quite so brittle.

. . .

Middle East and North Africa Regional Overview

As 2014 drew to a close, the world reflected on a year that was catastrophic for millions of people across the Middle East and North Africa; a year that saw unceasing armed conflict and horrendous abuses in Syria and Iraq, civilians in Gaza bearing the brunt of the deadliest round of fighting so far between Israel and Hamas, and Libya come increasingly to resemble a failed state caught up in incipient civil war. Yemen too remained a deeply divided society whose central authorities faced a Shi'a insurgency in the north, a vocal movement for secession in the south, and continuing insurgency in the southwest.

With the year in view, the heady hopes for change that drove the popular uprisings that shook the Arab-speaking world in 2011 and saw longstanding rulers ousted in Tunisia, Egypt, Libya, and Yemen appeared a distant memory. The exception was Tunisia, where new parliamentary elections passed off smoothly in November and the authorities took at least some steps to pursue those responsible for the legacy of gross violations of human rights. Egypt, by contrast, gave far less cause for optimism. There, the military general who led the ousting of the country's first post-uprising president in 2013 assumed the presidency after elections and maintained a wave of repression that targeted not only the Muslim Brotherhood and its allies, but political activists of many other stripes as well as media workers and human rights activists, with thousands imprisoned and hundreds sentenced to death. In the Gulf, authorities in Bahrain, Saudi Arabia, and the United Arab Emirates (UAE) were unrelenting in their efforts to stifle dissent and stamp out any sign of opposition to those holding power, confident that their main allies among the western democracies were unlikely to demur.

2014 also saw human savagery meted out by armed groups engaged in the armed conflicts in Syria and Iraq, notably the group calling itself Islamic State (IS, formerly ISIS). In Syria, fighters of IS and other armed groups controlled large areas of the country, including much of the region containing Aleppo, Syria's largest city, and imposed "punishments" including public killings, amputations and floggings for what it considered transgressions of its version of Islamic law. IS also gained ascendancy in the Sunni heartlands of Iraq, conducting a reign of terror in which the group summarily executed hundreds of captured government soldiers, members of minorities, Shi'a Muslims and others, including Sunni tribesmen who opposed them. IS also targeted religious and ethnic minorities, driving out Christians and forcing thousands of Yezidis and other minority groups from their homes and lands. IS forces gunned down Yezidi men and boys in execution-style killings, and abducted hundreds of Yezidi women and girls into slavery, forcing many to become "wives" of IS fighters, who included thousands of foreign volunteers from Europe, North America, Australia, North Africa, the Gulf and elsewhere.

Unlike many of those who perpetrate unlawful killings but seek to commit their crimes in secret, IS was brutally brazen about its actions. It ensured that its own cameramen were on hand to film some of its most egregious acts, including the beheadings of journalists, aid workers, and captured Lebanese and Iraqi soldiers. It then publicized the slaughter in polished but grimly macabre videos that were uploaded onto the internet as propaganda, hostage-bargaining and recruitment tools.

The rapid military advances achieved by IS in Syria and Iraq, combined with its summary killings of western hostages and others, led the USA to forge an anti-IS alliance in September that came to number more than 60 states, including Bahrain, Jordan, Saudi Arabia and the UAE, which then launched air strikes against IS positions and other non-state armed groups, causing civilian deaths and injuries. Elsewhere, US forces continued to mount drone and other attacks against al-Qa'ida affiliates in Yemen, as the struggle between governments and non-state armed groups took on an increasingly supranational aspect. Meanwhile, Russia continued to shield the Syrian government at the UN while transferring arms and munitions to feed its war effort without regard to the war crimes and other serious violations that the Syrian authorities committed.

IS abuses, and the publicity and sense of political crisis that they evoked, threatened for a time to obscure the unremitting and large-scale brutality of Syrian government forces as they fought to retain control of areas they held and to recapture areas from armed groups with seemingly total disregard for the lives of civilians and their obligations under international humanitarian law. Government forces carried out indiscriminate attacks on areas in which civilians were sheltering using an array of heavy weapons, including barrel bombs, and tank and artillery fire; maintained indefinite sieges that denied civilians access to food, water and medical supplies; and attacked hospitals and medical workers. They also continued to detain large numbers of critics and suspected opponents, subjecting many to torture and appalling conditions, and committed unlawful killings. In Iraq, the government's response to IS's advance was to stiffen the security forces with pro-government Shi'a militias and let them loose on Sunni communities seen as anti-government or sympathetic to IS, while mounting indiscriminate air attacks on Mosul and other centres held by IS forces.

As in most modern-day conflicts, civilians again paid the heaviest price in the fighting, as warring forces ignored their obligations to spare civilians. In the 50-day conflict between Israel and Hamas and Palestinian armed groups in Gaza, the scale of destruction, damage, death and injury to Palestinian civilians, homes and infrastructure was appalling. Israeli forces carried out attacks on inhabited homes, in some cases killing entire families, and on medical facilities and schools. Homes and civilian infrastructure were deliberately destroyed. In Gaza more than 2,000 Palestinians were killed, some 1,500 of whom were identified as civilians, including over 500 children. Hamas and Palestinian armed groups fired thousands of indiscriminate rockets and mortar rounds into civilian areas of Israel, killing six civilians, including one child. Hamas gunmen also summarily executed at least 23 Palestinians they accused of collaborating with Israel, including untried detainees, after removing them from prison. Both sides committed war crimes and other serious rights abuses with impunity during the conflict, repeating an all too familiar pattern from earlier years. Israel's air, sea and land blockade of Gaza, in force continuously since 2007, exacerbated the devastating impact of the 50-day conflict, severely hindered reconstruction efforts, and amounted to collective punishment—a crime under international law—of Gaza's 1.8 million inhabitants.

The political and other tensions at play across the Middle East and North Africa in 2014 reached their most extreme form in the countries torn by armed conflict, but throughout the region as a whole there were institutional and other weaknesses that both helped fuel those tensions and prevented their ready alleviation. These included a general lack of tolerance by governments and some non-state armed groups to criticism or dissent; weak or non-existent legislative bodies that could act as a check on or counter-weight to abuses by executive authorities; an absence of judicial independence and the subordination of criminal justice systems to the will of the executive; and a failure of accountability, including with respect to states' obligations under international law.

AMNESTY INTERNATIONAL is a global movement of people fighting injustice and promoting human rights.

EXPLORING THE ISSUE

Is the International Community Making Effective Progress in Securing Global Human Rights?

Critical Thinking and Reflection

1. Why was the international community reluctant to involve itself in human rights abuses in other countries for most of the last 500 years?
2. Should pre-1945 global leaders be held accountable for failure of the international community to make much progress in the promotion and protection of human rights?
3. Is it fair to place blame on the United Nations for the failure to protect human rights when the agency has no real enforcement mechanism?
4. Why would certain individuals today assert that their religious and national culture result in their looking at human rights differently?
5. Is Amnesty International able to be objective in its assessment of human rights problems worldwide, given that its very existence depends on such problems?
6. Do those who are seeking human rights rely too heavily on international organizations when national sovereignty is still operative?
7. Does the Amnesty International report over-emphasize the effect of the ravages of war on human rights?

Is There Common Ground?

Given that the list of specific human rights spelled out in the array of international documents is quite extensive, it is not surprising that the current track record of progress is viewed as mixed. And as a consequence, for the mainstream international human rights actors—be they governmental actors charged with promoting and protecting human rights, or nongovernmental advocacy groups who act as watchdogs of such actors—the typical message appears to present a cautionary tale. That is, on the one hand, each acknowledges success of certain steps taken and/or on specific human rights, while on the other hand, each posits a range of challenges that hinder progress. The difference between those formally charged with promotion and protection of human rights and those watchdogs who monitor the situation or those informal practitioners who take it upon themselves to try to solve the problem is one of degree rather than of kind. That is, the former bodies emphasize successes while lamenting impediments, while the latter group highlights abuses while acknowledging some successes. The YES and NO selections are examples of this common ground.

Additional Resources

Beltz, Charles R., *The Idea of Human Rights* (Oxford University Press, 2011)

This book examines the idea of human rights from a practical and advocacy approach, looking at the history and practice of human rights.

Beltz, Charles R. and Goodin, Robert E. (eds.), *Global Basic Rights* (Oxford University Press, 2011)

This edited book of important contributors focuses on some of the most difficult theoretical and practical questions relating to human rights.

Coleman, Paul, Koren, Elyssa and Miranda-Flefil, Laura, *The Global Human Rights Landscape: A Short Guide to Understanding the International Organizations and the Opportunities for Engagement* (Kairos Publications, 2014)

This book focuses on the international organizations that address human rights issues.

Donnelly, Jack, *International Human Rights*, 3rd Edition (Westview Press, 2013)

This book is a good introduction to the many issues associated with global human rights.

Goodhart, Michael, Human Rights: Politics and Practice, 2nd Edition (Oxford University Press, 2013)

This text approaches the study of human rights from the perspective of its political aspects.

Grove, Chris, "To Build a Global Movement to Make Human Rights and Social Justice a Reality for All," International Journal on Human Rights (vol. 11, June–December 2014)

The article argues that human rights represent both a relevant language and an effective framework for social change.

Hafner-Burton, Emilie M., *Making Human Rights a Reality* (Princeton University Press, 2013)

The book argues that those countries with human rights successes should take the lead worldwide for human rights protection.

Hopgood, Stephen, "Challenges to the Global Human Rights Regime: Are Human Rights Still an Effective Language for Social Change?" *International Journal on Human Rights* (vol. 11, June–December 2014)

The article examines recent challenges to the human rights regime.

Human Rights Watch, *World Report 2012: Events of 2011* (Seven Stories Press, 2012)

This 22nd annual *World Report* describes the human rights situation in over 90 countries and territories worldwide.

Ishay, Micheline, *The History of Human Rights: From Ancient Times to the Globalization Era, with a New Preface*, 2nd Revised Edition (University of California Press, 2008)

This is a well-written synthesis of the historical struggle since ancient times for human rights.

Lauren, Paul Gordon, *The Evolution of International Human Rights: Visions Seen*, 3rd Edition (University of Pennsylvania Press, 2011)

This book describes how the international community moved from centuries of indifference and active abuse into a current state of concern about human rights abuse.

Leite Gonçalves, Guilherme and Casta, Ségio, "The Global Constitution of Human Rights: Overcoming Contemporary Injustice or Juridifying Asymmetries?" *Current Sociology* (vol. 64, March 2016)

The article describes the expansion of actors and structures involved in extending human rights.

Mertus, Julie, *The United Nations and Human Rights: A Guide for a New Era*, 2nd Edition (Routledge, 2009)

This is a comprehensive guide to the role of the United Nations in the advancement of global human rights.

Minkler, Lanse (ed.), *The State of Economic and Social Human Rights: A Global Overview* (Cambridge University Press, 2013)

This book of readings from a wide range of disciplines examines questions about economic and social rights around the world.

Neier, Aryeh, *The International Human Rights Movement: A History* (Princeton University Press, 2012)

The book chronicles the history of international efforts against human rights abuses from the seventeenth century to current struggles.

Pegram, Tom, "Global Human Rights Governance and Orchestration: National Human Rights Institutions as Intermediaries," *European Journal of International Relations* (vol. 21, September 2015)

The article focuses on persistent compliance gaps in the enforcement of human rights.

Petrasek, David, "Global Trends and the Future of Human Rights Advocacy," *International Journal on Human Rights* (vol. 11, June–December 2014)

The article examines many trends that are likely to impact future efforts to secure human rights.

Ruggie, John Gerard, *Just Business: Multinational Corporations and Human Rights* (W. W. Norton & Company, 2013)

This book address a set of guiding principles about multinational corporations and human rights.

Weston, Burns H. and Grear, Anna, *Human Rights in the World Community: Issues and Action*, 4th Edition (University of Pennsylvania Press, 2016)

This text examines the broad spectrum of human rights including the processes and problems of implementation.

Internet References . . .

Amnesty International

www.amnesty.org

Human Rights Watch

www.hrw.org

United Nations

www.ohchr.org

United States Institute for Peace

www.usip.org

Selected, Edited, and with Issue Framing Material by:
James E. Harf, *Maryville University*
and
Mark Owen Lombardi, *Maryville University*

ISSUE

Do Adequate Strategies Exist to Combat Human Trafficking?

YES: Tierney Sneed, from "How Big Data Battles Human Trafficking," *U.S. News & World Report* (2015)

NO: United Nations Office on Drugs and Crime, from "Global Report of Trafficking in Persons: 2014," *United Nations Publications* (2014)

Learning Outcomes

After reading this issue, you will be able to:

- Gain an understanding of the nature of human trafficking and its underlying conditions.
- Discuss major characteristics of human trafficking patterns around the globe as highlighted in the 2014 report.
- Understand how modern technology and the web is being used in the fight against human trafficking.
- Understand that more progress has been made in creating awareness among governments around the world in the problem of human trafficking than on obtaining convictions for related crimes.

ISSUE SUMMARY

YES: Tierney Sneed's article details how new technologies are being used to address the problem of human trafficking.

NO: The 2014 United Nations Office on Drugs and Crime report spells out the magnitude of the problem with the compilation of major data collected about human trafficking.

In a typical week some American television show will sensationalize the sex or forced labor trafficking problem somewhere in the world and efforts of law enforcement agencies to combat the consequences of such activity. This is not surprising as the subject matter catches the eye of a significant number of TV viewers. And the problem is a pervasive one throughout the globe. Around 21 million individuals worldwide are in forced labor. An estimated 1.2 million children are involved. From 600,000 to 800,000 are taken across international borders each year, most of whom are women and children. And approximately $32 billion in annual profits fuels the problem.

What do we mean by human trafficking? It is defined by the United Nations (UN) as "the recruitment, transportation, transfer, harbouring or receipt of persons, by means of the threat or use of force or other forms of coercion, of abduction, of fraud, of deception, of the abuse of power or of a position of vulnerability or of the giving or receiving of payments or benefits to achieve the consent of a person having control over another person, for the purpose of exploitation" (*Trafficking in Persons— Global Patterns*, United Nations Office on Drug and Crime, April 2006). Exploitation may take any one of several forms: prostitution, forced labor, slavery, or other forms of servitude. The U.S. Department of State divides the types of human trafficking into seven categories: sex trafficking, child sex trafficking, forced labor, bonded labor or debt bond, domestic servitude, forced child labor, and unlawful recruitment and use of child soldiers.

Although slavery has been with us since ancient times, the existence of human trafficking across national borders, particularly involving major distances, is a relatively new escalation of a problem that in the past was addressed as a domestic issue, if addressed at all. The first evidence of modern international slavery occurred in 1877 at a meeting of the International Abolitionist Federation at Geneva. There, a report discussed dozens of women being sent to Austria and Hungary under the pretext of work as governesses for work in brothels. A decade later, the issue arose in London through a newspaper account, leading to a public outcry. In 1899, the first international congress to address the issue of white slave traffic was held, with 120 nations represented. International legislation soon followed. With the creation of the United Nations after World War II, the UN took responsibility for enforcing the agreement. But the issue does not appear to be high on the agenda of either the UN or its member states for much of the rest of the twentieth century.

It is not until a dramatic expansion occurred as communism was falling throughout the later half of the 1980s that the issue began to garner public attention. Louise Shelley (*Human Trafficking: A Global Perspective*, Cambridge University Press, 2010) suggests that this expansion was related to globalization, with its emphasis on "free markets, free trade, greater economic competition, and a decline in state intervention in the economy." The end of the cold war also played a major role as organized crime in the former Soviet republics and former East European communist countries discovered how lucrative human trafficking could be. Loosened controls in these countries led to exploitation of national legal systems as criminals operated across national boundaries with relative impunity. Added to this mix was the absence of a coordinated international attempt to combat the issue. Finally, there is the increased demand in both the labor and the sex areas. And the easy movement across national boundaries only exacerbates the problem.

Shelley sees globalization as the instrument through which human trafficking flourishes. The increased volume of international cargo means that inspections often go wanting, allowing smuggled individuals to easily cross national boundaries. Modern communication has been a major contributor, as websites blatantly advertise sex tourism, arranged relationships, and pornography, and e-mails and cell phones are standard operating procedures. Globalization has also meant decreased border controls throughout the developed world. Economic factors have played a role. First, it was the difficult transition from communism to capitalism that led the losers in this process to find other means of employment, with human trafficking as a likely result. Then, it was the downturn in the global economy in 2008 that played a strong role. Shelley also points to the dramatic increase on "grand corruption," building on decades of small-scale corruption

throughout the world. Finally, political factors have been major contributors to human trafficking. The end of the cold war not only loosened state control over every human activity, it also led to a dramatic increase in local conflicts, creating more victims who tried to flee the violence. In turn, this has led to a global condition of statelessness, as increasingly people are citizens of no country and live in limbo, with difficulty in finding employment and with no legal protection of a national government. Add the dramatic population increase in poorer countries and increased urbanization, and you find a large number of individuals, especially women, desperately searching for a better life.

People are abducted or "recruited" in the country of origin, transferred through a standard network to another region of the globe, and then exploited in the destination country. If at any point exploitation is interrupted or ceases, victims can be rescued and might receive support from the country of destination. Victims might be repatriated to their country of origin or, less likely, relocated to a third country. Too often, victims are treated as illegal migrants and treated accordingly. The United Nations estimates that 127 countries act as countries of origin, whereas 137 countries serve as countries of destination. Profits are estimated by the United Nations to be $7 billion per year, with between 700,000 and 4 million new victims annually.

When one hears of human trafficking, one usually thinks of sexual exploitation rather than of forced labor. This is not surprising as not only are individual victim stories more compelling, the former type of exploitation represents the more frequent topic of dialogue among policymakers and is also the more frequent occurrence as reported to the United Nations by a three-to-one margin. Trafficking for sexual exploitation accounts for 58 percent of all cases while trafficking for forced labor is 36 percent of all cases. The latter percentage is increasing, however. Trafficking for sexual exploitation occurs most frequently in Europe, Central Asia, and the Americas, while trafficking for forced labor occurs more often in Africa and the Middle East. With respect to victims, about 60 percent are women and 27 percent are children. Two out of every three child victims are girls. Victims from the 2007 and 2010 time period were from 118 countries and 136 nationalities. It is not surprising that most women and female children are exploited sexually, while most male adults and children are subjected to forced labor. Sexual exploitation is more typically found in Central and Southeastern Europe. Former Soviet republics serve as a huge source of origin. Africa ranks high as a region of victim origin as well, although most end up in forced labor rather than in sexual exploitation. Asia is a region of both origin and destination. Countries at the top of the list include Thailand, Japan, India, Taiwan, and Pakistan. The same UN study found that nationals of Asia and Europe represent the bulk of traffickers. And most traffickers who are arrested are nationals of the country where the arrest occurred.

Human trafficking has been part of the global landscape for centuries. What is different today is the magnitude and scope of the trafficking and the extent to which organized crime is involved in facilitating such nefarious activity. And yet the global community is still only in the position of trying to identify the nature and extent of the problem, let alone ascertaining how to deal with it. In April 2006, the United Nations Office on Drugs and Crime released a report on the human trafficking problem. Titled *Trafficking in Persons: Global Patterns* (United Nations Office on Drugs and Crime, April 2006), its message was clear. The starting point for addressing the problem is the implementation of the Protocol to Prevent, Suppress and Punish Trafficking in Persons, Especially Women and Children. National governments are called upon to take a leading role in (1) the prevention of trafficking, (2) prosecution of violators, and (3) protection of victims.

Consider the task of prevention. Nations are expected to establish comprehensive policies and programs to prevent and combat trafficking, including research, information, and media campaigns. Nations must attempt to alleviate the vulnerability of people, especially women and children. They must take steps to discourage demand for victims. Nations must also prevent transportation opportunities for traffickers. Finally, they must exchange information and increase cooperation among border control agencies. The UN report also suggests several steps with respect to prosecution. The first step is to "ensure the integrity and security of travel and identity documents"

and thus prevent their misuse. Domestic laws must be enacted making human trafficking a criminal offense, and these laws must apply to victims of both genders and all ages. Penalties must be adequate to the crime. Finally, victims must be protected and possibly compensated.

The third role, clearly an alternative approach, outlined in the UN report, focuses on protection of victims. This represents an alternative to the previous international focus, which essentially ignored the plight of the victims. Specifically, victims must be able to achieve "physical, psychological and social recovery." The physical safety of victims is also paramount. The final step relates to the future home of victims, whether they want to remain in the location where found or whether they wish to return home.

In the YES selection, Tierney Sneed describes how new technologies are being utilized to fight human trafficking. In the NO selection, the 2014 United Nations Office on Drugs and Crime report suggests that while much progress has been made in creating great awareness of the problem of human trafficking among national governments throughout much of the globe, only limited progress has been made in convictions of those guilty of trafficking. Until the world sees a major upswing in convictions, determining the amount of progress becomes a matter of viewing the glass as "half-full" or "half-empty."

For example, the report reveals that between 2010 and 2012, 40 percent of the world's countries showed less than 10 convictions per year. Some 15 percent did not even report one single conviction. Yet the number of detected victims has increased every year. The report concludes that "without robust criminal justice responses, human trafficking will remain a low-risk, high-profit activity for criminals."

YES ↵

Tierney Sneed

How Big Data Battles Human Trafficking

From services for victims to prosecuting offenders, new technologies are being utilized to address exploitation.

The Polaris Project, an anti-human trafficking organization based in the District of Columbia, for seven years has operated a national hotline for victims to call when they're in desperate need of help. The hotline has received more than 75,000 calls, and specialists manning it previously had to navigate between a caller's location on Google Maps, a Microsoft Word document detailing 215 protocols for handling specific situations and a list of nearly 3,000 resources—including shelters, legal services and local law enforcement agencies—available to victims through Polaris' network.

"It was a very complicated process, and in the calls where that matters most—where you're helping someone in crisis right now and she's calling from her hotel room and worried her pimp is going to come back—you really don't want to spend 10 minutes trying to find the right person to help her," says Jennifer Kimball, director of Polaris' data analysis program. "You want to be able to respond immediately."

A solution came after a December 2012 Google Ideas conference, where Polaris' executive director, Bradley Myles, crossed paths with representatives of Palantir Technologies, a data analysis firm whose biggest clients include the CIA, the U.S. military and major banks. Palantir's philanthropy arm was seeking "organizations dealing with complex data challenges and a great mission so that we could help enable them do what they do even better, and Polaris is an exemplar of that," says Peter Austin, philanthropic engineer at Palantir. The organizations soon became partners.

Within a few weeks, Palantir had created software that streamlined Polaris' network of resources into a single dashboard for call specialists to use.

"What Palantir does is sit on all those different data sets—the location information, it has those protocols, it has all 2,700 resources—and the call specialist says, 'This is where my crisis is happening,'" Kimball says. "You'll basically click a button and it will show you, 'OK, here is the way to respond in that area.'"

The dashboard doesn't just aggregate data according to geographical location, but also culls other relevant information—age, immigration status, language needs, shelter requirements—to help a specialist deliver assistance specific to a victim's situation.

. . .

"It takes a process that can be very complex and makes it instantaneous," Kimball says.

The partnership between a well-respected human rights organization and a Silicon Valley startup is one of many examples of big data and analysis being used in the fight against human trafficking. As the world becomes ever more connected, the Web is becoming both a dangerous outlet for traffickers seeking to exploit vulnerable populations and an archive of information that activists and officials can use to disrupt predatory networks.

"We're starting to see a lot more interdisciplinary involvement across fields, especially from engineering," says Theresa Leigh, a working group member of the Program on Human Trafficking and Modern Day Slavery at Harvard University's Kennedy School of Government. She notes that partnerships between anti-trafficking organizations and tech giants like Facebook and Google are in part a result of how those services have been abused by traffickers.

Across the world, social media is used as a tool to lure minors into sex trafficking. A Chicago-area man, for example, was accused of recruiting minors over Facebook, posting their pictures in an online ad for escort services and ensnaring them in prostitution.

"It's one of the few issues that everyone has problems with, and so we try to band together to do something about it," Leigh says. "Everybody is doing their part, and we're making much more ground than we did five, 10 years ago, because the data just didn't exist back then."

In addition to using data technology to make its call lines more efficient, Polaris' data team has used the information collected from phone calls, texts or emails with victims to better understand the larger trends among trafficking circuits.

"It's a very large data set, and it really shows us the trends of human trafficking and where and how trafficking is occurring in the United States. Over the past several years, what we've really come to realize in operating the hotline is we're able to help people in these one-off situations, but human trafficking is often a more organized, larger network," Kimball says. "What we've been doing with my team is starting to take those trends as a foundation to really dive deeper and to really build out these maps or analyses."

Such analyses have revealed insights that help Polaris better gear its outreach campaigns to victims. For instance, by identifying truck stops that have become hot spots for forced prostitution, advocates know where to advertise the hotline, can instruct truck stop operators on how to spot the warning signs of trafficking and can urge them to increase security. Activists also can promote law enforcement approaches that are backed up by hard data to better address the problem. Being able to identify the scope of sex exploitation occurring around an event like the Super Bowl, for example, can help determine how much law enforcement should be present, as well as other preventive measures.

"We are actually pursuing it from a preventative standpoint, being able to see patterns with big data that we haven't seen before," Leigh says. "We can have the individual data points to be able to justify policy and decisions to move forward. We can see what works in law enforcement and what doesn't work."

For labor trafficking, this means identifying destinations of demand, chokepoints in the recruitment pipeline and high-risk industry sectors like hospitality and traveling sales.

"The power of big data for social impact is really predicting the crisis," says Stefan Heeke, executive director of SumAll.org, a nonprofit arm of a data analysis firm that has studied trafficking trends and used its findings in public awareness campaigns.

Other groups examine the point at which vulnerable populations—those in impoverished, conflict- or disaster-prone areas—are entering pipelines that may lead to forced labor. By studying workers' interactions with recruiting companies, one can "start identifying information about those organizations, who they're serving and what the other end of that situation looks like," says Ryan Paterson, founder and CEO of technology company IST Research. His company has worked with various organizations to

understand commercial sex exploitation and is now analyzing data associated with forced and bonded labor.

Much of the information being studied is open-source—meaning it's Internet data available to the public—such as Craigslist ads, Google Search trends and exchanges on social media.

Launched by the Defense Department in February, the Memex program at the Defense Advanced Research Projects Agency enables partners, including anti-trafficking groups and law enforcement agencies, to use sophisticated search tools to cull publicly available information embedded in escort ads and other online content that could be connected to sex exploitation, and analyze it for trends and insights.

Before partnering with Memex, many organizations were depending on basic search engines like Google and Yahoo to collect links that would help them understand certain dynamics of human trafficking, DARPA program manager Christopher White says.

"That's just not comprehensive and it misses a lot. There's a lot of modern techniques that have come out in the last five or six years of big data that are really powerful here," White says. By using Memex, he says, groups can better grasp the scope of a network that could be trafficking women.

"Is this a mom-and-pop massage parlor? Is it an entrepreneurial young woman or boy who is making money? Or is this an organization that has a lot of people, a lot of locations, and they're moving people around?" White says.

Examining financial data—the flow of money through credit card companies, bank transfers and money wires—also has been crucial for law enforcement. As traffickers have become more clever in how they hide their profits, law enforcement, in partnership with major financial institutions, has had to refine the tools used to identify transactions connected to forced labor.

"Sex trafficking, like many other organized crimes, leaves classic electronic footprints," Manhattan District Attorney Cyrus Vance Jr. says. "Unless you are working systematically to find those footprints and clues, you are not going to find them."

In 2014, a working group that included the anti-trafficking conference Trust Women, Thomson Reuters Foundation CEO Monique Villa and Vance released a white paper with the help of financial institutions and anti-trafficking organizations that outlined the signs of trafficking operations.

"We were looking to figure out how the trafficking rings moved their money, if we could identify transactions and customer traits that might have a correlation to red flags for trafficking indicators for our investigators to pursue," says Barry Koch, senior vice president and chief compliance officer at Western Union, who worked on the paper. American Express, Bank of America, Barclays, and Citigroup were also among the companies participating. The paper names a variety of financial practices, as well as high-risk industries, that taken together suggest someone is running a forced labor operation.

For example, a nail salon that processes numerous $100 charges after 11 p.m. would warrant alert, Koch says, based on case studies done by the working group.

"We built models and utilized technology to identify transactions that had red flag-indicators, which we then investigated and in many instances we shared our results with law enforcement, which is required under certain regulations," Koch says. "We did it consistent to the right to financial privacy. These are not just exercises in data dumping."

The information financial data can provide is of extreme value to law enforcement and victim advocates. For one, it can corroborate survivors' stories, as trauma can impact their memories and being forced to relive their experiences can further harm them. It also helps investigators go after not just the trafficker who interacted with the victim, but also up the chain of command to bigger players operating complex networks.

Once they're identified, data also can help punish offenders. Survivors are often unwilling to testify, sometimes fearing humiliation, deportation or retaliation against their families. While prosecutors may not be able to convict suspects specifically of trafficking, they can at least tie them to related financial crimes, which also can carry heavy punishments.

In 2013 for instance, Vance's office achieved the conviction of a father and son team who ferried women between Manhattan and Pennsylvania, presenting at trial financial evidence related to the sham entertainment business the pair used to conceal their operation.

In that case, women involved spoke favorably in court about the offenders, who were acquitted of sex trafficking charges.

"This is the challenge of these cases, because they actually believed they were receiving love and affection; they just had no normal human understanding of what love and affection is," Vance says. "It had all the classic characteristics of domestic violence and abused women."

But the pair also was charged with promoting prostitution and money laundering, and ended up in prison.

"It achieved a just outcome, if not the perfect outcome," Vance says. "If we didn't have all that electronic data, if we hadn't done all the work we did that enabled us to put the case in court without [the women's] direct testimony, we would have had no case at all."

Under certain laws, financial institutions are required to turn over information like credit card charges to law enforcement. To obtain such records, authorities still must go through established legal processes like court orders and grand jury subpoenas.

"While the data and how it is captured and kept has changed radically from when I was an assistant in the 1980s, the manner by which we get access to it really hasn't," Vance says.

Nevertheless, the use of data collected on a massive scale also has led to some concerns about privacy issues.

"It's more of a philosophical question: Now that we know there are certain things we can do with the data, the question is, 'Should we do it?' At what point are civil liberties really being violated?" Leigh says. "No one wants to breach that line."

With the Polaris hotline, Kimball says consent is obtained from victims before certain information is shared, and Palantir—which has a team of civil liberty and privacy engineers—has constructed controls to limit access to victim data to only certain parties.

. . .

"Some people are, let's just say, reserved about the question of the reach at which Google and Palantir can have on your life," Leigh says. "But I would say for the most part working with them, they've been pretty fantastic."

Other programs, like Memex, have stuck with using Internet data that's accessible to the general public.

"There isn't an expectation of privacy and that's why we've drawn a bright line for this program to only look at publicly available information," White says.

But here still, questions arise.

"There is a lot of thoughtfulness that has to go into the handling and understanding of that space—especially when you get into the darker and deep Web, where even though it's public information, it's a little more sensitive," White says. "It's a research area and the issues are not entirely well-known."

There's yet another challenge embedded in the data-centric approach to fighting human trafficking, though: As methods to crack down on traffickers evolve, so do the traffickers.

"These are adaptive adversaries, so once you change a tactic, their response is they're going to change their tactic as well," Leigh says.

TIERNEY SNEED is a reporter for *U.S. News & World Report*, focusing on culture and social issues.

**United Nations Office
on Drugs and Crime**

Global Report on Trafficking in Persons 2014

Core Results

- Data coverage: 2010–2012 (or more recent).
- Victims of 152 different citizenships have been identified in 124 countries across the world.
- At least 510 trafficking flows have been detected.
- Some 64 per cent of convicted traffickers are citizens of the convicting country.
- Some 72 per cent of convicted traffickers are men, and 28 per cent are women.
- 49 per cent of detected victims are adult women.
- 33 per cent of detected victims are children, which is a 5 per cent increase compared to the 2007–2010 period.

The data collection has revealed wide regional difference with regard to the forms of exploitation (see figure).

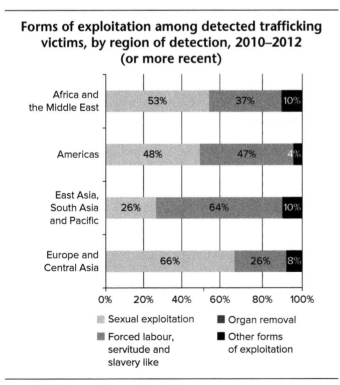

Forms of exploitation among detected trafficking victims, by region of detection, 2010–2012 (or more recent)

Source: UNODC elaboration on national data.

Executive Summary

1. Trafficking in Persons Happens Everywhere

The crime of trafficking in persons affects virtually every country in every region of the world. Between 2010 and 2012, victims with 152 different citizenships were identified in 124 countries across the globe. Moreover, trafficking flows—imaginary lines that connect the same origin country and destination country of at least five detected victims—criss-cross the world. UNODC has identified at least 510 flows. These are minimum figures as they are based on official data reported by national authorities. These official figures represent only the visible part of the trafficking phenomenon and the actual figures are likely to be far higher.

Most trafficking flows are intraregional, meaning that the origin and the destination of the trafficked victim is within the same region; often also within the same subregion. For this reason, it is difficult to identify major global trafficking hubs. Victims tend to be trafficked from poor countries to more affluent ones (relative to the origin country) within the region.

Transregional trafficking flows are mainly detected in the rich countries of the Middle East, Western Europe and North America. These flows often involve victims from the 'global south'; mainly East and South Asia and Sub-Saharan Africa. Statistics show a correlation between the affluence (GDP) of the destination country and the share of victims trafficked there from other regions. Richer countries attract victims from a variety of origins, including from other continents, whereas less affluent countries are mainly affected by domestic or subregional trafficking flows.

Main destination areas of transregional trafficking flows and their significant origins, 2010–2012

The arrows show the flows that represent 5% and above of the total victims detected in destination subregions

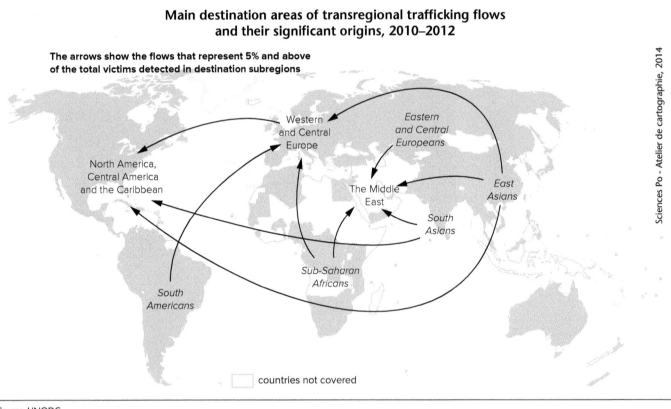

Sciences Po - Atelier de cartographie, 2014

☐ countries not covered

Source: UNODC.

2. A Transnational Crime That Often Involves Domestic Offenders and Limited Geographical Reach

Most victims of trafficking in persons are foreigners in the country where they are identified as victims. In other words, these victims—more than 6 in 10 of all victims—have been trafficked across at least one national border. That said, many trafficking cases involve limited geographic movement as they tend to take place within a subregion (often between neighbouring countries). Domestic trafficking is also widely detected, and for one in three trafficking cases, the exploitation takes place in the victim's country of citizenship.

A majority of the convicted traffickers, however, are citizens of the country of conviction. These traffickers were convicted of involvement in domestic as well as transnational trafficking schemes.

Dividing countries into those that are more typical origin countries and those that are more typical destinations for trafficking in persons reveals that origin countries convict almost only their own citizens. Destination

countries, on the other hand, convict both their own citizens and foreigners.

Moreover, there is a correlation between the citizenships of the victims and the traffickers involved in cross-border trafficking. This correlation indicates that the offenders often traffic fellow citizens abroad.

3. Increased Detection of Trafficking in Persons for Purposes Other Than Sexual Exploitation

While a majority of trafficking victims are subjected to sexual exploitation, other forms of exploitation are increasingly detected. Trafficking for forced labour—a broad category which includes, for example, manufacturing, cleaning, construction, catering, restaurants, domestic work and textile production—has increased steadily in recent years. Some 40 per cent of the victims detected between 2010 and 2012 were trafficked for forced labour.

Trafficking for exploitation that is neither sexual nor forced labour is also increasing. Some of these forms, such

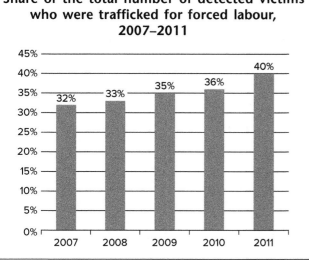

Share of the total number of detected victims who were trafficked for forced labour, 2007–2011

Source: UNODC elaboration on national data.

Forms of exploitation among detected trafficking victims, by region of detection, 2010–2012 (or more recent)

Africa and the Middle East: 53% | 37% | 10%
Americas: 48% | 47% | 4%
East Asia, South Asia and Pacific: 26% | 64% | 10%
Europe and Central Asia: 66% | 26% | 8%

Legend:
- Sexual exploitation
- Forced labour, servitude and slavery like
- Organ removal
- Other forms of exploitation

Source: UNODC elaboration on national data.

as trafficking of children for armed combat, or for petty crime or forced begging, can be significant problems in some locations, although they are still relatively limited from a global point of view.

There are considerable regional differences with regard to forms of exploitation. While trafficking for sexual exploitation is the main form detected in Europe and Central Asia, in East Asia and the Pacific, it is forced labour. In the Americas the two types are detected in near equal proportions.

4. Women Are Significantly Involved in Trafficking in Persons, Both as Victims and as Offenders

For nearly all crimes, male offenders vastly outnumber females. On average, some 10–15 per cent of convicted offenders are women. For trafficking in persons, however, even though males still comprise the vast majority, the share of women offenders is nearly 30 per cent.

Moreover, approximately half of all detected trafficking victims are adult women. Although this share has been declining significantly in recent years, it has been partially offset by the increasing detection of victims who are girls.

Women comprise the vast majority of the detected victims who were trafficked for sexual exploitation. Looking at victims trafficked for forced labour, while men comprise a significant majority, women make up nearly one third of detected victims. In some regions, particularly in Asia, most of the victims of trafficking for forced labour were women.

5. Detected Child Trafficking Is Increasing

Since UNODC started to collect information on the age profile of detected trafficking victims, the share of children among the detected victims has been increasing. Globally, children now comprise nearly one third of all detected trafficking victims. Out of every three child victims, two are girls and one is a boy.

The global figure obscures significant regional differences. In some areas, child trafficking is the major trafficking-related concern. In Africa and the Middle East, for example, children comprise a majority of the detected victims. In Europe and Central Asia, however, children are vastly outnumbered by adults (mainly women).

6. More Than 2 Billion People Are not Protected as Required by the United Nations Trafficking in Persons Protocol

More than 90 per cent of countries among those covered by UNODC criminalize trafficking in persons. Many countries have passed new or updated legislation since the entry into force of the United Nations Protocol against Trafficking in Persons in 2003.

Although this legislative progress is remarkable, much work remains. Nine countries still lack legislation altogether, whereas 18 others have partial legislation that covers only some victims or certain forms of exploitation. Some of these countries are large and densely populated, which means that more than 2 billion people lack the full protection of the Trafficking in Persons Protocol.

7. Impunity Prevails

In spite of the legislative progress mentioned above, there are still very few convictions for trafficking in persons. Only 4 in 10 countries reported having 10 or more yearly convictions, with nearly 15 per cent having no convictions at all.

The global picture of the criminal justice response has remained largely stable in recent years. Fewer countries are reporting increases in the numbers of convictions which remain very low. This may reflect the difficulties of the criminal justice systems to appropriately respond to trafficking in persons.

8. Organized Crime Involvement: Towards a Typology

Criminals committing trafficking in persons offences can act alone, with a partner or in different types of groups and networks. Human trafficking can be easily conducted by single individuals with a limited organization in place.

Criminalization of trafficking in persons with a specific offence, shares and numbers of countries, 2003–2014

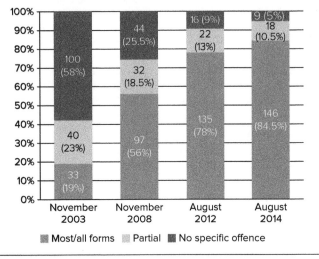

Source: UNODC elaboration on national data.

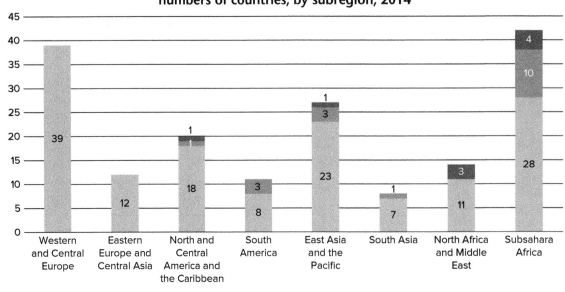

Source: UNODC elaboration on national data.

Typology on the Organization of Trafficking in Persons

Small Local Operations	Medium Subregional Operations	Large Transregional Operations
Domestic or short-distance trafficking flows.	Trafficking flows within the subregion or neighboring subregions.	Long distance trafficking flows involving different regions.
One or few traffickers.	Small group of traffickers.	Traffickers involved in organized crime.
Small number of victims.	More than one victim.	Large number of victims.
Intimate partner exploitation.	Some investments and some profits depending on the number of victims.	High investments and high profits.
Limited investment and profits.	Border crossings with or without travel documents.	Border crossings always require travel documents.
No travel documents needed for border crossings.	Some organization needed depending on the border crossings and number of victims.	Sophisticated organization needed to move large number of victims long distance.
No or very limited organization required.		Endurance of the operation.

This is particularly true if the crime involves only a few victims who are exploited locally. But trafficking operations can also be complex and involve many offenders, which is often the case for transregional trafficking flows.

Offenders may traffic their victims across regions to more affluent countries in order to increase their profits. However, doing so increases their costs as well as the risks of law enforcement detection. It also requires more organization, particularly when there are several victims. Cross-border trafficking flows—subregional and transregional—are more often connected to organized crime. Complex trafficking flows can be more easily sustained by large and well-organized criminal groups.

The transnational nature of the flows, the victimization of more persons at the same time, and the endurance in conducting the criminal activity are all indicators of the level of organization of the trafficking network behind the flow. On this basis, a typology including three different trafficking types is emerging. The trafficking types have some typical characteristics; however, as always, typologies are based on categorization to better explain and understand different aspects of trafficking. 'Pure' trafficking types may not exist since there is always some overlap between different types.

UNITED NATIONS OFFICE ON DRUGS AND CRIME is a global leader in the fight against illicit drugs and international crime.

EXPLORING THE ISSUE

Do Adequate Strategies Exist to Combat Human Trafficking?

Critical Thinking and Reflection

1. Why is greater attention now being paid to human trafficking?
2. Can anything be done to lower the demand for human traffic services?
3. Do you believe that governments are serious about combating human trafficking?
4. Do you consider progress made in combating human trafficking to be one of the glass being "half-full" or "half-empty"?
5. Are victims ignored in government strategies for dealing with human trafficking?
6. Can modern technology, including the web, be an effective tool in the fight against human trafficking?

Is There Common Ground?

The major area of agreement is on the recognition by governments throughout the world that there has been a major increase in human trafficking as a consequence of the end of the cold war and the rise of globalization. There is also agreement that coordinated efforts among governments at all levels—local, national, and global—are critical to successfully addressing the issue. Increasingly, governments agree that large data generating and analyzing capabilities can be used effectively in combatting trafficking.

Additional Resources

Aronowitz, Alexis A., *Human Trafficking, Human Misery: The Global Trade in Human Beings*, Reprint Edition (Scarecrow Press, 2013)

This book is a concise introduction of the subject of human trafficking.

Burke, Mary C., *Human Trafficking: Interdisciplinary Perspectives* (Routledge, 2013)

This classroom text examines the problem of human trafficking and how it is being addressed today.

Draglewicz, editor, *Global Human Trafficking: Critical Issues and Contexts* (Routledge, 2014)

This text examines recent studies of human trafficking from an interdisciplinary perspective.

Efrat, Asif, "Global Efforts against Human Trafficking: The Misguided Conflation of Sex, Labor, and Organ Trafficking," *International Studies Perspectives* (vol. 17, February 2016)

This articles examines global efforts against certain aspects of human trafficking.

"Everywhere in (Supply) Chains," *Economist* (vol. 414, March 14, 2015)

The article reports on the work of the global freedom network advocacy coalition in addressing human trafficking.

Foot, Kirsten, *Collaborating against Human Trafficking* (Rowman & Littlefield, 2015)

This book focuses on what is working against human trafficking.

Havocscope, *Prostitution: Prices and Statistics of Global Sex Trade* (Havocscope, 2015)

This short book provides an extensive list of data about prostitution around the globe.

Rao, Smriti and Presenti, Christina, "Understanding Human Trafficking Origin: A Cross-Country Empirical Analysis," *Feminist Economics* (vol. 18, April 2012)

The article focuses on the difficulty of differentiating between trafficking and migration.

United States Department of State, *Trafficking in Persons Report 2015* (Department of State, 2015)

The U.S. government report provides a comprehensive overview of the broad set of issues related to human trafficking around the globe.

Internet References . . .

Coalition Against Trafficking in Women

www.catwinternational.org

humantrafficking.org

www.humantrafficking.org

The Human Trafficking Project

www.traffickingptroject.org

Terres des homes

www.childtrafficking.org

United Nations

www.ungift.org

Selected, Edited, and with Issue Framing Material by:
James E. Harf, *Maryville University*
and
Mark Owen Lombardi, *Maryville University*

ISSUE

Should the United States and the West Address the Syrian Refugee Crisis by Allowing Them to Migrate to the West?

YES: Michael Ignatieff, from "The United States and the Syrian Refugee Crisis: A Plan of Action," White Paper, Harvard Kennedy School Shorenstein Center on Media, Politics and Public Policy (2016)

NO: Andrew C. McCarthy, from "The Controversy Over Syrian Refugees Misses the Question We Should Be Asking," *National Review* (2015)

Learning Outcomes

After reading this issue, you will be able to:

- Provide basic information about the number of Syrian refugees in various locations.
- Outline the history of the Syrian civil war and its consequences for its citizens.
- Understand the calls for the richer countries of the world to take a greater lead in accepting Syrian refugees.
- Understand the rationale behind the growing reluctance of Western countries to assume a larger responsibility for accepting refugees.
- Describe how a larger role for the United States in accepting refugees has a number of national interest potential payoffs.

ISSUE SUMMARY

YES: Michael Ignatieff, Edward R. Murrow Professor of Press, Politics and Public Policy at Harvard University, argues that the United States should help its major European allies by offering generosity, vision, and optimism to assist them in the resettlement of refugees.

NO: Andrew C. McCarthy suggests that even if the vetting process is perfect, there are two reasons why allowing a massive influx of refugees into the United States is "a calamity." The first is the vetting for terrorism ignores the real challenge, that of Islamic supremacism, of which violent terrorism is only a subset. The second is that the United States ignores the dynamics of jihadism, which suggests that there are individuals being admitted who are "apt to become violent jihadists."

Ahmad al-Abboud and his family arrived in Kansas City, Missouri on April 7, 2016 (*New York Post*, April 7, 2016). As Syrian refugees, they had been living in Jordan for three years after fleeing Syria to escape the ravages of the prolonged civil war raging in that country for a number of years. The al-Abboud family was part of the first wave of about 1000 Syrian refugees from Jordan to be settled in America since October 2015. This is one-tenth of the number set by President Barack Obama as his target by September 30, 2016. And the 10,000 represented a floor, not a ceiling, according to an American Embassy official in Amman, Jordan. But by late spring 2016, news accounts throughout the United States and Europe were headlining the fact that the United States was far behind in meeting the President's objective of at least 10,000 new

Syrian refugees to America (the United States typically accepts between 50,000 and 75,000 total refugees per year, and accepted 70,000 in 2015 fiscal year, about 1,500 from Syria). Jordan has been host to about 635,000 Syrian refugees, which in turn is part of the 4.7 million that have registered with the UN relief organization. It is further estimated that the 600,000+ official refugees represent only half the total number of Syrian refugees in Jordan. At the same time, over 350,000 Syrian refugees applied for asylum in a European Union country during 2015. Germany received the largest number of refugees, with Hungary in second place during this time frame. The al-Abboud family's path to freedom in America began with a journey replicated hundreds of thousands of times as Syrian individuals and families make the decision to risk the perils of fleeing to hopefully a far better place after seeing their country bombed and friends killed. They endure tremendous hardships along the way, documented virtually every night on American and European television.

The story of the Syrian refugee crisis can be traced back to late 2010 to Tunisia where a self-immolation of a street vendor inspired a movement that spread throughout Tunisia across North Africa and the Middle East in what became known as the Arab Spring. By March 2011, inspired by the initial Tunisian protest and the resultant protest movement throughout the broader region, teenage boys in southern Syria painted revolutionary slogans on a school wall, such as "The people want the fall of the regime." The perpetrators were arrested and tortured, which led to protests throughout the country. The Syrian government, in turn, responded with bullets leading to the deaths of some demonstrators. As expected, demonstrations grew in size and spread throughout the country, with demands that Syrian President Bashir al-Assad step down. Protests turned into episodic battles between government forces and opposition forces in various cities throughout Syria, and soon the country was engulfed in a major civil war. By 2012 violence reached Damascus, its capital. The war was a complex one, as highlighted in *The Atlantic* (Kathy Gilsinan, "The Confused Person's Guide to the Syrian Civil War," November 15, 2015). It has been part civil war between Syrian's people and their government, part religious war between Assad's minority sect aligned with Shiites fighters against Sunni rebel groups; and even part proxy war by Russia and Iran against the United States and its allies. It has led to several hundred thousand being killed and half the population displaced, not to mention the rise of ISIS. And this murky situation becomes even more fuzzy as foes are sometimes friends against a third party. Gilsinan puts the number of major rebel groups at 13 and the number of minor ones at 1300. And other countries are also involved with many joining the U.S.-led bombing efforts against ISIS and many more serving as points of entry for those who want to join the cause of ISIS and points of exit for those fleeing the horrors of war-torn Syria. Finally, a UN inquiry has found evidence of war crimes—murder, torture, rape, nerve gas, enforced disappearance—as well as other atrocities such as the lack of adequate food, water, health services, and other humanitarian assistance. Parties have agreed to cease fires from time to time, but the lack of enforcement has rendered them ineffective and short-lived.

In sum, the five-year old protracted and indecisive Syrian civil war with all of its terrible consequences has resulted in what author after author and reporter after reporter have called one of the greatest humanitarian disasters in history. Over six million Syrians, roughly half the country's population, have been displaced. Of this group, between four and five million have fled the country and are registered as refugees in neighboring countries, principally Turkey, Lebanon, Jordan, Iraq, Egypt, and some countries of North Africa. Three-fourths of these refugees are women and children, and only about one in eight live in a formal refugee camp. The majority of refugees are found in Jordan and Lebanon, where the lack of sufficient resources and infrastructure has doomed them to a precarious existence.

Clearly, the countries close by Syria that have served as the recipient country for refugees of the Syrian civil war have been overburdened by the influx of these displaced persons. And as more and more victims cross their borders, calls have increasingly been made to the United States and the richer developed countries of the European Union (and the Gulf Arab states) to assume a much larger share of the burden. This cry has come despite the fact that European countries have been much more open than the United States to assume a major share of the burden. In 2015 alone, more than 1.1 million Syrian refugees made their way to Europe. Many settled on the fringes rather than the center of Europe, with Turkey being a key recipient country.

There now appears to be reluctance on the part of many major countries in Europe to take on a larger burden. The horrific terrorist attacks in Paris in November 2015 had many consequences but one unfortunate result, both in Europe and America, was a growing belief that future terrorists could be masquerading as Syrian refugees. Add to it the San Bernardino killings, and calls in America for limiting Muslims have escalated. Thus, it is not just the economic burdens imposed by an increased pool of Syrian refugees that has led to opposition. The increased potential for homeland terrorist attacks perpetrated by individuals posing as Syrian refugees has resulted in a new red flag being thrown into the picture, particularly by individuals who philosophically are more likely to be xenophobic or nativist in nature. That is, they tend to have a fear of foreigners in the first place. And within the United States as well as in some European countries, this fear of or lack of compassion for non-natives is embraced by certain political blocs and thus become part of political battles waged among competing interests in many democratic countries of the West.

Yet, the United States has been a country of immigrants since its birth. It is at the core of who America is and who Americans are. The Statue of Liberty is not just a gift from France. It has been a living legacy of how the country has reached out to and welcomed during all of its existence those who wish to make America their new home for whatever reason. Over three million refugees have come to America in the past 40 years. In the past 60 years displaced persons from two civil wars, Hungary in 1956 and Cuba in the 1960s, resulted in hundreds of thousands of refugees being accepted. And America's guilt over Vietnam led to the country's opening its arms once again. But this was in the pre-9/11 days before homeland terrorism was a major threat. And while the United States is on record as wanting to admit large numbers in the coming years, disquieting opposition noises are being heard in many quarters. This opposition backlash has emerged despite the fact that many others make the moral argument that the Syrian problem is a consequence, albeit maybe an indirect one, of America's involvement in Iraq and elsewhere in the Middle East, and thus the country has a moral obligation to open its arms to those adversely affected by the consequences of this intervention.

But both American xenophobia and political opportunism have led to vigorous debate within the United States over the wisdom of embracing or even allowing refugees from war-torn Syria to enter the country. The backlash because of the Paris terrorist attacks is real but many believe it to be misguided. The latter argue that the more comprehensive and vigorous vetting process for Syrians, taking from 18 to 24 months, makes the system much more likely to catch any terrorist imposters. Syrian refugees undergo the most intense screening of anyone coming to the United States. Furthermore, the Paris terrorists were not Syrian refugees, who likely have the same fear of radical Islamic terrorism as do Americans. Syrian refugees are fleeing the exact terror that groups like ISIS perpetrate. And as Lauren Gamino et al. suggest (*The Guardian*, November 19, 2015), fear that an influx of Syrian refugees will "overrun the country" is also misguided. The number of Syrian refugees for the most recent three-year period totaled only 0.0007 per cent of America's population.

Opposition based on political considerations appears to exist as well. Thirty-one American governors have publicly voiced opposition to Syrian refugees being resettled in their states. All but one of the 31 is Republican. All GOP Presidential candidates who entered the 2016 race stated that they opposed admitting Muslim Syrian refugees, with many of them opposing Syrians of any religion. And Democratic candidate Bernie Sanders supported only the 10,000 threshold level proposed by the President as a floor level. The GOP-dominated Congress with the support of enough Democrats focused on enhancing barriers in the vetting process as well. American government officials were not alone in their increased opposition as the American public in the aftermath of the Paris terrorist attacks voiced a higher level of opposition.

In the YES selection, Michael Ignatieff of the Harvard Kennedy School argues that the United States should help its major European allies by offering generosity, vision, and optimism to assist them in the resettlement of refugees. His piece, a collaboration between Kennedy School faculty and students, suggests that it is in America's national interest to help its European allies successfully address the refugee crisis. In support of this thesis, he offers five positive outcomes. First, the United States would "reaffirm its historic leadership in refugee resettlement." Second, it "would demonstrate that refugee settlement will not endanger national security." Third, it would "send a powerful message to counter jihadi extremists' portrayal of the United States." Fourth, it would "support and stabilize European allies against resurgent anti-immigrant and anti-American populism." And fifth, it would "support and stabilize Middle Eastern front line states: Turkey, Jordan, and Lebanon." In the NO selection, Andrew C. McCarthy, a policy fellow at the National Review Institute, makes eight points in support of his position that the controversy over Syrian refugees is among the most deceitful public debates in recent memory.

YES ⤶

Michael Ignatieff

The United States and the Syrian Refugee Crisis: A Plan of Action

A Plan of Action

The Western world is witnessing the largest forced migration of peoples since World War II. America's closest ally in Europe, Germany, has opened its frontiers to admit over a million refugees from Syria, Afghanistan, Iraq, and Eritrea, while Italy has been struggling to cope with a flood of migrants and refugees from failed states and conflict zones in Africa. Greece has seen nearly eight hundred thousand refugees and migrants cross its borders in a single year.

The refugee and migration crisis is much more than a humanitarian drama. It is also a strategic challenge for the United States. Since 1945 Europe has been America's major strategic ally and most important trading partner. American engagement and support has helped Europe consolidate peace and prosperity on the continent. The United States will be weakened if Europe comes out of the refugee crisis weakened and divided. Thus far, while Europe has buckled under the crisis, America has remained a bystander.

This paper—a collaboration between Harvard Kennedy School faculty and students—argues that it is in America's national interest to help Europe manage and overcome this crisis by lending strong political support to its major European allies, particularly Germany, and by re-asserting its leadership role in refugee resettlement and integration. We propose a plan of action that renews American leadership and supports Europe while strengthening the national security of the United States.

Any refugee policy of the United States must strengthen, not weaken the security of its own citizens. In the wake of the terrorist attacks in San Bernardino, Paris, the Sinai, Beirut, Ankara, Bamako and Ouagadougou, a public debate has erupted over whether the U.S. should take any Syrian refugees. Republican Presidential candidates have declared that the security of American citizens must prevail over America's long-standing commitments to resettle refugees. Thirty governors, mostly Republican but also including some Democrats, have vowed to bar Syrian refugees from settling in their states. Congress is moving forward on bills that would make it significantly more difficult to accept refugees. President Obama has vowed to veto these measures and has stood by his plan to resettle 10,000 Syrian refugees on top of America's annual 70,000 quota from different lands. He has argued that America can keep faith with its commitment to Syrian refugees without jeopardizing the safety of American citizens.

This debate is a test of American commitment to the international refugee conventions. It is also a moment of truth for U.S. policy in the battle against jihadi extremism. In our view, the question is whether the U.S. will allow its refugee policies to be dictated by fear or by hope. We believe the U.S. must stand with its European and Middle Eastern allies to provide shelter and hope for families fleeing conflict in the Middle East. By doing so, U.S. refugee policies will refute jihadi messages of hate and division. We propose security measures that will allow the United States to accomplish these goals without compromising the security of American citizens.

We believe that by responding with generosity, vision and optimism, the refugee crisis offers the United States a historic opportunity to:

1. Reaffirm its historic leadership in refugee resettlement.
2. Demonstrate that refugee resettlement will not endanger national security.
3. Send a powerful message to counter jihadi extremists' portrayal of the United States.
4. Support and stabilize European allies against resurgent anti-immigrant and anti-American populism.
5. Support and stabilize Middle Eastern front line states: Turkey, Jordan, and Lebanon.

Our specific policy recommendations are that the U.S. should:

1. Surge resettlement in 2016 for 23,000 UNHCR Syrian refugees through U.S. military installations at Fort Dix.
2. Select UNHCR vetted refugees and repatriate them by air directly from camps in Jordan, Lebanon and Turkey.
3. Increase U.S. processing facilities in Lebanon, Jordan, and Turkey to resettle a further 40,000 refugees deemed vulnerable and in need of resettlement by the UNHCR.
4. Mandate full Federal funding for 8 months of integration and resettlement payments to Syrian refugees in American communities.
5. Increase U.S. assistance to UNHCR and WFP to stabilize and improve conditions in refugee camps in front line states.
6. Use all U.S. leverage and influence with Iran, Saudi Arabia, and Russia to negotiate a stand-in place cease-fire in Syria that would permit the eventual return of refugees.

In our view, these policies would affirm America's best historical traditions, confirm its humanitarian commitments to desperate people and support its strategic objectives in the fight against jihadi extremism.

U.S. policy so far has not met these objectives. Since the civil war began in 2011, the U.S. has taken in fewer than 2,000 refugees. The President's commitment to resettle 10,000 refugees is laudable, but it fails to meet the scale of the problem and fails to seize the opportunity for leadership that the refugee crisis presents.

While the U.S. has provided the lion's share of existing financial support to the international agencies—UNHCR and WFP—that provide relief in the camps, these agencies remain substantially underfunded. Deteriorating camp conditions and overcrowding helped precipitate the refugee exodus of 2015. As long as conditions in the camps in the front line states do not improve, refugee flows will continue. Refugee camps are also incubators and recruitment centers for jihadi extremism. To contain jihadi penetration of the refugees, it is important both to stabilize and improve conditions in the camps and also to provide hope for those who are desperate to leave and start a new life elsewhere.

In 2014, the UNHCR designated 130,000 Syrians in refugee camps in need of resettlement by 2016. The U.S. has traditionally resettled at least half of UNHCR- designated refugees. We believe the U.S. should fulfill this role and take in 65,000 Syrian refugees. Taking this number would relieve the pressure on the front line states and

send a message of solidarity to the European states struggling to cope with the refugee influx on their own. Refugee resettlement in the U.S., therefore, plays a critical role in strengthening and stabilizing critical American allies in Europe and the Middle East.

No refugee policy is viable if it compromises the security of Americans. Existing refugee screening processes are rigorous and effective. Of the 784,000 refugees that America has taken in since 9/11, fewer than ten have been charged with terrorist-related offenses and none have committed attacks. This record of safe refugee admission can be maintained and strengthened, especially if the refugees we propose to admit are repatriated directly to U.S. military installations and kept there until the vetting process is complete. In this report, we propose additional reforms of the admission and vetting process to increase the security it provides to Americans.

Some Americans question why Syrian refugees should be resettled here, but the fact is that there are no viable alternatives. The existing refugee camps in the Middle East are overcrowded and underfunded. The President has considered and rejected safe zones that could harbor displaced civilians inside Syria. Safe zones require air cover and ground troops. A safe zone is not safe without perimeter protection by combat capable ground troops and continuous air cover. No country has stepped forward to provide these ground troops, and the available ground forces—Kurdish fighters and Sunni militias—are unsuitable for the mission of civilian protection. Meanwhile the Syrian civil war grinds on, rendering refugee return currently impossible.

Nor can the U.S. safely assume that Europe can continue to absorb indefinite numbers of fleeing refugees. Sooner rather than later, Germany and other countries will find themselves unable to provide further assistance. When Europe closes its doors, pressure will increase on other countries, especially the United States, to step in and provide an alternative. If the refugees lose all hope of a better life, if they feel they have been abandoned, some of them will be easy targets for radicalization and terrorist recruitment. Keeping doors of refuge open for Syrian refugees is critical if the West is to prevail against jihadi extremism.

If the United States remains a bystander in the refugee crisis, existing strains in the U.S.-European alliance will grow and the disunity and instability of Europe will continue to increase, jeopardizing American and European unity of action in the face of Russian pressure in Ukraine and elsewhere. It is time for the United States to use its refugee policy to support Chancellor Merkel and other

European leaders. Doing so will reinforce these leaders, strengthen the Western alliance and help prevent anti-American, anti-Muslim and anti-immigrant voices gaining power in Europe.

Nor can the U.S. continue to look to the front line states—Turkey, Lebanon, and Jordan—to handle the refugee problem. They are all at capacity and further refugee flows will destabilize the fragile political order of all three. Taking 65,000 refugees will allow the U.S. to encourage other allies to take refugees; it will send a strong message of support to its front line allies; and it will assert a common front against jihadi propagandists who would like nothing more than to stop Western countries from providing refuge for civilians fleeing their murderous caliphate.

The most important dimension of refugee policy is strategic communication in the U.S. battle with jihadi extremism. The leaders of the Islamic State (IS) are masters of strategic disinformation. They want to convince Western publics that refugees fleeing barrel bombs and IS terror pose a security threat to states that give them refuge. It serves the strategic interests of terrorists if Western democracies begin to close their doors to desperate people. In this context, it is vital that U.S. refugee policy directly rebuts IS' strategies of disinformation. It is in the U.S. national interest to demonstrate that it can accept refugees and, in doing so, strengthen rather than weaken the security of its citizens in the battle against jihadi extremism.

Implementing the Plan

1. A Resettlement Surge

Over the next six months, the United States Government should transport 23,000 Syrian refugees from existing refugee camps in Turkey and Jordan to Joint Base McGuire-Dix-Lakehurst (MDL) for rapid screening and resettlement into American cities. The purpose of this temporary resettlement surge would be to quickly work through the backlog of Syrian refugees referred to the U.S. by UNHCR over the last several years that have not been resettled due to delays in U.S. screening. Only refugees already screened and accorded refugee status by UNHCR will be brought to America for processing at the base. Families, orphans, and victims of torture and recent combat in Syria will receive priority.

This operation will follow the example of Operation Provide Refuge in 1999, when over 4,000 Kosovar refugees were brought into the U.S., screened and resettled within one month. Like Operation Provide Refuge, multiple government agencies will participate to ensure rapid screening in a secure but humane environment. The U.S. military will be responsible for securing the operations and providing logistical and medical support. Each department that participates in security screening will have delegations at MDL under the leadership of the Department of Homeland Security.

Refugees will be airlifted directly from Incirlik Air Base in Southern Turkey to MDL. At MDL, the U.S. military will set up facilities for both the refugees and the U.S. government employees that will process them. As in Operation Provide Refuge, the refugees will stay in the barracks and all the entrances and exits to the base will be secured by the military. Food and medical care will be provided through MDL facilities.

The refugees will undergo all standard security and medical screening, but the process will be expedited because all the relevant U.S. government actors will be centralized in one place.

While refugees are being processed they will receive ESL lessons and cultural orientation from NGOs. Placement with a sponsoring organization will also be determined during this time.

After security and medical screening is complete, the refugee will be transported to the communities where they will be resettled and the sponsoring organization will take over responsibility for their integration.

The cost of a resettlement surge is difficult to estimate. The Canadian government is currently in the process of resettling 25,000 refugees in a similar manner to what is proposed here. Like our proposal, the refugees are being flown directly from camps in the Middle East to Canadian facilities where they are being screened and processed. A recently leaked budget estimate for the total cost of the Canadian resettlement was $826 million over the next six years, with $600 million in the first year. The per refugee cost is therefore approximately $33,000. This includes the cost of transporting the refugees, as well as the screening costs and all housing, food and education required to fully integrate them into society over a number of years.

The President has the power to authorize refugee admissions, but Public Law 96-212 (1980) requires him to designate the measure as a response to an 'emergency refugee situation' and then demonstrate to Congress that the Syrian situation is such an emergency. It will be important for the President to mobilize public support to secure Congressional support. In Chapter 5, we identify the constituencies and organizations that he will have to rally in order to maintain public and Congressional support.

Forceful action by the President to take more Syrians will provide immediate short-term relief to the countries bordering Syria that are struggling to deal with the refugee flow. It may also reduce refugee movement into Europe, assisting European leaders feeling domestic pressure to bar further refugees. Most importantly, it will give the U.S. standing to engage on refugee issues with other countries and the legitimacy to press for further resettlement and aid.

Screening refugees rapidly and in a controlled environment like Fort Dix is a more effective way to prevent any dangerous individuals from entering the country. A faster process is a more secure process. When a refugee passes a security check, U.S. security agencies are making a determination that this person does not pose a threat at the time of the investigation. If the resettlement process continues after that investigation for more than a year, the usefulness of that determination is reduced. Under the existing system, that forces the U.S. to run multiple, redundant checks. This proposed surge is a more secure and efficient alternative, because when a refugee passes security screening, he or she would be resettled within days, with no risk of radicalization in refugee camps.

2. Establish Additional Resettlement Support Centers in Europe

In addition to the surge resettlement through Fort Dix, the United States government should establish Resettlement Support Center (RSC) facilities in Athens, Greece and Munich, Germany to process approximately 40,000 additional refugees as close to their point of arrival in Europe as possible. This would supplement existing RSCs in Vienna and Istanbul. RSCs are the U.S. government hubs for all resettlement processing, including paperwork, security screening and medical checks. Greece and Germany receive the largest flow of refugees. Locating U.S. government capability there to screen refugees for resettlement in the United States would relieve pressure on our European allies and show that the U.S. stands shoulder to shoulder with their efforts to shelter those fleeing the conflict.

3. Streamline the Screening Process

The current screening process for refugees takes 18 to 24 months and involves multiple layers of medical and security screening, with built-in redundancy for checking and rechecking. Speeding up this process is important because refugees kept waiting in camps or in hostile foreign cities can easily be radicalized.

More processing should be done in parallel. For example, medical screening should begin at the same time as security screening so that lengthy medical tests have time to be completed. Medical screening should be contracted out to selected local clinics where the refugees can go directly. This will reduce the burden on U.S. government staff and the backlog of refugees waiting for medical clearance.

The current immigration vetting process is mostly paper-based, costly and slow. The U.S. government physically transfers paper files 6 times over thousands of miles to different processes centers within the U.S. and abroad to Embassies and Consulates. The process takes between 18 and 24 months. While this time frame has been touted as a strong security measure, it is the detail of security and medical checks and not the length of time that make the process secure. The time frame itself is reflective of inefficient administrative processes.

Several promising initiatives are under way which will enable the U.S. Customs and Immigration Service (USCIS) to enhance its capacity to process more refugee applications more quickly, all the while maintaining security integrity.

The following are policy recommendations that capitalize on these efforts:

1. **Deploy a "whole of government" approach for refugee visas:** Once one agency has determined that a case merits expedited processing, all agencies (Department of Homeland Security, Department of State and USCIS) should comply.
2. **Introduce new digital tools to speed the adjudication process:** USCIS and the Department of State are currently collaborating on a pilot program, the Modernized Immigrant Visa (MIV) Project, which digitizes the visa application and adjudication process. The MIV project is aimed at improving the visa applicant experience and increasing efficiencies in the adjudication process by digitizing as much of it as possible. A suite of applications, mainly belonging to USCIS and State, will more efficiently process and manage electronic immigrant records. The MIV pilot is being rolled out in Montreal, Buenos Aires, Rio de Janeiro, Frankfurt, Hong Kong and Sydney, with a wider launch in 2016. The U.S. could adapt the MIV tool for use with refugee populations in consular posts in Europe and the Middle East. The U.S. Digital Service, a team within the federal government that seeks to improve and simplify digital services, can create a cross-agency digital service team to support the implementation of the overall MIV pilot. The

U.S. Digital Service has a proven track record and is already seeking to assist USCIS in this project. They would be well positioned to oversee development of a refugee version of the MIV tool.

3. **Help refugees navigate the application process:** In 2015, the U.S. Digital Services and 18F, a consulting group within the General Services Administration, developed, a platform that allows users to access information about the immigration process and find immigration options. This tool could easily be enhanced to better respond to refugee needs by offering the location of the nearest U.S. embassy or UNHCR center capable of conferring refugee status.

4. Help Refugees Integrate Quickly

The U.S. already has a well-established partnership between federal, state and local agencies to assist refugee integration and resettlement. This existing set of partnerships and networks needs to be strengthened.

Resettlement agencies receive a stipend of $1,875 per refugee from the Department of State's Reception and Placement program as mandated by the Refugee Act of 1980. This money is given to these resettlement agencies to help refugees with airport pickup, initial rent, food, clothes, costs of agency staff salaries and other preliminary integration efforts. During the first eight months, local agencies also provide language and vocational training as well as job placement. In total around 300 agencies and organizations across the nation oversee refugee resettlement. In 2014, the Department of State spent $616.3 million on refugee resettlement inside the United States.

Under the Refugee Act of 1980 Syrian refugees will receive $420.00 per month for a two-person household for eight months with some additions for special cases. Through the office of Refugee Resettlement, Syrians will also benefit from a Refugee Cash Assistance program that will help subsidize their medical expenses until they are employed. The funding is immediately discontinued when a family finds employment income of more than $800 per month. Syrian refugees will also be required to repay the cost of travel to the United States.

While the current grant provided per refugee amounts to $1,875, research by refugee resettlement agencies has shown that the actual cost of initial resettlement is $3,492. In fact, federal funding currently accounts for only 39 percent of the total cost of refugee resettlement, with the remainder coming from private fund-raising.

Leaders of the State Department's budget committee in the Senate, Senator Lindsey Graham (R-S.C.) and Senator Patrick Leahy (D-Vt.), have proposed a funding model in the Middle East Refugee Emergency Supplemental Appropriations Act (S.2145, October 2015). This bill would provide an additional $1 billion in emergency funds to be used for refugee resettlement. Invoking an emergency requirement would exempt funds from discretionary spending limits and other budget enforcement rules. In return, the White House would need to report to Congress within 45 days on how it will use the money.

We recommend that federal funds for resettlement increase to meet the total needs of local agencies for the entire 8-month resettlement period.

To speed up integration, we recommend increased funding for the Department of Labor's Employment and Training Administration for refugees.

ETA has previously awarded grants to train refugee workers, in partnership with community organizations, to be able to acquire the necessary certifications, licenses and English language skills to pursue their professions in the U.S. **These grants should be offered to states for Syrian refugees to ensure their proper economic integration.** When new Americans can leverage and improve their skills, they are able to become successful entrepreneurs and self-sufficient members of society.

Refugees can also access U.S. Department of Education adult education and family literacy programs that provide basic English acquisition. **Specific to refugees is the Refugee Impact School Program, which should be extended to states that will be receiving Syrian refugees.** Administered by the Department of Education, it provides refugees with orientation, tutoring, after-school programming, parent-teacher conferences, interpretation assistance and additional information on school systems.

We endorse the recommendations of the White House Task Force on New Americans and we recommend their adoption for Syrian refugee resettlement, viz,

1. **Settlement Resources Information:** As soon as refugees arrive in the States, the Departments of State and DHS should identify opportunities to provide approved immigrant visa applicants and beneficiaries of an approved immigrant visa petition with information on critical settlement resources, including available English language learning opportunities.

2. **Identify Refugee Leaders Early:** Make citizenship more accessible by identifying and elevating community leaders who will raise awareness

about naturalization processes and the importance of civic engagement and who will engage with the broader community to highlight the needs of the refugee community.

3. **Increase funding for the Ethnic Community Self-Help Program:** This program provides support to refugee community-based organizations, cultural organizations and religious organizations that will facilitate the social integration processes for refugees.

MICHAEL IGNATIEFF is Edward R. Murrow Professor of Press, Politics and Public Policy at the Harvard Kennedy School. He previously served as leader of the Liberal Party in Canada and also served in the Parliament of Canada.

Andrew C. McCarthy   ➔ **NO**

The Controversy over Syrian Refugees Misses the Question We Should Be Asking

The jihad waged by radical Islam rips at France from within. The two mass-murder attacks this year that finally induced President Francois Hollande to concede a state of war are only what we see.

Unbound by any First Amendment, the French government exerts pressure on the media to suppress bad news. We do not hear much about the steady thrum of insurrection in the *banlieues:* the thousands of torched automobiles, the violence against police and other agents of the state, the pressure in Islamic enclaves to ignore the sovereignty of the Republic and conform to the rule of sharia.

What happens in France happens in Belgium. It happens in Sweden where much of Malmo, the third largest city, is controlled by Muslim immigrant gangs—emergency medical personnel attacked routinely enough that they will not respond to calls without police protection, and the police in turn unwilling to enter without back-up. Not long ago in Britain, a soldier was killed and nearly beheaded in broad daylight by jihadists known to the intelligence services; dozens of sharia courts now operate throughout the country, even as Muslim activists demand more accommodations. And it was in Germany, which green-lighted Europe's ongoing influx of Muslim migrants, that Turkey's Islamist strongman Recep Tayyip Erdogan proclaimed that pressuring Muslims to assimilate in their new Western countries is "a crime against humanity."

So how many of us look across the ocean at Europe and say, "Yeah, let's bring some of that here"?

None of us with any sense. Alas, "bring it here" is the order of the day in Washington, under the control of leftists bent on fundamentally transforming America (Muslims in America overwhelmingly support Democrats) and the progressive-lite GOP, which fears the "Islamophobia" smear nearly as much as the "racist" smear.

This, no doubt, is why what is described as the "controversy over Syrian refugees" is among the most deceitful public debates in recent memory—which, by Washington standards, is saying something.

Under a Carter administration scheme, the Refugee Admissions Program, the United States has admitted hundreds of thousands of aliens since 1980—and, as the Center for Immigration Studies explains, asylum petitions have surged since the mid-Nineties. If there is a refugee "crisis," it most certainly is no fault of ours: For example, the U.S. took in two-thirds of the world's refugees resettled in 2014, with Canada a distant second, admitting about 10 percent.

Those figures come from an invaluable briefing by Refugee Resettlement Watch, which illustrates that the *Syrian* component is but a fraction of what we must consider. Tens of thousands of what are called "refugees" have come to our shores from Muslim-majority countries. From Iraq alone, the number is 120,000 since 2007, notwithstanding the thousands of American lives and hundreds of billions of American taxpayer dollars sacrificed to make Iraq livable.

Many of the refugees are steered to our country by the United Nations High Commissioner on Refugees. Naturally, the UNHCR has a history of bashing Israel on behalf of Palestinian Islamists—indeed, it works closely with the U.N. Relief and Works Agency for Palestinian Refugees, one of Hamas's most notorious sympathizers. The UNHCR works in tandem with the State Department, which resettles the refugees throughout the U.S. with the assistance of lavishly compensated contractors (e.g., the U.S. Conference of Catholic Bishops, other Christian and Jewish outfits, and the U.S. Committee for Refugees and Immigrants)—often absent any meaningful consultation with the states in which Washington plants these assimilation-resistant imports.

Responsibility for vetting the immigrants rests with the Department of Homeland Security. As the ongoing controversy has illustrated, however, a background check is only as good as the available information about a person's background. In refugee pipelines like Syria, Iraq, Afghanistan, Somalia, and Sudan, such information is virtually nonexistent. (But don't worry, we can rest assured that the UNHCR is doing a fine job.)

Let's assume for fantasy's sake, though, that the vetting is perfect—that we have comprehensive, accurate information on each refugee's life up to the moment of admission. We would still have a calamity.

There are two reasons for this, and they are easily grasped by the mass of Americans outside the Beltway.

First, vetting only works *if you vet for the right thing*. Washington, in its delusional Islamophilia, vets only for ties to *terrorism*, which it defines as "violent extremism" in purblind denial of modern terrorism's Islamist ideological moorings. As the deteriorating situation in Europe manifests, our actual challenge is *Islamic supremacism*, of which jihadist terrorism is only a subset.

For nearly a quarter-century, our bipartisan governing class has labored mightily to suppress public discussion of the undeniable nexus between Islamic doctrine and terrorism. Consequently, many Americans are still in the dark about sharia, classical Islam's societal framework and legal code. We should long ago have recognized sharia as the bright line that separates authentic Muslim moderates, hungry for the West's culture of reason and individual liberty, from Islamic supremacists, resistant to Western assimilation and insistent on incremental accommodation of Muslim law and mores.

The promotion of constitutional principles and civic education has always been foundational to the American immigration and naturalization process. We fatally undermine this process by narrowly vetting for terrorism rather than sharia adherence.

Yes, I can already hear the slander: "You are betraying our commitment to religious liberty." Please. Even if there were anything colorable to this claim, we are talking about inquiring into the beliefs of aliens who want to enter our country, not citizens entitled to constitutional protections.

But the claim is not colorable in any event—it just underscores how willful blindness to our enemies' ideology has compromised our security. Only a small fraction of Islamic supremacism involves tenets that, in the West, should be regarded as inviolable religious conviction (e.g., the oneness of Allah, the belief that Mohammed is the final prophet, the obligation to pray five times daily). No one in America has any interest in interfering with that. For Muslims adherent to classical sharia, however, the rest of their belief system has nothing to do with religion (except as a veneer). It instead involves the organization of the state, comprehensive regulation of economic and social life, rules of military engagement, and imposition of a draconian criminal code.

Unlike the Judeo-Christian principles that informed America's founding, classical sharia does not abide a separation of spiritual from civic and political life. Therefore, to rationalize on religious-liberty grounds our conscious avoidance of Islamist ideology is to miss its thoroughgoing anti-constitutionalism.

Sharia rejects the touchstone of American democracy: the belief that the people have a right to govern themselves and chart their own destiny. In sharia governance, the people are subjects not citizens, and they are powerless to question, much less to change, Allah's law. Sharia systematically discriminates against women and non-Muslims. It is brutal in its treatment of apostates and homosexuals. It denies freedom of conscience, free expression, property rights, economic liberty, and due process of law. It licenses wars of aggression against infidels for the purpose of establishing sharia as the law of the land.

Sharia is also heavily favored by Muslims in majority-Muslim countries. Polling consistently tells us that upwards of two-thirds of Muslims in the countries from which we are accepting refugees believe sharia should be the governing system.

Thus, since we are vetting for terrorism rather than sharia-adherence, and since we know a significant number of Muslims are sharia-adherent, we are missing the certainty that we are importing an ever-larger population hostile to our society and our Constitution—a population that has been encouraged by influential Islamist scholars and leaders to form Muslim enclaves throughout the West.

This leads seamlessly to the second reason why the influx of refugees is calamitous. Not only are we vetting for the wrong thing, *we are ignoring the dynamics of jihadism*. The question is not whether we are admitting Muslims who currently have ties to terrorist organizations; it is whether we are admitting Muslims who are *apt to become violent jihadists after they settle here*.

The jihadism that most threatens Europe now, and that has been a growing problem in the United States for years, is the fifth-column variety. This is often referred to as "homegrown terrorism," but that is a misnomer. The ideology that ignites terrorism within our borders is not native: It is imported. Furthermore, it is ubiquitously available thanks to modern communications technology.

In assessing the dynamic in which ideological inspiration evolves into actual jihadist attacks, we find two necessary ingredients: (1) a mind that is hospitable to jihadism because it is already steeped in Islamic supremacism, and (2) a sharia-enclave environment that endorses jihadism and relentlessly portrays the West as corrupt and hostile.

Our current refugee policies promote both factors.

One last point worth considering: Washington's debate over refugee policy assumes an unmet American obligation to the world. It is as if we were not already doing and sacrificing far more than every other country combined. It is as if there were not dozens of Islamic countries, far closer than the United States to refugee hot-spots, to which it would be sensible to steer Muslim migrants.

Yet, there is nothing obligatory about any immigration policy, including asylum. There is no global right to come here. American immigration policy is supposed to serve the national interests of the United States. Right now, American immigration policy is serving the interests of immigrants at the expense of American national security and the financial security of distressed American workers.

Our nation is nearing $20 trillion in debt, still fighting in the Middle East, and facing the certain prospect of combat surges to quell the rising threat of jihadism. So why is Congress, under the firm control of Republicans, paying for immigration policies that exacerbate our peril?

ANDREW C. MCCARTHY is a columnist for the *National Review*. He served as Assistant US Attorney for the Southern District of New York where he led the successful prosecution of 12 defendants in the 1993 terrorist attack against the World Trade Center.

EXPLORING THE ISSUE

Should the United States and the West Address the Syrian Refugee Crisis by Allowing Them to Migrate to the West?

Critical Thinking and Reflection

1. Is there any foreseeable end to the civil war in Syria?
2. Do you believe that Middle East problems like the Syrian refugee crisis can be blamed on America's invasion of Iraq and the overthrow of its leader, Saddam Hussein?
3. Do you believe that America has a moral obligation to help solve the Syrian refugee crisis?
4. Do you believe that Europe because of its closer geographic proximity to Syria should assume a larger burden than the United States in solving the Syrian refugee problem?
5. Do you think that those on the political left who charge the political right with playing politics in its opposition to America's absorbing more Syrian refugees are themselves playing politics?

Is There Common Ground?

Despite differing viewpoints on how to address the Syrian refugee crisis, there is some common ground. Both sides understand the magnitude of the problem created by the prolonged civil war in Syria. Both understand how countries in proximity to Syria have endured a greater burden as a consequence. Both also accept that the wealthier countries have an economic obligation to help with refugee settlement issues. Both sides by and large also understand that the United States should take in some refugees (but maybe "not in my back yard"). Both sides also understand why people are more fearful in general of incoming non-Americans as a consequence of terrorist activity somewhere in the developed world (like the Paris attack).

Additional Resources

Bauer, Wolfgang, "Crossing the Sea with Syrians on the Exodus to Europe, And Other Stories," *Amazon Digital Services* (March 24, 2016)

This is a first-hand account of an undercover journalist's journey with Syrian refugees across the Mediterranean Sea to Europe.

Bin Talal, El Hassan, "Europe and the Future of International Refugee Policy," *Forced Migration Review* (no. 51, January 2016)

The article suggests long-term policies for EU leaders regarding the Syrian refugee crisis.

Clarke, Kevin, "Bishops' Delegation Reports Syrian Refugee Crisis at 'Tipping Point'," *America* (vol. 212, no. 10, March 23, 2015)

The articles describes the warning from the conference of Catholic bishops about the size of the Syrian refugee crisis.

Committee of Foreign Affairs, House of Representatives, "Examining the Syrian Refugee Crisis," *CreateSpace Independent Publishing Platform* (January 3, 2014)

This report highlights the refugee crisis as a result of the Syrian conflict.

Gabiam, Neill, "Humanitarianism, Development, and Security in the 21st Century: Lessons from the Syrian Refugee Crisis," *International Journal of Middle East Studies* (vol. 48, no. 2, May 2016)

The author discusses the concept of humanitarianism in the context of the Syrian refugee crisis.

Heisbourg, François, "The Strategic Implications of the Syrian Refugee Crisis," *Survival* (vol. 57, no. 6, December 2015)

The articles focuses on the fact there is no precedent for such a large flow of war refugees to Europe from the Middle East.

Peters, Michael A. and Besley, Tina, "The Refugee Crisis and the Right to Political Asylum," *Educational*

Should the United States and the West Address the Syrian Refugee Crisis by Allowing Them to Migrate to the West? by Harf and Lombardi

169

Philosophy & Theory (vol. 47, no. 13/14, December 2015)

The article describes the historical evolution of the right to political asylum since ancient times.

Saeed Al Mutar, Faisal, "The Syrian Refugee Crisis and the Need for Political Solutions," *Free Inquiry* (vol. 36, no. 2, February/March 2016)

The article addresses the political issues surrounding accepting refugees in Europe.

Subcommittee on Counterterrorism and Intelligence of the Committee on Homeland Security,

U.S. House of Representatives, "Admitting Syrian Refugees: The Intelligence Void and the Emerging Homeland Security Threat," *CreateSpace Independent Publishing Platform* (December 30, 2015)

The Congressional report examines the threats posed by Syrian refugees.

Thompson, Thomas, "Syrian Refugees: What-When-Why-How," *CreateSpace Independent Publishing Platform* (November 6, 2015)

This short volume provides basic information about many aspects of the Syrian refugee crisis.

Internet References . . .

Action Institute Power Blog

www.action.org/SyrianRefugee

Amnesty International

www.amnesty.org

European Union

www.syrianrefugees.eu

United Nations High Commissioner for Refugees

www.data.unhcr.org/syrianrefugees/region

United States Department of State

www.state.gov/

World Economic Forum

https://www.weforum.org/agenda/2015/11/europes-refugee-crisis-explained/

Selected, Edited, and with Issue Framing Material by:
James E. Harf, *Maryville University*
and
Mark Owen Lombardi, *Maryville University*

ISSUE

Is Global Income Inequality on the Rise in the International Community?

YES: **Oxfam Report**, from "An Economy for the 1%," *Oxfam* (2016)

NO: **Max Roser**, from "Income Inequality: Poverty Falling Faster than Ever but the 1% are Racing Ahead," *The Guardian* (2015)

Learning Outcomes
After reading this issue, you will be able to: • Understand the data related to income inequality both among and within states and regions. • Understand the complexity of economic development and wealth distribution. • Appreciate the differing viewpoints on wealth distribution and its impact on global development and the human experience. • Be able to determine your own views on income inequality and wealth distribution within a larger philosophical context.

ISSUE SUMMARY

YES: The Oxfam report issued this year provides data to characterize global inequality as a crisis where the wealthiest 62 people have amassed great wealth at the expense of billions of poor people.

NO: Roser argues that while a few have amassed great wealth, poverty as a whole is declining and tens of millions are moving into working and middle class positions.

The issue of income distribution and inequality has been of concern for centuries. Through varying eras of economic and systemic change, groups of individuals or classes of people have amassed great wealth. Whether through nobility/heredity, slavery, other forms of economic exploitation or through the benign means of entrepreneurship, invention or simply being first to market, wealth has developed, grown, and been concentrated in a relatively small group of people within and among states and societies. Some analysts and even religions through the centuries have assigned a host of factors to this development. These include everything from luck and hereditary to economic system, resource distribution, human exploitation, and societal values. Men such as Adam Smith in The Wealth of Nations, Karl Marx in Das Capital

and many others have addressed the issue from different political, social, and ideological perspectives.

In the world of globalization amidst the rapid changes in technology and dissemination of goods, services, and particularly capital at dizzying speed, the distribution of wealth has become a more pronounced and immediate issue. The sheer size and magnitude of the wealth that has been generated over the last fifty years through the massive expansion of global trade brings this issue front and center. Some see this wealth creation as part of a global tide that lifts all boats. They see the basic lot of most, but not all of the world citizens as being enhanced by wealth creation, global trade, and opening of markets. They point to data showing millions more people having access to a variety of goods and services that have elevated them from poverty to working- and middle-class status. Often, China,

India, Brazil, and Indonesia are among the states offered up to illustrate this point.

Others see this wealth expansion as skewed into the hands of a select few. They point to data within and among states, showing that wealth and resource control is in the hands of a few corporations and individuals, and they also point to the laws of wealth acquisition from the bond markets, trading regimes, and capital markets as skewed toward greater and greater inequality. In fact, many of the recent political contests within states in Europe, the United States, and Latin America have focused on these varying issues as cleavages within electoral contests. The 2016 U.S. presidential election, the recent British rejection of EU membership, and the elections in Canada and political crisis in Brazil are just a few examples.

In the YES article, the Oxfam report paints a scathing picture through statistical analysis of both the distribution of resources and wealth within global society and how the system allows for such maldistribution. Their simple conclusion if "the system is broken," it must be altered through concerted international action. In the NO article, Roser argues that while the top 1 percent have indeed profited greatly over the last generation, the fact remains that global poverty is in decline; millions of people are entering the middle class, and this important economic reality is missed when looking at inequality just between the top and the bottom only.

YES ⬅

Oxfam

An Economy for the 1%

How Privilege and Power in the Economy Drive Extreme Inequality and How This Can Be Stopped

The global inequality crisis is reaching new extremes. The richest 1% now have more wealth than the rest of the world combined. Power and privilege is being used to skew the economic system to increase the gap between the richest and the rest. A global network of tax havens further enables the richest individuals to hide $7.6 trillion. The fight against poverty will not be won until the inequality crisis is tackled.

An Economy for the 1%

The gap between rich and poor is reaching new extremes. Credit Suisse recently revealed that the richest 1% have now accumulated more wealth than the rest of the world put together.[1] This occurred a year earlier than Oxfam's much publicized prediction ahead of last year's World Economic Forum. Meanwhile, the wealth owned by the bottom half of humanity has fallen by a trillion dollars in the past five years. This is just the latest evidence that today we live in a world with levels of inequality we may not have seen for over a century.

'An Economy for the 1%' looks at how this has happened, and why, as well as setting out shocking new evidence of an inequality crisis that is out of control. Oxfam has calculated that:

- In 2015, just 62 individuals had the same wealth as 3.6 billion people—the bottom half of humanity. This figure is down from 388 individuals as recently as 2010.
- The wealth of the richest 62 people has risen by 45% in the five years since 2010—that's an increase of more than half a trillion dollars ($542bn), to $1.76 trillion.
- Meanwhile, the wealth of the bottom half fell by just over a trillion dollars in the same period—a drop of 38%.

- Since the turn of the century, the poorest half of the world's population has received just 1% of the total increase in global wealth, while half of that increase has gone to the top 1%.
- The average annual income of the poorest 10% of people in the world has risen by less than $3 each year in almost a quarter of a century. Their daily income has risen by less than a single cent every year.

Growing economic inequality is bad for us all—it undermines growth and social cohesion. Yet the consequences for the world's poorest people are particularly severe.

Apologists for the status quo claim that concern about inequality is driven by 'politics of envy.' They often cite the reduction in the number of people living in extreme poverty as proof that inequality is not a major problem. But this is to miss the point. As an organization that exists to tackle poverty, Oxfam is unequivocal in welcoming the fantastic progress that has helped to halve the number of people living below the extreme poverty line between 1990 and 2010. Yet had inequality within countries not grown during that period, an extra 200 million people would have escaped poverty. That could have risen to 700 million had poor people benefited more than the rich from economic growth.

There is no getting away from the fact that the big winners in our global economy are those at the top. Our economic system is heavily skewed in their favour, and arguably increasingly so. Far from trickling down, income and wealth are instead being sucked upwards at an alarming rate. Once there, an ever more elaborate system of tax havens and an industry of wealth managers ensure that it stays there, far from the reach of ordinary citizens and their governments. One recent estimate[3] is that $7.6 trillion of individual wealth—more than the

Global income growth that accrued to each decile 1988–2011: 46% of the total increase went to the top 10%[2]

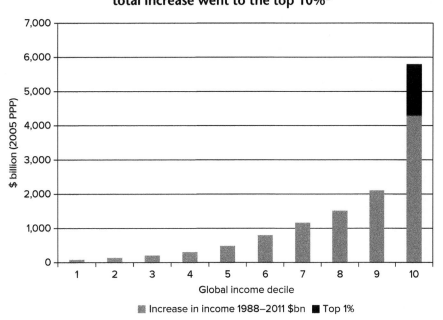

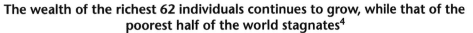

Increase in income 1988–2011 $bn   Top 1%

The wealth of the richest 62 individuals continues to grow, while that of the poorest half of the world stagnates[4]

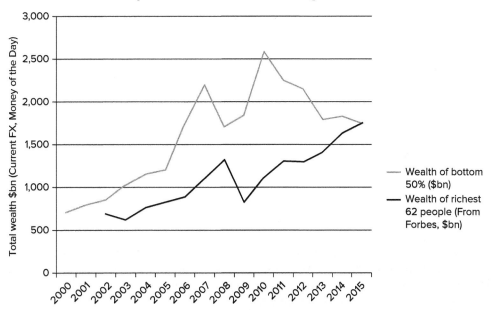

combined gross domestic product (GDP) of the UK and Germany—is currently held offshore.

Rising economic inequality also compounds existing inequalities. The International Monetary Fund (IMF) recently found that countries with higher income inequality also tend to have larger gaps between women and men in terms of health, education, labour market participation, and representation in institutions like parliaments.[5] The gender pay gap was also found to be higher in more unequal societies. It is worth noting that 53 of the world's richest 62 people are men.

Oxfam has also recently demonstrated that while the poorest people live in areas most vulnerable to climate change, the poorest half of the global population are responsible for only around 10% of total global emissions.[6] The average footprint of the richest 1% globally could be as much as 175 times that of the poorest 10%.

Instead of an economy that works for the prosperity of all, for future generations, and for the planet, we have instead created an economy for the 1%. So how has this happened, and why?

One of the key trends underlying this huge concentration of wealth and incomes is the increasing return to capital versus labour. In almost all rich countries and in most developing countries, the share of national income going to workers has been falling. This means workers are capturing less and less of the gains from growth. In contrast, the owners of capital have seen their capital consistently grow (through interest payments, dividends, or retained profits) faster than the rate the economy has been growing. Tax avoidance by the owners of capital, and governments reducing taxes on capital gains have further added to these returns. As Warren Buffett famously said, he pays a lower rate of tax than anyone in his office—including his cleaner and his secretary.

Within the world of work, the gap between the average worker and those at the top has been rapidly widening. While many workers have seen their wages stagnate, there has been a huge increase in salaries for those at the top. Oxfam's experience with women workers around the world, from Myanmar to Morocco, is that they are barely scraping by on poverty wages. Women make up the majority of the world's low-paid workers and are concentrated in the most precarious jobs. Meanwhile, chief executive salaries have rocketed. CEOs at the top US firms have seen their salaries increase by more than half (by 54.3%) since 2009, while ordinary wages have barely moved. The CEO of India's top information technology firm makes 416 times the salary of a typical employee there. Women hold just 24 of the CEO positions at Fortune 500 companies.

Across the global economy, in different sectors, firms and individuals often use their power and position to capture economic gain for themselves. Economic and policy changes over the past 30 years—including deregulation, privatization, financial secrecy and globalization, especially of finance—have supercharged the age-old ability of the rich and powerful to use their position to further concentrate their wealth. This policy agenda has been driven essentially by what George Soros called 'market fundamentalism'. It is this that lies at the heart of much of today's inequality crisis. As a result, the rewards enjoyed by the few are very often not representative of efficient or fair returns.

A powerful example of an economic system that is rigged to work in the interests of the powerful is the global spider's web of tax havens and the industry of tax avoidance, which has blossomed over recent decades. It has been given intellectual legitimacy by the dominant market fundamentalist world view that low taxes for rich individuals and companies are necessary to spur economic growth and are somehow good news for us all. The system is maintained by a highly paid, industrious bevy of professionals in the private banking, legal, accounting and investment industries.

It is the wealthiest individuals and companies—those who should be paying the most tax—who can afford to use these services and this global architecture to avoid paying what they owe. It also indirectly leads to governments outside tax havens lowering taxes on businesses and on the rich themselves in a relentless 'race to the bottom'.

As taxes go unpaid due to widespread avoidance, government budgets feel the pinch, which in turn leads to cuts in vital public services. It also means governments increasingly rely on indirect taxation, like VAT, which falls disproportionately on the poorest people. Tax avoidance is a problem that is rapidly getting worse.

- Oxfam analysed 200 companies, including the world's biggest and the World Economic Forum's strategic partners, and has found that 9 out of 10 companies analysed have a presence in at least one tax haven.
- In 2014, corporate investment in these tax havens was almost four times bigger than it was in 2001.

This global system of tax avoidance is sucking the life out of welfare states in the rich world. It also denies poor countries the resources they need to tackle poverty, put children in school and prevent their citizens dying from easily curable diseases.

Almost a third (30%) of rich Africans' wealth—a total of $500bn—is held offshore in tax havens. It is estimated that this costs African countries $14bn a year in lost tax revenues. This is enough money to pay for healthcare that could save the lives of 4 million children and employ enough teachers to get every African child into school.

Tax avoidance has rightly been described by the International Bar Association as an abuse of human rights[7] and by the President of the World Bank as 'a form of corruption that hurts the poor'. There will be no end to the inequality crisis until world leaders end the era of tax havens once and for all.

Companies working in oil, gas and other extractive industries are using their economic power in many different ways to secure their dominant position. This has a huge cost to the economy, and secures them profits far higher than the value they add to the economy. They lobby to secure government subsidies—tax breaks—to prevent the emergence of green alternatives. In Brazil and Mexico, indigenous peoples are disproportionately affected by the destruction of their traditional lands when forests are eroded for mining or intensive large-scale farming. When privatized—as happened in Russia after the fall of communism for example—huge fortunes are generated overnight for a small group of individuals.

The financial sector has grown most rapidly in recent decades, and now accounts for one in five billionaires. In this sector, the gap between salaries and rewards, and actual value added to the economy is larger than in any other. A recent study by the OECD[8] showed that countries with oversized financial sectors suffer from greater economic instability and higher inequality. Certainly, the public debt crisis caused by the financial crisis, bank bailouts and subsequent austerity policies has hurt the poorest people the most. The banking sector remains at the heart of the tax haven system; the majority of offshore wealth is managed by just 50 big banks.

In the garment sector, firms are consistently using their dominant position to insist on poverty wages. Between 2001 and 2011, wages for garment workers in most of the world's 15 leading apparel-exporting countries fell in real terms. The acceptability of paying women lower wages has been cited as a key factor in increasing profitability. The world turned its attention to the plight of workers in garment factories in Bangladesh in April 2013, when 1,134 workers were killed when the Rana Plaza factory collapsed. People are losing their lives as companies seek to maximize profits by avoiding necessary safety practices. Despite all the attention and rhetoric, buyers' short-term financial interests still dominate activities in this sector, as reports of inadequate fire and safety standards persist.

Inequality is also compounded by the power of companies to use monopoly and intellectual property to skew the market in their favour, forcing out competitors and driving up prices for ordinary people. Pharmaceutical companies spent more than $228m in 2014 on lobbying in Washington. When Thailand decided to issue a compulsory licence on a number of key medicines—a provision that gives governments the flexibility to produce drugs locally at a far lower price without the permission of the international patent holder—pharma successfully lobbied the US government to put Thailand on a list of countries that could be subject to trade sanctions.

All these are examples of how and why our current economic system—the economy for the 1%—is broken. It is failing the majority of people, and failing the planet. There is no dispute that today we are living through an inequality crisis—on that, the IMF, the OECD, the Pope and many others are all agreed. But the time has come to do something about it. Inequality is not inevitable. The current system did not come about by accident; it is the result of deliberate policy choices, of our leaders listening to the 1% and their supporters rather than acting in the interests of the majority. It is time to reject this broken economic model.

Our world is not short of wealth. It simply makes no economic sense—or indeed moral sense—to have so much in the hands of so few. Oxfam believes that humanity can do better than this, that we have the talent, the technology and the imagination to build a much better world. We have the chance to build a more human economy, where the interests of the majority are put first. A world where there is decent work for all, where women and men are equal, where tax havens are something people read about in history books, and where the richest pay their fair share to support a society that benefits everyone.

Notes

1 Credit Suisse (2015) 'Global Wealth Databook 2015'. Total net wealth at constant exchange rate (USD billion). http://publications.credit-suisse.com/tasks/render/file/index.cfm?fileid=C26E3824-E868-56E0-CCA04D4BB9B9ADD5

2 Source: Oxfam calculations based on Lakner-Milanovic World Panel Income Distribution (LM-WPID) database (2013). See Figure 1.

3 G. Zucman (2014) 'Taxing Across Borders: Tracking Personal Wealth and Corporate Profits', *Journal of Economic Perspectives*. http://gabriel-zucman.eu/files/Zucman2014JEP.pdf

4 Source: Oxfam calculations, see Figure 3.

5 C. Gonzales, S. Jain-Chandra, K. Kochhar, M. Newiak and T. Zeinullayev (2015) 'Catalyst for Change: Empowering Women and Tackling

Income Inequality'. IMF. http://www.imf.org/external/pubs/ft/sdn/2015/sdn1520.pdf

6 T. Gore (2015) 'Extreme Carbon Inequality: Why the Paris climate deal must put the poorest, lowest emitting and most vulnerable people first', Oxfam, http://oxf.am/Ze4e

7 M. Cohn (2013) 'Tax Avoidance Seen as a Human Rights Violation', *Accounting Today*. http://www.accountingtoday.com/news/Tax-Avoidance-Human-Rights-Violation-68312-1.html

8 OECD (2012) 'OECD Employment Outlook 2012', OECD Publishing. Chapter 3, 'Labour losing to capital: what explains the declining labour share?'. http://www.oecd.org/els/employmentoutlook-previouseditions.htm

OXFAM is an international confederation of 18 organizations working together with partners and local communities in more than 90 countries.

Max Roser

➡ **NO**

Income Inequality: Poverty Falling Faster Than Ever but the 1% Are Racing Ahead

Is the Gap Between Rich and Poor Widening? It's Not as Simple as That, Says Dr Max Roser, from the Institute for New Economic Thinking

How are the benefits of economic growth shared across society? Much of the current discussion assumes that income inequality is rising, painting a gloomy picture of the rich getting richer while the rest of the world lags further and further behind. But is it really all bad news?

The reality is complex, yet by looking at recent empirical data we can get a comprehensive picture of what is happening to the rich and the poor.

Let us start with the share of total income going to that much-maligned 1%. Reconstructed from income tax records, this measure gives us the advantage of more than a century of data from which to observe changes.

The blue line in the left-hand panel below shows the long-term trend in the US. Prior to the second world war, up to 18% of all income received by Americans went to the richest 1%. The share of the top 1% then dropped substantially, increasing again in the early 1980s until it returned to its 1939 level. This U-shaped long-term trend for top income share is not unique to the US; several other English-speaking countries shown in the left-hand panel below followed the same pattern. After a decline in the past, inequality is now on the rise.

However, it is not a universal phenomenon. The right-hand panel shows that in a number of equally rich European countries, and in Japan, things developed quite differently. Just like in the countries on the left, the income share of the rich reached a low point in the 1970s, but then, rather than bouncing back up to previous levels, it remained flat or increased only modestly, giving us an L-shape on the graph. Income inequality has decreased drastically since the beginning of the 20th century so that today, a much smaller share of total incomes is paid to the very rich.

One lesson to take away from this empirical research is that there is reason to believe we can do something about inequality. If there was a universal trend towards more inequality it would be in line with the notion that inequality is determined by global market forces and technological progress, where it is very hard (or for other reasons undesirable) to change the forces that lead to higher inequality. It is dangerous to believe that there is a unanimous trend to higher inequality, as this encourages the belief that growing inequality is inevitable.

The reality of different trends suggests that it is not global forces that shape the distribution of incomes, but the country-specific institutional and political framework. Therefore it is crucial to understand the institutional settings that allowed some countries to achieve economic growth without returning to the old levels of top income inequality. A big step in this direction is the forthcoming book from the inequality researcher Sir Tony Atkinson, in which he makes concrete proposals on how to reduce inequality, based on the insights from the periods in which inequality decreased.

The data on top income share does not measure the share of income that reaches the pockets of the rich, but the gross income before taxes are paid. Yes, the rich do avoid paying taxes, and top marginal income tax rates were higher in the past, but progressive taxation still does a great deal to narrow the gap between rich and poor: in the US, 37% of the total sum of income tax is paid by the top 1%, while less than 3% is paid by the bottom 50%. The redistribution means that the incomes of the poor are higher after taxes (because of transfer payments such as pensions, child benefits, and unemployment benefits) and the incomes of the rich are reduced after taxes (due to generally progressive income tax rates).

This difference between pre-redistribution market incomes and eventual disposable income is shown in the next chart. Redistribution through taxes and transfers reduces inequality considerably.

In this chart, inequality is measured with the Gini index, an inequality measure that not only looks at the top of the income distribution but captures the whole distribution.

We can see that market income inequality in the UK, the US, and France is fairly similar (Gini between .5 and .52)—but there are big differences in how much these countries reduce inequality by redistribution. Inequality in market and disposable income are steadily increasing in the US, and compared to similarly rich countries, the US redistributes comparatively less.

This data is based on household surveys, a shortcoming of which is that they under-report top incomes. Likewise, the shortcoming of the top income measure is that it is necessarily silent about what is happening within the bottom 99% of the distribution. Taking the top 1% chart and the market income v disposable income charts together allows us to understand how inequality has developed. In the UK, the pre-tax share of the top 1% has been rising continuously since the late 1970s, but disposable income for all earners followed a very different trajectory, with inequality increasing rapidly in the 1980s but not changing much since then. If anything, income inequality has actually fallen in the UK over the past 25 years. In summary, the incomes of the poor in the UK are growing as fast as the incomes of the rich, apart from the top 1%, whose incomes are racing away.

Before we turn to the global income distribution, I want to shift the focus from Europe and North America to South America. In all of the South American countries shown in the following chart, income inequality has fallen since 2000. It is shown in the chart for the bigger South American countries (and it can be studied for the other countries on OurWorldInData.org). Rapidly falling inequality in South America that lifts millions out of poverty is a huge success, and demonstrates once again that there is not one simple answer to whether inequality is rising or falling within individual economies.

We have looked at the changing income distribution within countries, but what about the global picture?

The chart below shows global income estimates in 1820, adjusted, as always, to take inflation and price level differences between countries into account. Two centuries ago, no country in the world had a life expectancy over 40, and even in relatively rich countries such as England and France, the poor were so malnourished and weak that they were effectively excluded from the labour force. The share of the world's population living in absolute poverty was estimated to be around 90%.

Over the next 150 years, some countries achieved economic growth while others remained poor. Europe and the European offshoots in North America and Oceania grew rapidly, but most of Asia, most of Latin America and all of Africa remained poor. The consequence of this was a hugely unequal world. The bimodal, "two-humped" blue line for 1970 shows the world income distribution of a planet clearly divided into rich and poor countries.

The world has changed since the 1970s. The circle of countries achieving economic growth now includes much of Asia, Latin America, and for the last two decades, Africa. The consequence of this is that global poverty is falling faster than ever before—the share of the global population living in poverty has decreased from more than 50% in 1981 to 17% in 2011.

Due to rapid growth in formerly poor countries, the world income distribution has changed dramatically. The bimodal distribution of very high inequality across countries in 1970 has changed into a unimodal distribution of lower inequality today. The latest research on this question shows that the Gini index for global inequality has fallen from 72.2 in 1988 to 70.5 in 2008 (the last year for which we have data).

There is no reason for complacency, and a long way to go to improve living standards for the worst-off globally, but we can take a clear and heartening message from the data: world income inequality and poverty are in decline.

Dr. Max Roser is a James Martin fellow at the Institute for New Economic Thinking at the Oxford Martin School, Oxford University. If you are interested in long-term trends of living standards around the world, follow him on Twitter, where he shares many data visualizations of long-term trends from his web publication Our World in Data.

EXPLORING THE ISSUE

Is Global Income Inequality on the Rise in the International Community?

Critical Thinking and Reflection

1. How do both authors use the data on inequality to make different cases?
2. What is more important, the distribution of income or the decline of poverty?
3. Are there systemic changes that the international system can make to mitigate inequality?
4. Should the primary goal be the elimination of inequality or the creation of greater wealth and opportunity and what are the implications of each position?

Is There Common Ground?

Most analysts agree that the current global economic system creates wealth and opportunity for more and more people across the globe, despite the varying types of governments and systems that people live under. The key question is what are the rules under which that wealth is created, obtained, controlled, and used, and does it benefit a larger segment of global society over time or not? Analysts on both sides agree that ignoring widening gaps between rich and poor only acerbates socio-political tensions and leads to demagogues, political disorder, and ultimately violence.

Additional Resources

Hickel, Jason, "Global Inequality May Be Worse Than We Think," *The Guardian* (April 7, 2016).

Lynch, Connor, "Stephen Hawking on the Future of Capitalism and Inequality," *Counterpunch* (October 15, 2015).

Milanovic, Branko, *Global Inequality: A New Approach for the Age of Globalization* (Harvard University Press, 2016).

Murray, Iain and Young, Ryan, "People, Not Ratios: Why the Debate over Income Inequality Asks the Wrong Questions," Competitive Enterprise Institute (May 25, 2016).

Internet References . . .

Competitive Enterprise Institute

cei.org

The Global Policy Institute

globalpolicy.org

The Institute of Policy Studies

inequality.org

Selected, Edited, and with Issue Framing Material by:
James E. Harf, *Maryville University*
and
Mark Owen Lombardi, *Maryville University*

ISSUE

Is Social Media Becoming the Most Powerful Force in International Politics?

YES: **Ritu Sharma**, from "Social Media as a Formidable Force for Change," *The Huffington Post* (2015)

NO: **Kathy Gilsinan**, from "Is ISIS's Social-Media Power Exaggerated?" *The Atlantic* (2015)

Learning Outcomes

After reading this issue, you will be able to:

- Understand the pervasiveness and growth of social media in the political and social world.
- Be able to evaluate it both as a tool for change in varying forms regardless of the outcomes.
- Understand that as a tool it can be used by governments, people, movements, religions, and any other social construct for any purpose.
- Reflect on how it has changed political and social movements as compared to years past and how it might shape future events.

ISSUE SUMMARY

YES: Ritu Sharma sees social media as a powerful mobilizing force and uses many examples to illustrate that point.

NO: Kathy Gilsinan looks at ISIS's use of social media and questions its power and influence, questioning the real reach of social media.

The revolution that is the Internet and now the advent of social media is as deep and far reaching as any communication revolution in history. The data analytics alone are staggering. Worldwide, there are hundreds of millions of users of mobile technology that links these users through a plethora of social networks from Facebook, twitter, and Instagram to many you may not have heard of. Since 2001 global population access to a cell phone network grew from 58 to 95 percent. Facebook's internet.org is now available to one billion people across 17 countries. Internet usage in India has grown 37 percent per year. In 2015, advertisers spent over 23 billion on social media marketing. Once a person is online, they use social media networking sites on average 76 percent of the time. And of the total world population of 7.395 billion people, approximately 2.307 are active social media users.

These statistics though fascinating merely reflect awareness, access, and usage. They do not grasp the depth of how social media is being used; who and what messages are being transmitted; what political and social causes are being promoted; what actions are being organized and initiated and of course what impact are they having. Today small groups of like-minded people around the globe can connect with each other, share information, mobilize political action, raise money, and directly impact policy and governmental actions without ever leaving their homes.

Two recent examples illustrate this growing power. The Arab spring of 2010–2016 despite its mixed record of success showed the power of political unrest mixed with social media. Protests in Tunisia, Egypt and elsewhere, which in the past would have led to crackdowns and perhaps modest reforms, brought down governments because social media was able to mobilize popular opinion both within countries and globally faster than government's ability to analyze and react.

The second example is its use by ISIS. The radical terrorist group known as ISIS (or sometimes referred to as ISIL or Daesh) has mobilized a seeming worldwide network of followers despite their being rooted in a remote portion of the Middle East deep in the deserts of Syria and Iraq. Since their emergence they have used social media and a sophisticated propaganda network through Internet sites like YouTube, Facebook, and others to message, recruit, and turn individuals into terrorist supporters and actors. The most recent horrific attacks in San Bernadine and Orlando are prime illustrations of this approach.

Social media is an evolving dynamic tool for communication, expression, mobilization, and action. It is now being utilized in thousands of ways by hundreds of groups for both good works (crowdsourcing for cancer research or to find lost children) and evil ends (recruiting suicide bombers or spreading gospels of hate). Its power and impact across the global landscape is dynamic and very much still under debate.

In the YES selection, Sharma makes the case that social media is potent, impactful, and limitless in how it can influence the global political and social landscape. In the NO selection, Gilsinan takes the example of ISIS and intimates that social media's influence may be overblown and that we may be enamored with a new method of communication and overstating how it can impact in this case the effect of a group like ISIS.

YES ←

Ritu Sharma

Social Media as a Formidable Force for Change

The power of social media is hard to dismiss. What once seemed like a trivial way to keep in touch with friends, sharing photos and jokes, has become a force for societal change, shining light on subjects previously unknown, deepening conversations and empowering citizens of the world to unite and effect change in a number of ways. Interestingly, social media as a medium for connecting, organizing and communicating is powering and spreading democracy far better than billions of dollars of aid or war in corners of world very resistant to such change.

Creating Visibility

The ability of the Internet, and social media particularly, to bring issues to bear is unique. Consider the still-too-common problem of domestic violence. Though former President Clinton passed the Violence Against Women Act in 1994, and changes have been made, incidents like the recent Ray Rice video have made it clear we still have a ways to go.

Enter social media, and the hashtag. The sharing of experiences marked by grassroots campaigns like #WhyIStayed is just one example of the people of the world refusing to sit idly by and let anyone fall into blame-the-victim habits by showing just how many have been affected; how surprising some of those people might be (Meredith Viera); and how resolute and united they are in demanding the attention of those in a position to enact change.

The subtext of every #WhyIStayed, every #YesAllWomen seems to be, "You think this isn't a widespread problem? Think again."

The Evidence Is Mounting

It would be easy to shrug off these campaigns as flashes in the pan, or much ado that comes to nothing in the end, but it wouldn't be true. These are not isolated incidents,

and they're not just noise. Recent history has given us a number of examples of hashtag campaigns that have led to change. Here are a few:

#Ferguson called for awareness of police brutality and the racial divide in Missouri. The story might not have come to many folks' attention save for the hashtag which caused people to follow events as they unfolded on social media, blogging, sharing and talking about it all.

#NetNeutrality continues to bring attention to the fight to keep the Internet free and open, without fast lanes for elite customers who can pay more, resulting in nearly 4 million comments on the open forum established by the FCC to help them decide on proposed Net Neutrality rules.

#IceBucketChallenge raised record-breaking funds for the ALSA ($115 million since July 29) thanks to the grassroots awareness campaign which went viral, proving that a bucket of ice water has the means to warm people's hearts and loosen their purse strings.

#ArabSpring brought attention to protests and unrest in countries from Tunisia to Syria, uniting those ready to take action in hopes of a new order.

> The people were ruled for decades in fear of regimes this year the fear was over taken by hope and dreams of change.. #ArabSpring
> 10:40 AM - 20 Oct 2011

#DelhiGangRape brought the culture of violence and rape in India into the spotlight, effecting changes in the law and in sexual education in the country. The events also inspired playwright Eve Ensler to start One Billion Rising, a global campaign dedicated to ending violence, and demanding change and justice for women.

While recent occurrences of gang rape in India show the problem is far from solved.

#WakeUpAkhilesh has become a rallying cry to hold Akhilesh Yadav, the chief minister of the northern Indian state of Uttar Pradesh, accountable to his constituents. The fight for change continues.

The Benefits of Social Media and Hashtag Unity

One of the most powerful aspects of social media is that it provides an environment and a medium for people to express themselves independently, and yet find community. This "hashtag unity," to coin a term, is as real and as powerful as a group of people physically gathered in the same space. It can educate, heal and provoke change by sheer strength of vocal numbers.

And it doesn't cost a thing. Consider the resources spent on the military, for the sake of bringing peace to war-ravaged countries, while we struggle at home to find funds for health care and education. Then consider that we may have found a better alternative to reform, thanks to this byproduct of Silicon Valley, able to solve some of our most pressing problems using social media.

We're already impacting the issues of health (from the perspective of terminal diseases like ALS) and overall well-being (impossible in an environment of domestic abuse), gender equality, sexual violence, civil rights. These issues being brought to light offer all of us the chance to educate and be educated. Quite unexpectedly, social media has become the strongest tool of democracy at our disposal.

It's not just noise. *It's having an effect.* The issues highlighted above each gained exposure through social media, but that exposure inspired people to take action. Those taking action shared their experiences, which were amplified by social networks spreading them in solidarity. There is action, attention and further action. So it's a virtuous circle.

Taking It Further—Dos and Don'ts

So what do we do next? What are our roles as nonprofits, as members of society?

I think it's important for us to seek out and boost the causes that mean the most to us. To shine that spotlight, to educate and be part of that hashtag unity. We should take action where we can, but sometimes simply guiding the action of others is enough. "Share this story, sign this petition, recognize that this is happening and be part of the chorus of voices demanding change."

It's not always in our power to do much more than that, but when enough of us band together, we can convince those with the power to create solutions to act.

There are caveats, of course (just as fire can keep you warm and cook your meals, it can also burn your house down). Social media can be a community, or it can be a mob—as community leaders, let's be sure we encourage our constituents to use it responsibly.

For example, we would never want to encourage or promote any acts of vigilante justice. We lose our power when we give in to revenge, not to mention losing credibility. These social media campaigns offer a self-correcting democratic method, where the story is no longer controlled by only a few people, or people trying to keep things quiet. Our job as citizens is to press for solutions by the proper authorities—not to take things into our own hands.

Social media alone will not solve all the world's problems—I know that. But without necessarily meaning to it has served as a very powerful tool in imparting democracy, education and justice, both at home and abroad.

So what if we put our heads together and did attempt to thoughtfully harness this power? Think of the ways we could educate each other. Think of the ways we could empower people to own their own activism, to truly be the change they wish to see. We can incorporate government and educate via these platforms, but we can also go beyond government, training others in advocacy and activism so they can help themselves.

Social media put an oft-ignored disease into the collective consciousness. It got laws changed in India. Yes, there is more to do—always. But as more of us take on the burden, the lighter it will be.

RITU SHARMA is a public speaker, consultant, event planner, facilitator, and thought leader. She is the Co-Founder and Executive Director of Social Media for Nonprofits, an organization committed to bringing quality and accessible social media education to nonprofits around the world, primarily through conferences that bring together the top thinkers and doers in the social media and nonprofit spaces.

Kathy Gilsinan  **NO**

Is ISIS's Social-Media Power Exaggerated?

The Group Is Famously Active on Twitter and Has Attracted Thousands of Foreign Fighters. But to What Extent Is One Related to the Other?

"Terrorist groups like al-Qaeda and ISIL deliberately target their propaganda in the hopes of reaching and brainwashing young Muslims, especially those who may be disillusioned or wrestling with their identity," President Obama said last week in remarks wrapping up a Washington summit on Countering Violent Extremism. "The high-quality videos, the online magazines, the use of social media, terrorist Twitter accounts—it's all designed to target today's young people online, in cyberspace."

The remarks reflected what's become something of a truism as the media routinely reports on ISIS's "slick" propaganda apparatus, Western recruits becoming radicalized through social media, and the U.S. government's sluggishness—or outright ineptitude—in fighting back on the Internet. The State Department has a Center for Strategic Counterterrorism Communications with a team dedicated to countering "terrorist propaganda and misinformation about the United States across a wide variety of interactive digital environments," which, it admits on the department's website, "had previously been ceded to extremists." That office, as *The New York Times* recently reported, is slated for expansion. The online information war was a focus of last week's summit.

But what if ISIS's much-hyped social-media juggernaut isn't as important as all of these measures suggest?

"We know it has the potential to influence, but exactly how and at what levels are quite unknown," Anthony Lemieux, an associate professor of communication at Georgia State University, wrote in an email. Lemieux is researching that very question, but in the meantime it's difficult to find a reliable estimate of how many ISIS fighters have been radicalized and recruited primarily through social media. Max Abrahms, a political-science professor and terrorism specialist at Northeastern University, suspects the number is lower than many people believe. "There are other groups"—such as Boko Haram in

Nigeria—"that have rapidly expanded their membership size in the absence of social media," he pointed out to me. "Battlefield success is a better predictor" of group size than is social-media activity, Abrahms said. If, as some contend, ISIS's battlefield momentum has already stalled, its recruitment could suffer even as its social-media activity remains constant.

In tandem with its military successes, ISIS has also likely benefited from an influx of foreign fighters to Syria that predates the group's blitzkrieg in the summer of 2014. A record number of foreigners had already joined a variety of Syrian rebel groups by mid-2013, a full year before ISIS captured the Iraqi city of Mosul and began consolidating territory across the Syria-Iraq border. At the time, Thomas Hegghammer, a senior research fellow at the Norwegian Defence Research Establishment, acknowledged the role of social media in "the scale and speed of the mobilization." But, he continued, "this does not mean that social media in and of itself drives recruitment." Citing poorly policed borders and ease of travel to Syria, Hegghammer theorized: "The bottom line is that record numbers of foreign fighters are going to Syria because they can." Since then, ISIS's victories, among other factors, have enabled the organization to eclipse other rebel groups in terms of recruitment.

Western policymakers are quite reasonably preoccupied with ISIS's recruitment of jihadists from Europe and the United States. But by far the biggest suppliers of the Islamic State's foreign fighters are Middle Eastern and North African countries, particularly Tunisia and Saudi Arabia, where broadband access lags behind access rates in the West. Among those who are online, according to a Soufan Group study of foreign fighters in Syria, potential recruits in the Levant and the Gulf "are interconnected within self-selected bubbles, and are isolated from anything outside." This implies both that social media helps ISIS amplify its message among closed groups that are already receptive to it, and that there are limits to how far that message can spread beyond those circles.

ISIS does have an enthusiastic base of supporters on English-language Twitter. But there's a major difference between retweeting beheading videos or trolling the State Department and actually going to fight for the Islamic State. (For example, one vocal online propagandist for ISIS was revealed last December to be a corporate executive living in Bangalore, India.) According to official estimates cited in *The New York Times*, 150 people have "traveled, or tried to travel, to fight in Syria from the United States." It's not clear how many people actually succeeded in doing so. But it's hard to imagine social media being the decisive factor in the decision to leave, say, suburban Colorado for Syrian battlefields, even if social media proved helpful in planning the trip. The causal relationship could just as easily work in the opposite direction: People engage with ISIS through social media because they're already radicalized, rather than getting radicalized through social media.

Perhaps it's the novelty of ISIS's social-media operation, or its status as a major source of information about an otherwise largely opaque enemy and war zone, that make the group's tweets seem excessively threatening. The recent report that ISIS produces a large volume of social-media chatter—90,000 "tweets and other social media responses every day," by the *Times*'s estimate, which may be conservative—doesn't actually say much about the role those communications play in terrorist recruitment.

As Brian Jenkins of the Rand Corporation noted in a 2011 study of "homegrown" extremists in the United States, reaching potential recruits online "does not mean radicalizing, and radicalizing does not mean recruitment to violent jihad." He also noted that while the Internet could "serve as a source of inspiration . . . it may also become a substitute for action, allowing would-be terrorists to engage in vicarious terrorism while avoiding the risks of real action." Similarly, J.M. Berger, a Brookings analyst who studies extremists' use of social media, observed in 2011, "There is a tremendous amount of radical activity online. Very little of that activity will translate into real-world threats." Writing in *The Atlantic*, Berger also pointed out that even the volume of this activity may be less than meets the eye, since ISIS uses a number of techniques, including automated tweets and organized efforts to trend certain hashtags, to give the impression of a larger network of online support.

ISIS is surely a formidable force. Offline, the group is now estimated to be 30,000-strong or bigger. But the roots of its expansion probably don't lie on the virtual battlefield. More likely, they're on the real one.

KATHY GILSINAN is a senior editor at *The Atlantic*, where she oversees the Global section.

EXPLORING THE ISSUE

Is Social Media Becoming the Most Powerful Force in International Politics?

Critical Thinking and Reflection

1. What are the dimensions of social media that make it a powerful force?
2. What actors seem most adept at using it to further their political goals?
3. Is social media an inherent good or evil or is it simply a tool and its character is shaped by the actor using it?
4. Can governments control the impact of social media on their own citizens?

Is There Common Ground?

Determining the relative power of social media when compared to other forces in global politics is certainly debatable and a moving target. The evolution of technology to communicate, disseminate, shield, and secure is rapidly changing on an almost weekly basis. What most analysts can agree on is that social media has empowered millions of more people and groups within the international system to engage in both good works and crime and violence. This explosion of access for millions has made the international system more volatile, more fluid, and potentially more dangerous.

Additional Resources

Finsterlin, Marc, "Is India's Social Media Already a Potent Force?" *Aquarius* (March 23, 2013)

This article describes how social media shapes Indian life and politics.

Luis Moreno, "How Social Media Could Revolutionize Third World Cities," *The Atlantic* (November 13, 2012)

The author discusses how the Internet will change the way cities are run and how it empowers inhabitants.

Pew Research Center, "Emerging Nations Embrace Internet, Mobile Technology" (February 13, 2014)

Data shows mobile technology is present across the world.

Shirky, Clay, "The Political Power of Social Media: Technology, the Public Sphere, and Political Change," *Foreign Affairs* (January/February, 2011)

The article provides a balanced argument for the power of social media used effectively.

Toyama, Kentaro, "Malcolm Gladwell Is Right: Facebook, Social Media and the Real Story of Political Change," *Salon* (June 6, 2015)

The author argues that social media gets too much credit for change movements.

Internet References . . .

The Harmony Institute

Harmony-Institute.org

Web trends

webtrends.org

Wired Magazine

wired.com

Selected, Edited, and with Issue Framing Material by:
James E. Harf, *Maryville University*
and
Mark Owen Lombardi, *Maryville University*

ISSUE

Are Cyber-groups Terrorists or a Potential Force for Good?

YES: **David Auerbach**, from "The Sony Hackers are Terrorists," *BITWISE* (2014)

NO: **Evan Schuman**, from "Anonymous Just Might Make All the Difference in Attacking ISIS," *Computerworld* (2015)

Learning Outcomes

After reading this issue, you will be able to:

- Appreciate and understand the enormous power behind the use of cyber terrorism.
- Understand that cyber power can be projected by individuals, small groups, and states with equal force.
- Appreciate that this strategy can take on many forms and be employed in a variety of ways to undermine groups, organizations, and societies.
- Speculate as to how it will impact your life and the lives of millions in the coming years.

ISSUE SUMMARY

YES: Auerbach views hackers like the recent Sony hack as terrorism because of the economic and social impact.

NO: Schuman points out that hacker groups like Anonymous may indeed be forces for good in terms of combating real terrorists like ISIS.

Cybercrime, identity theft, cyber-scams, cyber terrorism, and of course cyber security are all becoming part of the daily lives of organizations, businesses, and governments, and of course tens of millions of people around the globe. The Internet empowers people with mobility, ease, functionality, and fingertip access to goods, services, information, and wants. It also exposes those same people to security challenges that jeopardize personal information, trade secrets, military data, business strategy, and countless other forms of information that can bring down peoples credit, businesses, and even governments.

This new reality has led to the emergence of individuals, groups, organizations, and agencies within states that use cyber skills to access information, steal resources, and undermine competitors either for financial, philosophical, or political reasons. The rise of cyber actors in this Internet space has been significant and profound. In 2015, over

165 million personal records were exposed and that was 38 percent higher than the previous year. Fifty-two percent of businesses believe they will be hacked this year. There are on average 100,000 hacks per minute globally.

Many of these new actors appear sinister and secretive by their very nature. They hack into systems, compromise data, and use it for their own political, social, or economic agenda. Since most of us do not fully understand how they accomplish these hacks, we fear them and their capability, given that so much of our lives are now accessible on the Internet.

Over the last 20 years these groups have engaged in activities that promote everything from anarchy, social justice, and economic destabilization to Islamic fundamentalism, cyber espionage, and governmental downfall. The following questions arise: are these cyber actors a new form of terrorism? Are they creating a new battlefield in terrorizing citizen's populations for the purpose of achieving some political end? Or are they potentially actors who are policing the international cyberspace, holding power and authority accountable and making sure state actors are operating by some set of recognized international law? These are some of the questions that have been raised by the actions of people like Edward Snowden and his hacks of the U.S. intelligence community. It's also the question that is asked by groups like Anonymous when they hack into a government server or engage in social protest or when they turn their sights on ISIS in the wake of the Charlie Hebdo attacks.

In the new territory of cyberspace, the rules are ill-defined, the issues are fluid and the methods of attack can be ambiguous. Groups like Anonymous for example can use some of the cyber skills to attack government agencies, reveal secure information, or engage in civil disobedience. They also can turn their skills against groups like ISIS to combat their own violent agenda. States are also operating in this space and some with great effectiveness. It is said that U.S., Soviet, Chinese, Syrian, Iranian, and Israeli cyber capabilities have engineered attacks against one another or other states to pursue their own political interests.

Our two authors view this issue from very different perspectives. Auerbach sees the very nature of cyber-attacks as extra-legal and indeed as clear negatives in the world of international economic, political, and social exchange. No matter what issue or cause one wraps themselves in, the nature of the action makes it wrong, and therefore terrorism in some particular form and a threat to global stability. Schuman sees the act of cyber-attack as almost amoral, and rather than target, rationale and cause being the determining factors in its being equated with terrorism.

YES ↵

David Auerbach

The Sony Hackers Are Terrorists

We've Never Seen a Cyberattack Like This One. Here's Why It's so Frightening

The Sony Pictures hack is important, and the Sony Pictures hack is terrifying.

In a series of cyberattacks that were first noticed on Nov. 24, a mysterious group calling itself the Guardians of Peace stole and subsequently leaked personal and medical information from every Sony Pictures employee, revealed scads of confidential internal information, left the company technologically crippled, and issued vague demands that "our request be met." That last one increasingly appears to center on The Interview, a James Franco–Seth Rogen comedy about killing North Korean leader Kim Jong-un, which Sony is now offering theaters the choice of not showing. (As of today, the four largest movie-theater chains say they won't screen the movie. Update, Dec. 17, 5:06 p.m.: After those exhibitors decided not to show The Interview, Sony Pictures announced that it had "decided not to move forward with the planned December 25 theatrical release.") We don't yet know exactly who is behind this cyberassault. (Update, Dec. 17, 7:52 p.m.: The New York Times is now reporting that "American intelligence officials have concluded that the North Korean government was 'centrally involved' in the recent attacks.") What's clear, however, is that it represents a wake-up call that's been coming for a long time.

The Sony hack isn't important because of its technological sophistication, which is impressive but probably not particularly innovative. While neither Sony nor the FBI has released the exact details, so far there is little to suggest that this was some brilliant, unprecedented maneuver on the order of the NSA's still-astounding StuxNet, a virus which managed to sneak its way into the isolated nuclear facilities of Iran and sabotage them. What's remarkable is the sheer destruction leveled at Sony and its employees. For perhaps the first time, a major American company really did suffer a worst-case cyberassault scenario.

As someone who suffered through and reviled the hysteria of the post-9/11 era, I want to stress that most hackers, from script kiddies to the members of Anonymous, are not terrorists. The Guardians of Peace are different. With yesterday's threat of violence against theaters showing The Interview—"The world will be full of fear. Remember the 11th of September 2001."—I don't know what else to call them.

Consider most of the high-profile hacks of recent years, like the theft of millions of credit card numbers from Target, or The Fappening's stolen celebrity nudes from Apple accounts, or, indeed, the theft of 77 million Sony PlayStation accounts in 2011. All of these were costly, damaging thefts of private information, but they were fundamentally thefts. Not this time. While tabloid rags are salivating over the juicy Hollywood gossip and Aaron Sorkin is writing impassioned polemics against revealing stolen information, these hackers, whoever they are, genuinely do deserve to be termed cyberterrorists. Many attacks are for financial gain or revelation of valuable or salacious information. The latter is a factor here, but the overriding aim seems to have been to damage Sony Pictures and its employees to the point at which they could barely even function. To my knowledge, there has never before been a cyberattack of this scale. The Guardians of Peace didn't just steal 100 TB (an ungodly amount) of sensitive data, they also used "wiper malware" to more or less destroy Sony's internal systems, leaving its entire infrastructure crippled. Just consider what Kevin Roose of Fusion has reported:

Sony Pictures' network subsequently went down for two days, forcing employees to use personal e-mail accounts, work from home, and in some cases, resort to paper and pencil to do their work . . . "It's just business as usual, if the year was 2002," one Sony TV staffer wrote to me in a Facebook message. "[There are] lots of PAs having to run jump-drives back and forth all over the place, and hand delivering hard copies of files and scripts."

Or what another insider told the Wrap: "Every PC in the company is useless and all of the content files have either been stolen or destroyed or locked away. . . . The IT department has absolutely no idea what hit them." Since Sony's security "department" is apparently the Three Musketeers plus managerial overhead—"Three information security analysts are overseen by three managers, three directors, one executive director and one senior-vice president," according to Fusion—I don't blame them, though I do blame Sony Pictures. The studio's security appears to be little better than Sony Playstation's was in 2011, and probably worse.

This is the real story. Sony Pictures' systems were not just compromised but obliterated, with the company now sent back to what's comparably the technological Stone Age. Because of the centrality of IT infrastructure to every aspect of a company's functions, it's not even clear whether Sony has the ability to pay people accurately at the moment, as its payroll system has been reportedly destroyed. In this, the attack resembles two other wiper incidents, as reported by Kaspersky's Kurt Baumgartner: the 2012 "Shamoon" attack against Saudi Aramco, and the 2013 DarkSeoul attacks against South Korean banks and broadcasters. Those events skirted the line of cyber-terrorism without quite crossing it. And while this attack is particularly damaging to Sony's rank and file, the hack itself poses no threat to people's lives or critical infrastructure. But by so effectively creating a climate of fear and making threats of actual violence, the Guardians of Peace have raised the specter of genuine cyberterroristic acts to come. These acts aren't scary because they're ingenious, but because they could be easily replicated by anyone with the right resources and enough malice.

Sony, in contrast, has played up the technical sophistication of the attack, which is both an overstatement and a distraction. FBI Assistant Director James Demarest said that 90 percent of systems couldn't have withstood the Guardians of Peace's attack, but that's not really saying much. As we've seen in so many cases, the average state of cyber security is rather weak. I take Demarest's number to mean that Google, Apple, Microsoft, the federal government, and other companies with serious security expertise could have easily withstood the attack, but companies closer to the average haven't yet insulated themselves against whatever particular vectors the Guardians used to compromise Sony. That's the scary part: In terms of security, Sony Pictures wasn't terrible, but just average. It's likely that comparable amounts of damage could have been inflicted on many companies via the same vectors of attack.

That doesn't mean we should panic. Again, a good security system could have prevented the hack, but, compared to most previous attacks, this is a whole new ballgame. It's not about money or humiliation. It's about fear and wanton destruction, possibly with the intent of causing Sony and others to accede to the demand to "Stop immediately showing the movie of terrorism." The threat of the Guardians is serious, and they herald a world in which a bad security story may not just mean the loss of user trust and revenue, but the obliteration of a company's ability to function. So as the media fiddles while Sony burns, wallowing in the stolen emails and pointing fingers, there's much we all need to learn about our own security vulnerabilities and longstanding inadequacies. We need to do it now.

David Auerbach is a writer and software engineer based in New York.

Evan Schuman

 **NO**

Anonymous Just Might Make All the Difference in Attacking ISIS

The Hacking Group's Activities Have Always Seemed Dubious, but in This Case, Success Will Be Quite Welcome

In the wake of the horrific attacks in Paris, military operations against ISIS terrorist strongholds have increased. When someone hits you, it's natural to hit back. But can you win by killing an enemy that seeks death—and when those who are killed inspire desperate others to replace them?

Along with the news that the French had launched air strikes against ISIS positions came the word that the cyber-revenge group calling itself Anonymous has declared war on ISIS. I never thought I would say this, but Anonymous might be our savior. Attacking ISIS militarily is necessary, but the group has always exerted its influence through social media, using it for both fundraising and recruitment. Both activities are essential to ISIS' continuing existence and effectiveness. The weaponry it uses in its terror campaign is expensive, and when every successful operation ends in death or the arrest of all participants, recruitment is critical.

Which makes Anonymous' involvement intriguing. ISIS, you should be very scared. Anonymous' official Twitter area for Operation Paris has already reported that "more than 3,824 Twitter accounts pro-ISIS are now down," a number that has been steadily climbing. That's on top of tens of thousands of pro-ISIS Twitter accounts that Anonymous took down before the Paris attacks, along with a large number of donation pages for the group—housed on the dark Web—that Anonymous closed, according to a report in The Atlantic.

Anonymous' key message on Twitter is: "Make no mistake: #Anonymous is at war with #Daesh. We won't stop opposing #IslamicState. We're also better hackers." That last sentence is key. What Anonymous is talking about is a cyberwar against ISIS, one that is not restricted by any of the laws that could hamper a cyberattack undertaken by U.S., French or Russian governments. And, to Anonymous' credit, they are indeed better hackers.

This could be very effective. While governments bomb and soldiers attack on the ground, it's essential that ISIS' recruitment and funding be killed. If DDoS and other shutdown tactics make social media useless to the terror group, it will find it far more difficult to fund terror and recruit replacements.

Let's take a look at Anonymous' prospects for weakening ISIS in these areas.

Funding

ISIS does get some funding from governments, but that's generally done quietly. Governments are subjected to International pressure to make ISIS a pariah, so most funding from nation states is kept secret. But Anonymous is impressively good at revealing secrets, so its activities could make it much harder for evil-oriented governments to back ISIS without paying a price.

ISIS also makes money by selling items it steals, but that funding is limited.

Mostly, ISIS relies on small donations from supporters, after they've been indoctrinated. The Anonymous campaign can be effective in this area in two ways. First, by repeatedly shutting down huge numbers of donation sites, Anonymous will make it far harder for people who want to donate to do so. Second, by shutting down propaganda sites, it can reduce the number of people who feel the desire to donate.

By the way, this has to be done in conjunction with military attacks. As those attacks happen, ISIS scum (sorry, but that's the harshest language *Computerworld* will let me use) will use them to recruit people and beg for money. Anonymous shutdowns will make it far harder to get those messages out and for anyone who does see such messages to respond.

Recruitment

In this area, Anonymous' cyber warfare can be far more effective than bombs and bullets. Recruitment requires three elements: Get messages to potential supporters and sympathizers; convince them to support you; give them the means to connect with you. (Note: Attempts to infiltrate these operations are another great tactic because, if nothing else, they force ISIS to be ultra-suspicious of anyone too eager to join.)

Social media and the dark Web are the only practical ways—beyond word-of-mouth, which is far too slow and inefficient for a global operation—ISIS can do this. Its leaders have learned social media well. If Anonymous can keep ahead of ISIS in shutting down propaganda, recruitment and fundraising sites and feeds of all kinds, its cyberwar, in conjunction with standard military attacks, can work.

So, though I have never had kind words for Anonymous in the past, I am truly hopeful that in this endeavor it will find success. When we see atrocities like what happened last week in Paris, it's natural for many of us to feel helpless. But the coders and other technically oriented folk who have joined forces with Anonymous have a chance to make a difference.

EVAN SCHUMAN has covered IT issues for a lot longer than he'll ever admit. The founding editor of retail technology site StorefrontBacktalk, he has been a columnist for CBSNews.com, *RetailWeek*, *Computerworld*, and *eWeek* and his byline has appeared in titles ranging from *Business-Week*, *VentureBeat*, and *Fortune* to *The New York Times*, *USA Today*, *Reuters*, *The Philadelphia Inquirer*, *The Baltimore Sun*, *The Detroit News*, and *The Atlanta Journal-Constitution*.

EXPLORING THE ISSUE

Are Cyber-groups Terrorists or a Potential Force for Good?

Critical Thinking and Reflection

1. What is the nature of a cyber-attack and is it inherently wrong and illegal or open to interpretation?
2. Are cyber-attacking groups terrorists in terms of the nature of the attack or its overall impact?
3. Can states be cyber terrorists?
4. Is cyber terrorism the new frontline in the war on terror or merely a more benign distraction?

Is There Common Ground?

There is likely to be disagreement for some time on whether cyber-groups are by definition terrorists or not. Just as conventional terrorist groups have always argued your point of view determines whether one is a terrorist. ("One man's terrorist is another man's freedom fighter.") One thing is clear, cyber-attacks and cyber-groups are on the rise and there is no sign that this form of behavior or the groups that employ it will subside in the foreseeable future.

Additional Resources

Hayward, John, "Cyberterrorism Is the Next Big Threat, Says Former CIA Chief," *Breitbart* (May 20, 2015)

Former Chief of counterintelligence argues that cyber terrorism is the singular threat we all face.

Holden, Dan, "Is Cyberterrorism the New Normal?" *Wired* (January, 2015)

The lingering reality of cyber terrorism.

Love, Dylan, "8 Things That Anonymous, The Hacker 'Terrorist' Group, has done for Good," *Business Insider* (April 27, 2013)

This article blurs the line between cyber terrorism and "Freedom fighter."

Morrone, Stefan, "Anonymous vs ISIS: Vigilante Justice in the War Against Terrorism," Institute for Ethics and Emerging Technologies (November 24, 2015)

The author argues that hacker groups can be allies against ISIS.

Stewart, Scott, "The Coming Age of Cyberterrorism," *Stratfor* (October 22, 2015)

The author discusses how cyber terrorism is on the rise and here to stay.

Internet References . . .

The Brookings Institute

brookings.edu

HACKMAGEDDON

hackmageddon.com

Terrorism Analysts

terrorismanalysts.com

Terrorism Research

terrorism-research.com

Unit 3

UNIT

The New Global Security Dilemma

*W*ith the end of the cold war, the concept of security was freed from the constraints of bipolar power politics and a purely state-centric focus. No longer were issues framed simply in terms of the United States–Soviet Union conflict (Vietnam, Afghanistan, and the Middle East), but rather analyzed in more complex ways related to issues of ethnicity, religious fundamentalism, cultural division, nuclear proliferation, and new asymmetrical forms of conflict driven not the least of which by technological evolution. What is clear over the ensuing 25 years is that the global security agenda is becoming more complicated and the issues more numerous as new actor, issues, modes of communication, warfare, and ideologies take center stage.

This new security agenda poses difficult challenges for state and non-state actors alike in terms of how to respond, when or if they should use force, how they can survive challenges, and whether their ways of promoting a more peaceful world are working. Issues from the role of social media to cyberwarfare to nuclear proliferation in Iran and North Korea to the emergence of China and a resurgent Russia and of course the growth of terrorism in the Islamic world now hold sway over the discourse. When examined within the accelerating forces of globalization and technology, however, these challenges take on new and at times more frightening manifestations.

Selected, Edited, and with Issue Framing Material by:
James E. Harf, *Maryville University*
and
Mark Owen Lombardi, *Maryville University*

ISSUE

Is Cyber Warfare the Future of War?

YES: **David Gewirtz**, from "Why the Next World War will be a Cyberwar First, and a Shooting War Second," *ZDNet Government* (2015)

NO: **Thomas Rid**, from "Cyberwar and Peace: Hacking Can Reduce Real-World Violence," *Foreign Affairs* (2013)

Learning Outcomes
After reading this issue, you will be able to:
• Describe the elements that would make up a cyberwar.
• Examine the pros and cons of why a country would want to wage a largely cyberwar.
• Speculate as to whether cyberwar will replace conventional war in the future, and if so, explain why or why not.

ISSUE SUMMARY

YES: Gewirtz articulates the view that cyber warfare will be a crucial and important element of the next and perhaps all future wars.

NO: Rid, however, is skeptical of the notion that cyberwar is the wave of the future and in fact may act as an effective deterrent to real violence.

Drone strikes, cyber hacks, computer viruses, and robot soldiers: these are some of the tools being used today by a few states and some non-state actors in the prosecution of war. These new ways of waging conflict against adversaries are surgical, effective, and often deadly, and greatly decrease the casualties on the attacking side. They are also direct manifestations of the computer, technological revolution that is upon us and accelerating on a daily basis.

For millennia, war has been waged by human beings directly. It involved the killing of another group's soldiers and the occupation of territory. Historically, it has always been impacted by technological advances, from the horse bringing mobility; to the lance and later gunpowder; to rail and later airpower which expanded the theaters of war and of course the most significant advancement, the invention and use of nuclear weaponry. At every level of "advancement" soldiers have been required to engage in attacks and occupy territory. In the computer age and particularly in the twenty-first century a new form of warfare has emerged which is now called cyberwar. In this new environment, states and in selected areas, non-state actors can attack power grids, water systems, information and financial records. They can use robotics such as drones to kill or bomb and they can even construct cybernetic soldiers to engage in selected attacks.

Today the United States uses cyberwar in its fight against terrorism and ISIS. The sophistication of the technology used in cyberwar makes aspects of its use difficult for only a few states. Other states use cyber-attacks to diminish, disrupt, and degrade their adversaries' economic and social systems. The ability to hack systems and use cyber-attacks is relatively simple and therefore poorer states like Syria, North Korea, and Iran are just as able to employ such tactics as wealthy states such as the United States, China, Britain, and Russia.

Analysts such as P.W. Singer and others have written about the future use of robotic soldiers, DNA signature-guided explosives, remote and undetectable weaponry, and other such "futuristic" devices. Each of these inventions expands the options for waging war while in some ways mitigating some of the previous negative effects of traditional warfare.

However, as we are learning with drone strikes and cyber-attacks, cyberwar has a variety of effects that have heretofore been under-examined. Does it make war easier to wage and therefore more likely? Does it violate aspects of international law built around clear state boundaries? Does it inhibit individual freedoms and protections long honored by some societies? Can it empower actors who before were deterred from confronting directly states with enormous military advantages, such as the United States.

In the YES selection, Gewirtz contends that cyber warfare and its expansion of options will definitely hold sway for all actors as the preferred means of waging war. The positives will surely outweigh the negatives in his analysis. In the NO selection, Rid recognizes the impact of cyber weapons on warfare but contends that we overestimate its ultimate impact and may in fact mitigate violence because of the sheer lethality of such weapons.

YES ↵

David Gewirtz

Why the Next World War will be a Cyberwar First, and a Shooting War Second

Everything we do revolves around the Internet. Older technologies are finding themselves eclipsed by their Internet-based substitute solutions.

Even technologies historically unrelated to networking (like medical instruments) are finding themselves part of the Internet, whether as a way to simply update firmware, or using the network to keep track of telemetry and develop advanced analytics.

Whether we're talking about social networking, financial systems, communications systems, journalism, data storage, industrial control, or even government security—it is all part of the Internet.

That makes the world a very, very dangerous place.

Historically, wars are fought over territory or ideology, treasure or tradition, access or anger. When a war begins, the initial aggressor wants something, whether to own a critical path to the sea or strategic oil fields, or "merely" to cause damage and build support among certain constituencies.

At first, the defender defends, protecting whatever has been attacked. Over time, however, the defender also seeks strategic benefit, to not only cause damage in return, but to gain footholds that will lead to an end to hostilities, a point of leverage for negotiation, or outright conquest.

Shooting wars are very expensive and very risky. Tremendous amounts of material must be produced and transported, soldiers and sailors must be put into harm's way, and incredible logistics and supply chain operations must be set up and managed on a nationwide (or multinational level).

Cyberwar is cheap. The weapons are often co-opted computers run by the victims being targeted. Startup costs are minimal. Individual personnel risk is minimal. It's even possible to conduct a cyberwar without the victims knowing (or at least being able to prove) who their attackers are.

Cyberwar can be brutal, anonymous—and profitable.

But the damage done by a cyberwar can be huge, especially economically. Let's follow that idea for a moment.

One of the big reasons the U.S. won the Cold War (and scored highly in many of its other conflicts) is because it had the economic power to produce goods for war, whether capital ships or food for troops. A economically strong nation can invest in weapons R&D, creating a technological generation gap in terms of leverage and per-capita effectiveness compared to weaker nations.

But cyberwar can lay economic waste to a nation. Worse, the more technologically powerful a nation is, the more technologically dependent that nation becomes. Cyberwar can level the playing field, forcing highly connected nations to thrash, to jump at every digital shadow while attackers can co-opt the very resources of the defending nation to force-multiply their attacks.

Sony is still cleaning up after the hack that exposed many confidential aspects of its relationship with stars and producers. Target and Home Depot lost millions of credit cards.

The Snowden theft, while not the result of an outside hack, shows the economic cost of a national security breach nearly $47 billion. Cyberwar can also cause damage to physical systems, ranging from electric power stations to smart automobiles.

And when a breach can steal deeply confidential information of a government's most trusted employees, nothing remains safe or secret. The U.S. Office of Personnel Management was unwittingly funneling America's personnel data to its hackers for more than a year. Can you imagine?

We think China was responsible for the OPM hack. Despite the gargantuan nation's equally gargantuan investments in America (or, perhaps, because of them), China has been accused of many of the most effective and persistent penetrations perpetrated by any nation.

Providing additional reason to worry, Russia and China have recently inked an agreement where they agreed to not launch cyberattacks against each other. They have also agreed to share cyberwarfare and cyberdefense technology, creating an Asian axis of power that can split the world in half.

On the other side of the geopolitical spectrum are the American NSA and British GCHQ, two organizations who share signals intelligence and—if the screaming is to be believed—spy as much upon their own citizens as enemies of the state.

It is important to note that the destabilization of Allied intelligence can be traced to Edward Snowden, who ran to and is currently living in Russia after stealing a vast trove of American state secrets. Ask yourself who gained from the Snowden affair. Was it America? No. Was it Snowden? Not really. Was it Russia? You betcha.

China, of course, supplies us with most of our computer gear. Every iPhone and every Android phone, nearly all our servers, laptop computers, routers—heck, the entire technological core of American communications—has come from China. The same China that has been actively involved in breaching American interests at all levels.

Russia and China. Again and again and again.

In the center of all this is the main body of Europe, where the last two incendiary world wars were fostered and fought.

Nations fall when they are economically unstable. Greece is seeing the writing on the wall right now. It is but one of many weak European Union members. Other EU members are former Soviet states who look eastward towards Putin's Russia with a mixture of fear and inevitability.

This time, Germany isn't the instigator of unrest, but instead finds itself caught in the middle—subject to spying by and active in spying on its allies—the only nearly-super power of the EU.

Here's How the Coming World Cyberwar Will Play Out

An enemy (or even a supposed "friendly" nation) decides it needs the strategic upper hand. After years of breaches, it has deep access to nearly every powerful government and business figure in the United States. Blackmail provides access into command and control and financial systems.

Financial systems are hit and we suffer a recession worse than the Great Recession of 2008–2009. Our budget for just about everything (as well as our will) craters. Industrial systems (especially those that might post a physical or economic threat to our attacker) are hit next. They are shut down or damaged in the way Stuxnet took out centrifuges in Iran.

Every step America takes to respond is anticipated by the enemy—because the enemy has a direct pipeline to every important piece of communication America produces, and that's because the enemy has stolen enough information to corrupt an army of Snowdens.

While this is all going on, the American public is blissfully in the dark. Citizens just get angrier and angrier at the leadership for allowing a recession to take hold, and for allowing more and more foreigners to take American jobs.

Europe, which has always relied on America to keep it propped-up in the worst of times, will be on its own. Russia will press in from the north east. ISIS will continue to explode in the Middle East. China will keep up its careful dance as it grows into the world's leading economic power.

India, second in size only to China and a technological hotbed itself, remains a wild card, physically surrounded by Europe, the Middle East, China, and Russia. India continues to live in conflict with Pakistan, and with Pakistan both unstable and nuclear-tipped, Indo-Pak, too, is on the precipice.

A world war is about huge nations spanning huge geographic territories fighting to rewrite the map of world power. Russia, China, ISIS (which calls itself the Islamic State), India, Pakistan, the US, the UK, and all of the strong and weak members of the EU: we certainly have the cast of characters for another global conflict.

I could keep going (and, heck, one day I might game the full scenario). But you can see how this works. If enemy nations can diminish our economic power, can spy on our strategic discussions, and can turn some of our key workers, they can take us out of the battle—without firing a single shot.

We are heading down this path now. I worry that we do not have the national or political will to turn the tide back in our favor. This is what keeps me up at night.

DAVID GEWIRTZ is an author, US policy advisor and computer scientist. He is one of America's foremost cyber-security experts, and is a top expert on saving and creating jobs. He is also director of the US Strategic Perspective Institute as well as the founder of ZATZ Publishing.

Thomas Rid   ➔ **NO**

Cyberwar and Peace: Hacking Can Reduce Real-World Violence

Cyberwar Is Coming!" declared the title of a seminal 1993 article by the RAND Corporation analysts John Arquilla and David Ronfeldt, who argued that the nascent Internet would fundamentally transform warfare. The idea seemed fanciful at the time, and it took more than a decade for members of the U.S. national security establishment to catch on. But once they did, a chorus of voices resounded in the mass media, proclaiming the dawn of the era of cyberwar and warning of its terrifying potential. In February 2011, then CIA Director Leon Panetta warned Congress that "the next Pearl Harbor could very well be a cyberattack." And in late 2012, Mike McConnell, who had served as director of national intelligence under President George W. Bush, warned darkly that the United States could not "wait for the cyber equivalent of the collapse of the World Trade Centers."

Yet the hype about everything "cyber" has obscured three basic truths: cyberwar has never happened in the past, it is not occurring in the present, and it is highly unlikely that it will disturb the future. Indeed, rather than heralding a new era of violent conflict, so far the cyber-era has been defined by the opposite trend: a computer-enabled assault on political violence. Cyberattacks diminish rather than accentuate political violence by making it easier for states, groups, and individuals to engage in two kinds of aggression that do not rise to the level of war: sabotage and espionage. Weaponized computer code and computer-based sabotage operations make it possible to carry out highly targeted attacks on an adversary's technical systems without directly and physically harming human operators and managers. Computer-assisted attacks make it possible to steal data without placing operatives in dangerous environments, thus reducing the level of personal and political risk.

These developments represent important changes in the nature of political violence, but they also highlight limitations inherent in cyberweapons that greatly curtail the utility of cyberattacks. Those limitations seem to make it difficult to use cyberweapons for anything other than one-off, hard-to-repeat sabotage operations of questionable strategic value that might even prove counterproductive. And cyber-espionage often requires improving traditional spycraft techniques and relying even more heavily on human intelligence. Taken together, these factors call into question the very idea that computer-assisted attacks will usher in a profoundly new era.

The Thin Case for Cyberwar

One reason discussions about cyberwar have become disconnected from reality is that many commentators fail to grapple with a basic question: What counts as warfare? Carl von Clausewitz, the nineteenth-century Prussian military theorist, still offers the most concise answer to that question. Clausewitz identified three main criteria that any aggressive or defensive action must meet in order to qualify as an act of war. First, and most simply, all acts of war are violent or potentially violent. Second, an act of war is always instrumental: physical violence or the threat of force is a means to compel the enemy to accept the attacker's will. Finally, to qualify as an act of war, an attack must have some kind of political goal or intention. For that reason, acts of war must be attributable to one side at some point during a confrontation.

No known cyberattack has met all three of those criteria; indeed, very few have met even one. Consider three incidents that today's Cassandras frequently point to as evidence that warfare has entered a new era. The first of these, a massive pipeline explosion in the Soviet Union in June 1982, would count as the most violent cyberattack to date—if it actually happened. According to a 2004 book by Thomas Reed, who was serving as a staffer on the U.S. National Security Council at the time of the alleged incident, a covert U.S. operation used rigged software to engineer a massive explosion in the Urengoy-Surgut-Chelyabinsk pipeline, which connected Siberian natural gas fields to Europe. Reed claims that the CIA managed to insert malicious code into the software that controlled the pipeline's pumps and valves. The rigged valves supposedly resulted in an explosion that, according to Reed, the U.S. Air Force rated at three kilotons, equivalent to the force of a small nuclear device.

But aside from Reed's account, there is hardly any evidence to prove that any such thing happened, and plenty of reasons to doubt that it did. After Reed published his book, Vasily Pchelintsev, who was reportedly the KGB head of the region when the explosion was supposed to have taken place, denied the story. He surmised that Reed might have been referring to a harmless explosion that happened not in June but on a warm April day that year, caused by pipes shifting in the thawing ground of the tundra. Moreover, no Soviet media reports from 1982 confirm that Reed's explosion took place, although the Soviet media regularly reported on accidents and pipeline explosions at the time. What's more, given the technologies available to the United States at that time, it would have been very difficult to hide malicious software of the kind Reed describes from its Soviet users.

Another incident often related by promoters of the concept of cyberwar occurred in Estonia in 2007. After Estonian authorities decided to move a Soviet-era memorial to Russian soldiers who died in World War II from the center of Tallinn to the city's outskirts, outraged Russian-speaking Estonians launched violent riots that threatened to paralyze the city. The riots were accompanied by cyber-assaults, which began as crude disruptions but became more sophisticated after a few days, culminating in a "denial of service" attack. Hackers hijacked up to 85,000 computers and used them to overwhelm 58 Estonian websites, including that of the country's largest bank, which the attacks rendered useless for a few hours.

Estonia's defense minister and the country's top diplomat pointed their fingers at the Kremlin, but they were unable to muster any evidence. For its part, the Russian government denied any involvement. In the wake of the incident, Estonia's prime minister, Andrus Ansip, likened the attack to an act of war. "What's the difference between a blockade of harbors or airports of sovereign states and the blockade of government institutions and newspaper websites?" he asked. It was a rhetorical question, but the answer is important: unlike a naval blockade, the disruption of websites is not violent—indeed, not even potentially violent. The choice of targets also seemed unconnected to the presumed tactical objective of forcing the government to reverse its decision on the memorial. And unlike a naval blockade, the attacks remained anonymous, without political backing, and thus unattributable.

A year later, a third major event entered the cyber-Cassandras' repertoire. In August 2008, the Georgian army attacked separatists in the province of South Ossetia. Russia backed the separatists and responded militarily. The prior month, in what might have been the first time that an independent cyberattack was launched in coordination with a conventional military operation, unknown attackers had begun a campaign of cyber-sabotage, defacing prominent Georgian websites, including those of the country's national bank and the Ministry of Foreign Affairs, and launching denial-of-service attacks against the websites of Georgia's parliament, its largest commercial bank, and Georgian news outlets. The Georgian government blamed the Kremlin, just as the Estonians had done. But Russia again denied sponsoring the attacks, and a NATO investigation later found "no conclusive proof" of who had carried them out.

The attack set off increasingly familiar alarm bells within American media and the U.S. national security establishment. "The July attack may have been a dress rehearsal for an all-out cyberwar," an article in *The New York Times* declared. Richard Clarke, a former White House cybersecurity czar, warned that the worst was yet to come: the Georgian attack did not "begin to reveal what the Russian military and intelligence agencies could do if they were truly on the attack in cyberspace." Yet the actual effects of these nonviolent events were quite mild. The main damage they caused was to the Georgian government's ability to communicate internationally, thus preventing it from getting out its message at a critical moment. But even if the attackers intended this effect, it proved short-lived: within four days after military confrontations had begun in earnest, the Georgian Foreign Ministry had set up an account on Google's blog-hosting service. This move helped the government keep open a channel to the public and the news media. What the Internet took away, the Internet returned.

Earlier this year, the Pentagon announced that it would boost the staff of its Cyber Command from 900 to 4,900 people, most of whom would focus on offensive operations. William Lynn, formerly the Pentagon's second-in-command, responded to critics of the move by assuring the public that the Department of Defense would not militarize cyberspace. "Indeed," Lynn said, "establishing robust cyberdefenses no more militarizes cyberspace than having a navy militarizes the ocean."

In a sense, Lynn is right: cyberspace has not been militarized, precisely because the U.S. government, along with many other governments, has not actually established robust cyberdefenses. Defending against cyber-sabotage means hardening computer systems, especially those that control critical infrastructure. But such systems remain staggeringly vulnerable. Defending against cyber-espionage means avoiding the large-scale theft of sensitive data from companies and government agencies. But as illustrated by the recent leaks of classified information regarding the National Security Agency's domestic surveillance, Western intelligence agencies are only now beginning to understand digital counterespionage and the proper role of human informants in a digitized threat environment.

What has been militarized is the debate about cyber-attacks, which is dominated by the terminology of warfare. What appears as harmless inconsistency—constantly warning of cyberwar's dangers while neglecting to protect against them—masks a knotty causal relationship: for a number of reasons, loose talk of cyberwar tends to overhype the offensive potential of cyberattacks and diminish the importance of defenses. First, it encourages the false idea that two states exist: cyberwar and cyberpeace. In fact, the threat of a cyberattack is ever present and will not go away. Second, when U.S. military officials hype cyberwar, it leads the public to believe that the Pentagon is in charge of dealing with the threat. In fact, companies and individuals need to take responsibility for their own security. And finally, advocates of the concept of cyberwar often suggest that the best defense is a good offense. That is not the case: consider, for example, that designing the next Stuxnet will not make the U.S. energy grid any safer from digital attacks. To avoid further distorting the issue, the debate over cyberattacks must exit the realm of myth.

THOMAS RID is a Reader in War Studies at King's College London. His most recent book is *Cyber War Will Not Take Place* (Oxford University Press, 2013), from which this essay is adapted. Copyright © Oxford University Press, 2013.

EXPLORING THE ISSUE

Is Cyber Warfare the Future of War?

Critical Thinking and Reflection

1. Can cyberwar eliminate the need for conventional war?
2. Will cyberwar be a constant factor in the behavior of states and other actors if it is made as "painless" as it appears to be?
3. What are the political ramifications of such an evolution?
4. Will it make warfare more or less likely?
5. Does it eliminate some of the deleterious effects of war on its combatants?

Is There Common Ground?

"Futurists and purists" will debate the nature of warfare for some time in the twenty-first century as they did in the twentieth century with the advent of airpower and mechanized tanks. What all can agree upon is that technology has allowed warfare to develop high levels of precision that will shape and guide its prosecution for decades to come.

Additional Resources

Clarke, Richard A., *Cyber War: The Next Threat to National Security and What to Do about It* (Harper Collins, 2010).

Rand, "Cyber Warfare" Rand Corporation (June 2016).

Singer, P.W., "How the United States Can Win the Cyberwar of the Future," *Foreign Policy* (December 18, 2015).

Will, George, "The Destructive Threat of Cyber warfare," *The National Interest* (April 13, 2016).

Internet References . . .

Future Technology 500

www.futuretechnology500.com

Rand Corporation

www.rand.org/

Space Wars

www.spacewars.com

Selected, Edited, and with Issue Framing Material by:
James E. Harf, *Maryville University*
and
Mark Owen Lombardi, *Maryville University*

ISSUE

Are Russia and the United States in a New Cold War?

YES: **Andrej Krickovic and Yuval Weber,** from "Why a New Cold War with Russia is Inevitable," *The Brookings Institution* (2015)

NO: **Matthew Rojansky and Rachel S. Salzman,** from "Debunked: Why There Won't be Another Cold War," *The National Interest* (2015)

Learning Outcomes

After reading this issue, you will be able to:

- Understand the nature of the US–Russian relation in the aftermath of the cold war (1991–2008).
- Explore the issues that now seem to divide the two states.
- Analyze the role of Vladimir Putin in the fractured nature of the relationship today.
- Determine what policy areas lend themselves to cooperation.
- Speculate as to what a second cold war might mean for Europe and the United States.

ISSUE SUMMARY

YES: Krickovic and Weber argue that substantive differences between the United States and Russia on NATO membership, the status of Ukraine, among others will lead to further deterioration in the relationship.

NO: Rojansky and Salzman, however, see the social, economic, and political interconnections as too deep for another true cold war to develop.

For the better part of 50 years (1946–1991), the United States and the Soviet Union were the two global superpowers. Together they engaged in a cold war in which they competed economically, politically, militarily, and philosophically for hegemony on the world stage. In conflicts from Korea to the Middle East, Vietnam to Afghanistan these two states representing two distinct ideologies competed for control, influence, and power. At various times and through various stages, that competition was fierce, violent, and at times extremely dangerous. During the Korean War, the creation of the Berlin wall, the Cuban Missile Crisis, the Sino-Soviet war, the Vietnam War, Afghanistan, and the subsequent fall of the Berlin wall, the two states were pitted against one another and often the crises danced perilously close to nuclear confrontation. It

is important to note that during this period the United States and the USSR also cooperated on arms control treaties, non-proliferation agreements, and some selected mutual trade arrangements. Overall, many if not all of the global security issues that the world faced were placed within an East–West, cold war prism of thought and action, whether they fit that dichotomy or not.

With the collapse of the Soviet Union and its East European satellite states, a new era began that left the United States as the sole remaining superpower. American military and economic power was unchallenged on a global scale. This led to a sustained period of Russian–US cooperation on a host of economic, political, and security issues including the first Gulf War, the second invasion of Iraq, and a host of agreements related to issues as fractious as Iranian nuclear capability, North Korea, the

Israeli-Palestinian issue among others. This cooperation began to show signs of fracturing and change with the ascendency of Vladimir Putin in 1999–2000. In the first iteration of Putin's rule 1999–2008, communication and cooperation continued and despite some minor differences, the nature of the relationship was more cooperative than confrontational. Since Putin's reassertion of authority in 2008, issues have emerged in areas such as Georgia, Chechnya, Iran, Syria, Ukraine among others that have caused Kremlin watchers and others to speculate if the relationship has shifted back to rivalry and antagonism.

In the YES selection, Andrej Krickovic and Yuval Weber see these issues as festering and divisive sources of real conflicts between the West and the East, between the United States and Russia, and they do not see a time when Putin will back down from pressing his interests. According to them, this will lead to further deterioration and decline in the US–Russian relation that will at the very least look very much like a second cold war.

In the NO selection, Mathew Rojansky and Rachel S. Salzman view the economic and social relationships between their economies and structures as too deep, too intertwined, and interdependent to foster a second cold war. While they recognize that there will be tensions over areas such as Ukraine, there will not be the kind of intractable conflict that characterized the cold war of the latter half of the twentieth century.

YES ↵

Andrej Krickovic and Yuval Weber

Why a New Cold War
with Russia Is Inevitable

This is a critical moment in U.S.-Russia relations. The civil war in Ukraine is settling into a mutually hurtful stalemate; a workable nuclear deal with Iran has been concluded; and Russia is ramping up its presence in Syria, which increases the danger of confrontation with the United States but also opens up the potential for cooperation against the Islamic State (or ISIS). Before a more hawkish U.S. administration comes to power—and before anti-Americanism becomes further entrenched in Russia as evidenced by the latest Levada Center public polling data—perhaps there is an opportunity for Washington and Moscow to overcome their current impasse.

This is our hope. But theory and evidence point to a sobering conclusion: Neither side can make the concessions necessary to resolve their current differences and prevent relations from deteriorating even further.

Commitment Anxiety

The chief concern of those calling for negotiation between the United States and Russia is that while the current relationship is beset by a number of serious differences, the downside of an openly hostile relationship is even worse. These voices argue that without an updated European security framework to resolve some of the worst tensions (and implicitly to update the post-Cold War settlement), a new Cold War between the two camps will emerge.

The competition would not be as encompassing as before, but it would make cooperation on vital issues outside of Europe—including Iran, ISIS, and Syria—unsustainable and lead to an inherently more unstable international order. In turn, they advocate for a mutually acceptable framework for regional order to avoid future conflicts from arising in other areas of the post-Soviet space.

A larger "grand bargain" to regulate the structure of international and regional relations based on mutual accommodation is a worthy goal. Nevertheless, we are skeptical that such a grand bargain can be reached, because it would suffer from acute commitment problems. Russia would have to convince the United States and its allies that it would not push for even greater revisions to the status quo. The United States would have to demonstrate to Russia that it would stick to any bargain and not go back to the policies that threaten it.

Impossible Concessions

Among the insights of bargaining theory is that states can overcome commitment problems by accepting costly concessions that signal their resolve to abide by agreements. What hypothetical concessions could both sides make to make the grand bargain stick?

Russia could atone for its actions in Crimea—which Washington and other Western capitals see as a grave breach of international law and order—by either reversing the annexation or by using economic and other inducements to get Kiev to recognize the new status quo. The United States could address Russia's fears of encirclement by NATO, either by agreeing to the formation of a pan-European security organization with authority above NATO's (as Dmitri Medvedev proposed during his presidency). Or, it could formally abrogate NATO's right to enlarge its membership and recognize the neutrality of the post-Soviet states on Russia's Western borders.

While these concessions could conceivably make a bargain work, we believe that the domestic political costs of trying to implement such agreements would be too high for leaders of both sides. Any demand for the return of Crimea to Ukraine (as many Western voices have implored) is a non-starter for Russia. The Kremlin has invested so much in the "Return of Crimea" discourse that even minor concessions on this issue would shake the regime's legitimacy to its very foundations (and perhaps even threaten a nationalist revolt).

The alternative—inducing Ukraine to recognize de facto the loss of Crimea—is also problematic, with severe domestic blowback almost certain to befall any government in Kiev that made the deal. Moreover, given its current economic difficulties, Moscow may not be able to deliver the economic inducements necessary to win Kiev's compliance.

On the other side, any concession that would give Russia a de facto veto over NATO's policies would be rejected outright by the alliance's members. The more moderate option, the abrogation of NATO enlargement and the recognition of Russia's sphere of influence in the post-Soviet space, is unacceptable to American leaders and publics. It would validate a realpolitik understanding of international relations that is fundamentally at odds with their views of international relations—in which every state should be free to choose its alliances.

A Downward Slide?

As things stand now, neither side can make the concessions necessary to make a grand bargain work. As a result, both now find themselves sliding towards a new Cold War that neither really wants.

We hope that statesmen on both sides will prove us wrong by finding the courage and foresight necessary to overcome these commitment problems. But it is difficult to be optimistic given the current political climate, as talk in both capitals is dominated by the sort of Russia- and America-bashing, which prevents either side from developing an appreciation of the other's security concerns.

ANDREJ KRICKOVIC is an assistant professor, National Research University—Higher School of Economics in Moscow, Russia.

YUVAL WEBER is an assistant professor, National Research University—Higher School of Economics in Moscow, Russia.

Matthew Rojansky and Rachel S. Salzman   **NO**

Debunked: Why There Won't
Be Another Cold War

In the wake of the ongoing crisis in Ukraine, talk of a "New Cold War" is in vogue. Even experts who studied the Soviet Union and Russia from the depths of mutually assured destruction and détente to the fall of Communism now say that it will be decades again before "normal" relations between Russia and the West can resume. We disagree.

The "New Cold War" narrative goes something like this: Vladimir Putin's invasion of Ukraine has constituted an unpardonable violation of the European security order and of international law, which will poison relations with the United States and Europe to the point that any hopes of rapprochement would be dashed. Putin himself is determined to make contempt for Western values—and anti-Americanism especially—the centerpiece of his regime's official ideology. Further, he has successfully brainwashed the younger generation of Russians with these views. For the foreseeable future, we will at best achieve limited agreements on issues like arms control, only because of a grudging recognition that without them, we risk destroying each other and the world. It's the Cold War and détente all over again.

True, relations between Russia and the West have become severely strained, and the sources of that tension—principally the catastrophic situation in Ukraine—may not change any time soon. But we believe the "New Cold War" narrative misreads both Russia and the West, especially when it comes to the generation that has fully come of age after the collapse of the Soviet Union and the end of the original Cold War.

As members of that generation in the United States, both Russia watchers with experience working on second-track diplomacy and living and working in Russia, we suggest that it is a mistake to define the current conflict primarily through the lens of the previous one. The declaration of a "New Cold War" has come too soon, and the label does not fit.

In geopolitical terms, there are clear differences between the current conflict and the Cold War: the absence of a global ideological dimension to the conflict; the prevalence of tension in the post-Soviet space versus in other regions; and the much greater relative power of non-Western states (China, India, Brazil, and others) that have, so far, refused to take sides. As others have noted, such differences do not preclude a renewal of Cold War–type relations between the United States and Russia, with considerable spillover effects for Europe and other regions.

Yet what distinguishes the contemporary situation from the conflict that governed the second half of the twentieth century—and the reason this cannot be called a New Cold War—is the profound difference in interpersonal relations.

Far from a new iron curtain of mutual hostility and distrust descending between East and West, Russians and Americans can in most cases travel back and forth, interact freely with one another and seek to find common ground on the toughest issues through mutually respectful dialogue. During recent visits to Moscow, we each found Russian experts at all levels open to engagement; with the younger generation especially keen to offer helpful and insightful analysis of events in Russia, Ukraine and Russia's relations with the West.

Make no mistake: "helpful and insightful" is not a euphemism for "pro-Western." Just as anywhere else, attitudes among the rising generation of Russian experts run the gamut. But they are professional, intelligent and willing to engage with Americans, even if that engagement does not yield an immediate meeting of the minds. They also have (so far) been willing to speak openly and on the record without fear of official repercussions, a clear distinction from a time when even a banal conversation with foreigners was fraught with complications and outright risks.

And thanks to the Internet and international voice calling, our dialogues are able to continue with great ease—something that Cold Warriors never dreamed could be possible. Through social media, we exchange messages and links with Russian friends and colleagues on a daily—sometimes even hourly—basis. This is not just the usual trope about the Internet making the world smaller. On Facebook and Twitter, we see the images that Russian peers curate when their Russian friends are watching and without them necessarily thinking about the Americans in the room. But it also means we see each other's birthdays, children and family vacations. Our relations with our Russian counterparts are therefore fundamentally deeper and more complex than were those of the previous generation of American and Russian experts.

This is all to the good, because while such dialogue cannot alone deliver an end to the current conflict, it helps ensure that tensions will run less deep and for less long. Still, it takes a critical mass of expert voices on both sides to maintain such engagement and to have any impact on official opinion or policy in Moscow and Washington.

In this respect, the United States is unfortunately less well equipped than it should be. Ironically, the disappearance of Cold War–era U.S. government programs for supporting Russia-related scholarships and travel to Russia has decreased the number of young Americans wishing to join our field and the number of opportunities available to them. As a result, the voices of the Cold War generation still predominate.

Without a renewed appreciation of the importance of preparing American experts to work with and on Russia, the interpersonal connections that have made such a difference for us and our Russian colleagues will atrophy and disappear. Then we really will find ourselves in a new Cold War.

MATTHEW ROJANSKY is Director of the Kennan Institute at the Wilson Center in Washington DC.

RACHEL S. SALZMAN is a Doctoral Candidate in Russian and Eurasian Studies at the Johns Hopkins School of Advanced International Studies.

EXPLORING THE ISSUE

Are Russia and the United States in a New Cold War?

Critical Thinking and Reflection

1. What are the attributes of the first cold war that appear to be present in the relation between Russia and the United States today?
2. What are the areas of conflict and agreement that currently exist?
3. Is the tension between Russia and the United States a function of Putin's leadership or a deeper set of ideological issues?
4. Are there ways and areas where the relationship can be made more cooperative?

Is There Common Ground?

As the relationship between Russia and the United States continues to develop, there will undoubtedly be mixed signals as to its inherent negative or positive aspects. What most analysts can agree upon is that President Putin's control over the Russian state is the key determining factor in the quality and tenor of the US–Russian relation. As long as president Putin remains in control, areas of rivalry and conflict will emerge and shape the tenor of the relationship.

Additional Resources

Koshkin, Pavel, "What a new Cold War between Russia and the US Means for the World," *Russia Direct* (April 25, 2014)

The author talks about the state of the Russian–US relation.

Norton-Taylor, Richard, "Is a New Cold War with Russia Inevitable or is Terrorism the Real Enemy?" *The Guardian* (March 22, 2016)

The author argues that anti-terrorism can bring the United States and Russia together.

Pepper, John, "Beyond a New Cold War," *The Nation* (March 11, 2016)

Former CEO of Proctor & Gamble argues for more cooperation between Russia and the United States.

Stavridis, James, "Are We Entering a New Cold War?" *Foreign Policy* (February 17, 2016)

The author lays out a policy agenda for dealing with Russia.

Internet References . . .

Foreign Policy

foreignpolicy.org

The New Cold War

www.newcoldwar.org

Russia Direct

www.russia-direct.org

Selected, Edited, and with Issue Framing Material by:
James E. Harf, *Maryville University*
and
Mark Owen Lombardi, *Maryville University*

ISSUE

Can ISIS Be Defeated in the Near Future?

YES: Max Boot, from "How ISIS can be Defeated," *Newsweek* (2015)

NO: Aaron David Miller, from "5 Reasons the U.S. Cannot Defeat ISIS," *Real Clear World* (2015)

Learning Outcomes
After reading this issue, you will be able to:
• Describe the current tactics being employed against ISIS.
• Evaluate their effectiveness thus far.
• Determine what other approaches may be viable in attempts to defeat ISIS.
• Speculate as to what factors seem most important in the goal of defeating ISIS.

ISSUE SUMMARY

YES: Boot lays out a strong military strategy for the defeat of ISIS that relies on a coalition of ground forces.

NO: Miller lays out the currently existing factors that will prevent the defeat of ISIS.

ISIS has emerged out of the shifting Sunni political and military evolution in Iraq in the aftermath of the US invasion and Iraq's own civil war. Inspired by Abu Musab al-Zarqawi, the leader of the al Qaeda branch in Iraq that fought the American occupation of Iraq and the Shi'ite government of Prime Minister al-Maliki, ISIS is now a hardline political–military organization based in Syria and Iraq that is bent on resurrecting the Caliphate, an ancient concept in Sunni, Islamic culture revolving around the keepers of the "true" faith of Islam. This transformation from al-Qaeda terrorist affiliate to its own even more extreme Islamic fundamentalist organization occurred during the period between Zarqawi's death in 2006 and its re-emergence in 2010 in Iraq and Syria.

The goal of ISIS is to lead the Islamic world, control the interpretation of the Islamic faith, and apply a strict, revisionist interpretation of its tenants on all believers and non-believers alike. Its methods of enforcement are terror, beheadings, military attacks, and extreme violence as well as a sophisticated use of social media, propaganda,

and the Internet. ISIS is led by Abu Bakr al-Baghdadi and is supported by mostly former military officers of the Saddam Hussein regime. Based in al-Raqqa, Syria, it is Sunni based and led, and it controls the territory across eastern Syria and western Iraq to the outskirts of Baghdad.

With an estimated 30,000 active fighters and affiliates from Nigeria and Libya to Egypt, Yemen, and East Asia, it has captured Islamic fundamentalist momentum among Sunnis and has proclaimed Caliphate in Mosul, Iraq in 2014. Through very high-profile televised executions, propaganda, social media, and coordinated attacks in Europe and the United States, they have captured the attention of and fear from the Western world as well as moderate Sunnis and of course Shi'ite Muslims.

The United States along with selected Arab states, France, Britain, Russia, and others have been waging a campaign designed to roll back ISIS successes in Iraq and degrade and ultimately destroy ISIS within Syria. Thus far, this policy has had some success despite the high-profile ISIS attacks in Brussels, Paris, Nigeria, and Libya and the most recent lone wolf attacks in the United States, particularly San Bernardino and Orlando.

Max Boot argues that ISIS can be defeated but only with a coordinated ground invasion by well-trained U.S. and other world forces. He argues that air power alone will not defeat the group. Only by defeating ISIS on the ground can the organization and its ideology be undermined, discredited, and ultimately ended. Miller articulates the key factors that must be addressed in order to even hope to defeat ISIS and how some of these are far from being addressed in any meaningful way.

YES ↵

Max Boot

How ISIS can be Defeated

President Barack Obama's strategy in Syria and Iraq is not working. The president is hoping that limited air strikes, combined with U.S. support for local proxies—the peshmerga, the Iraqi security forces, the Sunni tribes and the Free Syrian Army—will "degrade and ultimately destroy" the Islamic State of Iraq and Syria (ISIS).

U.S. actions have not stopped ISIS from expanding its control into Iraq's Anbar Province and northern Syria. If the president is serious about dealing with ISIS, he will need to increase America's commitment in a measured way—to do more than what Washington is currently doing but substantially less than what it did in Iraq and Afghanistan in the past decade.

And although Obama will probably not need to send U.S. ground-combat forces to Iraq and Syria, he should not publicly rule out that option; taking the possibility of U.S. ground troops off the table reduces U.S. leverage and raises questions about its commitment.

A Big Threat

A reasonable goal for the United States would be neither to "degrade" ISIS (vague and insufficient) nor to "destroy" it (too ambitious for the present), but rather to "defeat" or "neutralize" it, ending its ability to control significant territory and reducing it to, at worst, a small terrorist group with limited reach.

This is what happened with ISIS's predecessor, Al-Qaeda in Iraq, during 2007 and 2008, before its rebirth amid the chaos of the Syrian civil war. It is possible to inflict a similar fate on ISIS, which, for all of its newfound strength, is less formidable and less organized than groups like Hezbollah and the Taliban, which operate with considerable state support from Iran and Pakistan, respectively.

Although not as potent a fighting force as Hezbollah or the Taliban, ISIS is an even bigger threat to the United States and its allies because it has attracted thousands of foreign fighters who could return to commit acts of terrorism in their homelands.

What It Will Take to Defeat ISIS

To defeat ISIS, the president needs to dispatch more aircraft, military advisers, and special operations forces, while loosening the restrictions under which they operate. The president also needs to do a better job of mobilizing support from Sunnis in Iraq and Syria, as well as from Turkey, by showing that he is intent on deposing not only ISIS but also the equally murderous Alawite regime in Damascus. Specific steps include:

Intensify air strikes. So far, the U.S. bombing campaign against ISIS has been remarkably restrained, as revealed by a comparison with the strikes against the Taliban and Al-Qaeda in Afghanistan after 9/11.

When the Taliban lost control of Afghanistan between October 7, 2001, and December 23, 2001—a period of 75 days—U.S. aircraft flew 6,500 strike sorties and dropped 17,500 munitions. By contrast, between August 8, 2014, and October 23, 2014—76 days—the United States conducted only 632 airstrikes and dropped only 1,700 munitions in Iraq and Syria. Such episodic and desultory bombing will not stop any determined military force, much less one as fanatical as ISIS.

Lift the prohibition on U.S. "boots on the ground." Obama has not allowed U.S. Special Forces and forward air controllers to embed themselves in the Free Syrian Army, Iraqi security forces, Kurdish peshmerga, or in Sunni tribes when they go into combat as he did with the Northern Alliance in Afghanistan. This lack of eyes on the ground makes it harder to call in air strikes and to improve the combat capacity of U.S. proxies. Experience shows that "combat advisers" fighting alongside indigenous troops are far more effective than trainers confined to large bases.

Increase the size of the U.S. force. Military requirements, not a priori numbers dreamed up in Washington, should shape the force eventually dispatched. The current force, even with the recent addition of 1,500 more troops for a total of 2,900, is inadequate. Estimates of necessary troop size range from 10,000 personnel (according to General Anthony Zinni, former head of Central Command) to 25,000 (according to military analysts Kim and Fred Kagan). The total number should include Special Forces teams and forward air controllers to partner with local forces as well as logistical, intelligence, security, and air contingents in support.

Work with all of Iraq's and Syria's moderate factions. The United States should work with the peshmerga, Sunni tribes, the Free Syrian Army and elements of the Iraqi security forces (ISF) that have not been overtaken by Iran's Quds Force, rather than simply supplying weapons to the ISF. Given Shiite militia infiltration, working exclusively through the ISF would risk empowering the Shiite sectarians whose attacks on Sunnis are ISIS's best recruiter.

The United States should directly assist Sunni tribes by establishing a small forward operating base in Anbar Province, and also increase support for and coordination with the Free Syrian Army. Current plans to train only 5,000 Syrian fighters next year need to be beefed up.

Send in the Joint Special Operations Command (JSOC). Between 2003 and 2010, JSOC—composed of units such as SEAL Team Six and Delta Force—became skilled at targeting the networks of Al-Qaeda in Iraq. Its success was largely due to its ability to gather intelligence by interrogating prisoners and scooping up computers and documents—something that bombing alone cannot accomplish. JSOC squadrons should once again be moved to the region (they could be stationed in Iraq proper, the Kurdistan Regional Government, Turkey and/or Jordan) to target high-level ISIS organizers.

Draw Turkey into the war. Obama should do what he can to increase Turkey's involvement in the anti-ISIS campaign. If the Turkish army were to roll across the frontier, it could push back ISIS and establish "safe zones" for more moderate Syrian opposition members.

Turkish President Recep Tayyip Erdogan has said that he will not join the fray without Washington's commitment to overthrowing Syrian President Bashar al-Assad, whom he rightly sees as the source of instability in Syria. Assuming Erdogan has honestly outlined his conditions for Turkish involvement in Syria, a greater U.S. commitment, demonstrated by a no-fly zone and airstrikes on Assad's forces, should be sufficient to entice Ankara to play a greater role.

Impose a no-fly zone over part or all of Syria. Even though U.S. aircraft are overflying Syria, they are not bombing Assad's forces. This has led to a widespread suspicion among Sunnis that the United States is now willing to keep Assad in power. More broadly, Sunnis fear that Obama is accommodating Assad's backers in Tehran to allow Iran to dominate Mesopotamia and the Levant. A no-fly zone over part or all of Syria would address these concerns and pave the way for greater Turkish involvement.

The United States should act to ensure that Assad does not take advantage of the anti-ISIS campaign to bomb opposition centers. Obama could announce that no Syrian aircraft will be allowed over designated "safe zones." Such a move would garner widespread support among Arab states, undercutting attempts to portray U.S. action as a war against the Muslim world.

There are legitimate concerns that overthrowing Assad now, before the Syrian opposition is ready to fill the vacuum, would be counterproductive and potentially pave the way for a jihadist takeover of all of Syria. But instituting a partial or even a complete no-fly zone would not lead to Assad's immediate ouster. It would, however, facilitate the moderate opposition's ability to organize an administration capable, with international help, of governing Syria once Assad finally goes.

Mobilize Sunni tribes. As long as the Sunni tribes of Iraq and Syria continue to tacitly support ISIS, or at least not to resist it, defeating ISIS will be almost impossible. But if the tribes turn against ISIS, as they did against Al-Qaeda in Iraq in 2007, a rapid reversal of fortunes is likely.

Galvanizing Sunni tribes into action will not be easy; Iraqi Sunnis feel that the United States betrayed them after the surge by leaving them under Shiite domination in Baghdad. The fact that Haidar al-Abadi replaced former Prime Minister Nouri al-Maliki in September is a good first step. But Abadi is also a Shiite from the same Dawa Party as Maliki, making it unlikely that Sunnis will fight ISIS if they once again find themselves subordinated to Shiite rule.

This concern could be allayed if the United States were to engineer a political deal to grant Sunnis autonomy within the Iraqi federal structure, similar to what the Kurdistan Regional Government already enjoys. To assuage Sunnis' fear of betrayal, the United States should pledge to indefinitely maintain advise-and-assist forces in Iraq—even without Baghdad's agreement, U.S. forces could at least remain in the Kurdish area.

Prepare now for nation-building. The United States should lay the groundwork for a post-conflict settlement in both Iraq and Syria that does not necessarily require keeping both political entities intact. In the Iraqi context, this means offering greater autonomy to the Sunnis and guaranteeing the Kurds that their hard-won gains will not be jeopardized; the United States should propose to permanently station troops in the Kurdistan Regional Government. This is not necessarily synonymous with Kurdish independence, but the United States should give serious consideration to dropping its longtime opposition to the creation of a Kurdish state or possibly even two—one in Syria and one in Iraq.

Social fragmentation in Syria will make postwar reconstruction difficult; after three years of civil war, it may not be possible to reconstitute the country as it previously existed. The U.S. goal should simply be to ensure that Syrian territory is not controlled by either Shiite or Sunni extremists. The postwar settlement in the former Yugoslavia, which involved the dispatch of international peacekeepers and administrators under the United Nations, European Union, and NATO mandates, could be a possible model.

The United States should push U.N. Special Envoy for Syria Staffan de Mistura to work in cooperation with the Arab League, the E.U., NATO, the United States and even Russia to create a post-Assad administration that can win the assent of Syria's sectarian communities. As Kenneth Pollack of the Brookings Institution has suggested, "[T]he U.S. should provide most of the muscle, the Gulf states most of the money, and the international community most of the know-how."

This is admittedly an ambitious goal. Neither Assad nor ISIS is in imminent danger of falling, and it will be challenging to impose any kind of order in Syria. But the United States should not repeat the mistake it made in Iraq and Libya of pushing for regime change absent a plan to fill the resulting vacuum. Admittedly even the best-laid plans can fail, but failure is guaranteed if no such plans are in place.

Down the "Slippery Slope"?

Critics will call this strategy too costly, alleging that it will push the United States down a "slippery slope" into another ground war in the Middle East. This approach will undoubtedly incur a greater financial cost (dispatching 10,000 troops for a year would cost $10 billion) and a higher risk of casualties among U.S. forces. But the present minimalist strategy has scant chance of success, and it risks backfiring—ISIS's prestige will be enhanced if it withstands half-hearted U.S. air strikes.

Left unchecked, ISIS could expand into Lebanon, Jordan, or Saudi Arabia. Greater American involvement could galvanize U.S. allies—the most important being Turkey and the Sunni tribes of Iraq and Syria—to commit more resources to the fight.

If this plan is not implemented, a major ground war involving U.S. troops becomes more probable, because the security situation will likely continue deteriorating. By contrast, this strategy, while incurring greater short-term risks, enhances the odds that ISIS will be defeated before Obama leaves office.

Max Boot is the Jeane J. Kirkpatrick senior fellow in national security studies at the Council on Foreign Relations. His latest book is *Invisible Armies: An Epic History of Guerrilla Warfare from Ancient Times to the Present.* This article first appeared on the Council on Foreign Relations website.

Aaron David Miller   **NO**

5 Reasons the U.S. Cannot Defeat ISIS

On Monday, U.S. President Barack Obama will sit down with Iraqi Prime Minister Haider al-Abadi to talk about the strategy to fight the Islamic State. The president will lay out what he wants Iraq to do, including making good on promises to empower Sunni militias and tribes. Indeed, there are many things the United States can do to counter the Islamic State: It can increase the number of special forces deployed in the region; assign U.S. troops as spotters and coordinators with forward-deployed Iraqi units; supply weapons directly to vetted Sunni militias; and increase airstrikes.

But what it cannot do is defeat the Islamic State and eliminate it from Iraq and Syria. Even if we finesse the problem and use Obama's clever turn of phrase, to "ultimately defeat" ISIS, as our goal, we had better get used to a very long war. Even with such a war, victory as conventionally defined may still be elusive. Here is why.

The Islamic State will die only when the Middle East is reborn: This will not happen for years to come, if indeed it ever does. The Islamic State, or more specifically its forerunner, al Qaeda in Iraq, rose as a Sunni insurgency in response to the U.S. invasion of Iraq and to Shiite regional dominance. The group was energized by Shiite triumphalism in Iraq and received a further boost from the rapid U.S. withdrawal from the country. Now it has surged largely as a result of regional dysfunction, and it succeeds in countries where no governance (Syria) or bad governance (Iraq) are the rule, not the exception. The Islamic State's spread to Yemen, Libya, and Sinai is fed by the expanse of empty, uncontrollable spaces, by access to weapons and money, and by the spread of a vicious Islamist ideology that speaks to the grievances of an embattled Sunni community searching for an identity around which to rally. Rooting out the organization would require transformational change in both Syria and Iraq. An important facet of that change would be the rise of good governance that empowers and includes Sunnis as well as Shia.

Defeating ISIS requires a Solution to the Syria Problem: ISIS is an Iraqi organization, and Iraq is where its aspirations lie. But Syria is where its putative caliphate has been established, and as a base for expansion it continues to hold promise. The Assad regime's brutal policies create potential ISIS recruits faster than the West can possibly train Sunnis to oppose the group. Further, most Sunnis want to fight Assad, not ISIS, and ISIS cooperates with the regime at times in order to weaken rival Sunni groups. In this confusion and chaos, ISIS thrives. Indeed, even if the civil war somehow ended, ISIS might well be the beneficiary. As the strongest power on the ground, it might expand further, even threatening to take its first major Arab capital—Damascus. Without a solution to Syria—and none is likely—there is no defeating ISIS.

There is no regional military force capable of defeating ISIS: The solution to ISIS is not a military one. Still, military force could stop ISIS gains and begin to lay the basis for the group's demise. But there is no force, nor combination of forces, willing or able to accomplish this objective. The notion of an Arab state coalition will remain a thought experiment, and the Iraqi military, as seen recently in Ramadi, is not up to the job. Political considerations—largely Shiite pushback—prevent the training and arming of Sunni tribes and militias. The Kurds are too weak, and their peshmerga too localized a force. Even Iran's Shiite militias would have a hard time defeating the Islamic State in Sunni-majority areas, and relying on Iran would threaten the already precarious balance between Sunni and Shiite Iraqis. A fully effective Iraqi national army, with the will and the capacity not just to retake territory but to hold it, would be the answer—but that for now seems a distant dream.

The United States lacks the will for this fight: Americans could defeat the Islamic State on the battlefield—certainly in Iraq, and probably in Syria, too. But the odds of this administration, or even one led by a Republican successor, being willing to make the necessary commitment to both battlefields, seem very small indeed. The American public and the U.S. Congress have grown risk-averse after years of investment in the Middle East that brought no tangible returns. Moreover, at times military force is simply an instrument to achieve sustainable political goals. There is simply no reason to believe that the political end state in Iraq or Syria would turn out any better than it did in Iraq or Afghanistan over the past decade, when the United States deployed tens of thousands of troops and spent trillions of dollars.

Lack of a mandate: The Obama administration turned its attention back toward Iraq after the Islamic State beheaded individual Americans and seemed ready to plan attacks against the United States. A Pew poll in February showed that while there is support for more assertive action against the Islamic State, there is also growing concern that the United States would become too deeply involved in Iraq and Syria. This would seem to give the administration the political space to do more against the Islamic State, but within certain limits. What would that mean? Perhaps more airstrikes, or a greater deployment of special forces positioned more centrally. There is no mandate to pursue anything like the kind of nationbuilding effort we have seen in Iraq and Afghanistan since 2001.

Trying to determine the right approach toward ISIS, and the right amount of resources to dedicate to the task, remains the central challenge for this administration and for its successor. Perhaps a significant terror attack in the United States would shift that balance toward a more aggressive strategy—but even then, the same constraints would apply. Fourteen years after 9/11, we have yet to defeat the terrorist derivatives that al Qaeda spawned, including the Islamic State. At best, we can degrade ISIS's capabilities; keep it on the defensive; hold the line against further takeovers of Iraqi territory; mobilize local allies against it; and most important, try to prevent and preempt its efforts to direct attacks on U.S. soil. But defeating ISIS is for now an unattainable objective—one to ponder during the long war to come.

AARON DAVID MILLER, a Vice President at the Woodrow Wilson Center, served as a Middle East negotiator, analyst, and adviser in Republican and Democratic administrations.

EXPLORING THE ISSUE

Can ISIS Be Defeated in the Near Future?

Critical Thinking and Reflection

1. Is ISIS a state or a terrorist organization and how does the answer to that question determine what its potential future may be?
2. What are the military, political considerations of an all-out effort to defeat and crush ISIS for the larger Muslim world?
3. Is ISIS a manifestation of another variant of al Qaeda or is it a whole new movement within the ongoing civil war between Sunni and Shi'ite Islam for control over its future?
4. Is defeating the ISIS state the same as defeating the virulent ideology that supports it?

Is There Common Ground?

Evaluating the military capability of ISIS and then crafting strategy to defeat ISIS has many ideas, variants, and positions. Some are more confrontational and direct while others are more focused on the use of Muslim proxy forces to eliminate ISIS. There is, however, one area of agreement and that is that ISIS as a movement and ideology must be defeated whatever the cost. If ISIS's brand of Islamic ideology were to take hold and grow, the human devastation and negative implications for Muslims and non-Muslims would be catastrophic both within the Middle East and throughout the world.

Additional Resources

Carpenter, Ted Galen, "We Defeat ISIS/ Then What?" *The National Interest* (December 31, 2015)

The author explores what will change when ISIS is defeated.

Coll, Steven, "ISIS after Paris," *The New Yorker* (November 30, 2015)

The author examines ISIS status in the aftermath of the attacks.

Ibrahim, Azeem, "Here's What Happens After ISIS is Defeated and It May Create a New Era of Global Jihadism," *Business Insider* (January 26, 2016)

The author examines what comes after ISIS.

Shapiro, Ben, "No the Islamic State Will not be Defeated-And if It is, We Still Lose," *Breitbart* (November 24, 2015)

The author criticizes U.S. handling of ISIS.

Internet References . . .

Aljazeera

www.aljazeera.com

The Federalist

www.thefederalist.com

Foreign Policy

http://foreignpolicy.com/

Selected, Edited, and with Issue Framing Material by:
James E. Harf, *Maryville University*
and
Mark Owen Lombardi, *Maryville University*

ISSUE

Is the Iran Nuclear Program Agreement Good for America and for the World?

YES: John Kerry, from "Remarks on Nuclear Agreement With Iran," National Constitutional Center, Philadelphia, Pennsylvania, U.S. Department of State (2015)

NO: David E. Sanger and Michael R. Gordon, from "Future Risks of an Iran Nuclear Deal," *The New York Times* (2015)

Learning Outcomes

After reading this issue, you will be able to:

- Describe the evolution of Iran's nuclear program since the 1950s.
- Describe the path of negotiations on the Iran nuclear program deal during the final years before an agreement was reached.
- Outline the main components of the Iran nuclear program agreement.
- Describe the battleground issues that are likely to define the discussion of the agreement over the next several years.
- Gain insight into the politics within the American political system over the Iran nuclear program deal.

ISSUE SUMMARY

YES: John Kerry, U.S. Secretary of State, whose team negotiated the Iran nuclear program agreement, lays out the rationale behind the support of the deal by many top nuclear scientists and other experts, and the reasons why he believes the world in general and the most relevant countries within that world will be safer as a consequence of the deal.

NO: David Sanger and Michael Gordon of *The New York Times* criticize the deal, focusing on the fact that in 15 years Iran will be free to produce massive quantities of uranium and thus be in a position to produce nuclear bombs quickly.

Iran's nuclear program has been of interest to the United States and other Western powers for a long time, even predating the ugly episode of American diplomats being held hostage in the late 1970s following the revolution that overthrew Iran's more secular leader, the Shah, in 1979. In fact, Iran had benefited from President Eisenhower's Atoms for Peace Program in the late 1950s, which allowed countries such as Iran to harness nuclear energy for peaceful domestic purposes. In the years following Eisenhower's overtures but before the 1979 revolution, the United States and other Western powers were interested in helping Iran harness nuclear power. Iran signed the Nuclear Non-Proliferation Treaty in 1968 and soon France and West Germany joined the United States in providing help, and this assistance grew over the next few years. But by the mid-1970s, the United States and other Western powers became concerned about the Shah's nuclear ambitions and support for nuclear development began to wane. Iran, in turn, became more secretive with regard to its nuclear activities. And following the revolution in 1979, Western assistance came to a stop.

Greg Bruno of the Council on Foreign Relations relates (Iran's Nuclear Program, March 10, 2010) that a combination of three factors brought Iran's nuclear program to a halt. The new supreme leader, Ayatollah Khomeini, opposed nuclear technology. The second factor was that the exodus of nuclear scientists among the broader exodus of the educated and privileged class of the earlier regime left a major void. Finally, Israel, suspecting an immediate nuclear threat from Iraq, destroyed the latter's nuclear facility in a 1981 bombing raid, a factor that gave the Iranians pause.

Within a few years, Iran's post-revolutionary government soon came to believe that it needed a robust nuclear program, and the death of Khomeini in 1989 paved the way for a renewed emphasis on nuclear energy. Bruno argues that the discovery of a previously unknown Iraqi nuclear program, as well as a growing U.S. presence in the region, caused Iran to more vigorously push its nuclear program. The Iranians came to view nuclear capability as an effective deterrent against others, particularly the United States, Iraq, and Israel. Within a couple of decades Iran developed a vast network of nuclear-related facilities—"uranium mines, enrichment plants, conversion sites, and research reactors." These activities had been in full swing for a while, even though suspicions had been raised early in the new century. Bruno writes that in 2002 a London-based Iranian opposition group revealed details about two such facilities. Another covert Iranian nuclear site was subsequently made public by the United States, Britain, and France. As a signatory to the Non-Proliferation Treaty, Iran has been subject to a variety of restrictions from the UN Security Council and other countries' unilateral sanctions due to its secretive nuclear work. The United States has led a vigorous effort over the years to make certain that the International Atomic Energy Agency (IAEA) was documenting Iranian non-compliance. The United States has itself imposed significant sanctions on Iran over the years since the early days of the post-revolutionary regime, including a total U.S. trade embargo.

In 2008, then-Senator Barack Obama campaigned on a platform with two prongs: first that the United States would increase the pressure on Iran to come into compliance with its international obligations, especially through additional sanctions, and second that the United States was open to pursuing diplomacy with Iran to see if the nuclear issue could be resolved peacefully. The Iranian nuclear program had grown dramatically in size and capability over the previous decades, and it was getting close to a point where Iran could have enough fissile material to build a nuclear weapon relatively soon. Within a month of President Obama assuming office in January 2009, Iran successfully launched its first satellite, raising concerns about a potential for a ballistic missile delivery system. In 2010, additional sanctions by the UN Security Council, the European Union, and the United States were imposed on Iran. Throughout the first years of the Obama Administration, the P5+1 (France, Russia, Germany, Britain, China, and the United States) met with Iran regularly to discuss the nuclear issue, but no agreement resulted. Iran continued moving its nuclear program forward, all the while suggesting that it was only for peaceful purposes, but the P1+5, the United Nations, and the IAEA all raised flags over Iran's suspected activities. As Iran's program grew, pressure from the international community grew as well. In September 2012, Israeli Prime Minister Benjamin Netanyahu used the occasion of an address before the UN General Assembly to draw what the Arms Control Association called a "redline for an Israeli attack on Israel" (Arms Control Association, Timeline of Nuclear Diplomacy with Iran, January 2016). This red line was defined as "Iran amassing enough uranium enriched to 20 percent," enough for one bomb. Iran was getting ever closer to this "red line." While P5+1 discussions with Iran had been occurring for years, a new threshold was crossed in July 2013 when Iran elected a more moderate President, Hassan Rouhani, who called for serious negotiations with the P5+1, in large part because the sanctions had crippled Iran's economy and Iran knew negotiations that dealt with the nuclear issues would be the only way to relieve the economic pressure. It was clear in this new round of talks that Iran had brought to the table a new seriousness of purpose, what U.S. negotiator Wendy Sherman called a candor missing in the previous rounds of negotiations. By the fall of 2013, the differences had been narrowed, according to U.S. Secretary of State John Kerry. In early 2014, a Joint Action Plan was implemented, which halted the progress of Iran's nuclear program and gave the negotiators time and space to hammer out a final agreement that prevented Iran from obtaining a nuclear weapon. Public pronouncements took on a more optimistic tone as the year ended and all parties began working toward an agreement, which the P5+1 and Iran did for the next 18 months.

In spring 2015 the U.S. Senate decided to get involved, arguing that the President must submit any agreement to the Congress for approval. In early March, Israeli Prime Minister Netanyahu made an unprecedented speech before the Congress, arguing strongly against any agreement with Iran. And in July the announcement came from the P5+1 and Iran that an agreement had been reached on a comprehensive deal. President Obama immediately proclaimed that "We have stopped the spread of nuclear weapons in this region." The IAEA announced a roadmap for its investigation of the past military aspects of Iran's nuclear program. President Obama, in turn, sent the agreement to Congress, which had 60 days to review the deal. Meanwhile, five eminent scientists and engineers on behalf of 24 other scientists and engineers sent a letter to the White House dated August 8, 2015 extolling the virtues of the agreement and congratulating the President on completing the deal. On September 2, the 34th US Senator came out in support of the agreement, thus presumably ending the Senate's ability to override a Presidential veto on its action. A few days later, the 42nd Senator voiced support, thus ending any formal Senate opposition to the agreement. Both houses of Congress nonetheless tried to stop the agreement through Congressional action, but they failed. In October, Iran's parliament approved the agreement.

Exactly what was agreed upon by the parties to the Iran nuclear program deal? Julian Boeger of *The Guardian* describes its components succinctly in a July 14, 2015 report. In sum, Iran accepted strict limits on and monitoring of its nuclear program in return for the lifting of extensive economic sanctions that had been put in place because of Iran's nuclear program. Specific points of the agreement are as follows: (1) Iran's installed centrifuges, which are used to enrich uranium, will be reduced by two-thirds; (2) its underground enrichment plant near Qom will be used only for non-military research; (3) its enriched uranium stockpile will be reduced by 98 percent; (4) there are limits on research and development on advanced centrifuges so that there could be no sudden upgrading of enrichment capacity that would lead to a breakout time from one year to a few weeks; (5) the reactor core at Arak would be removed and filled with concrete, and the reactor will be redesigned so that it produces no weapons-grade plutonium; (6) IAEA inspectors will have full access to all Iran's declared nuclear sites and will be able to visit non-declared sites; (7) Iran has agreed to a "road map" with IAEA to provide access to facilities and people suspected of prior involvement on warhead design; (8) the United States and the European Union agree to suspend or cancel economic and financial sanctions once the agreed-upon steps have been taken by Iran; and (9) the agreement will be

incorporated into a new UN Security Council resolution that will address the arms embargo, placing a five-year limit and an eight-year limit on the ban of missile technology. In January 2016, the deal took effect after Iran had completed all of its nuclear-related commitments, and sanctions were then lifted.

Battleground issues remain as the first-year anniversary of the implementation of the agreement approached. Robert Einhorn summarized them in "Debating the Iran Nuclear Deal" (August 2015). He describes six important areas of contention. (1) What happens to Iran's nuclear program after the deal's first decade? (2) How does the deal address concerns about the possible military dimensions of Iran's past nuclear work? (3) Is IAEA access to sensitive sites timely enough? (4) What is the significance of the restrictions on conventional arms transfers and ballistic missile activities? (5) What are the implications of sanctions relief? (6) What are the consequences of rejecting the deal? While the latter is a moot point for now, pending the upcoming American election, the other five issues have been the subject of much discussion during the entire process and in the months following the signing and implementation of the agreement.

In the YES selection, Secretary of State John Kerry in a speech at the National Constitution Center in Philadelphia less than two months after the agreement's signing describes how close Iran had been to developing a nuclear weapon and how the recently concluded agreement successfully pushes back or even eliminates that possibility in the future. The Secretary concludes that the agreement "will cement the support of the international community behind a plan to ensure that Iran does not ever acquire or possess a nuclear weapon … (and thus) will remove a looming threat from a uniquely fragile region, discourage others from trying to develop nuclear arms, make our citizens and our allies safer, and reassure the world that the hardest problems can be addressed successfully by diplomatic means." In the NO selection, David E. Sanger and Michael R. Gordon of *The New York Times* suggest that most of the constraints on Iran's actions end after 15 years and then Iran will be free to produce uranium "on an industrial scale."

YES ⤶

John Kerry

Remarks on Nuclear Agreement With Iran

Two months ago, in Vienna, the United States and five other nations—including permanent members of the UN Security Council—reached agreement with Iran on ensuring the peaceful nature of that country's nuclear program. As early as next week, Congress will begin voting on whether to support that plan. And the outcome will matter as much as any foreign policy decision in recent history. Like Senator Lugar, President Obama and I are convinced—beyond any reasonable doubt—that the framework that we have put forward will get the job done. And in that assessment, we have excellent company.

Last month, 29 of our nation's top nuclear physicists and Nobel Prize winners, scientists, from one end of our country to the other, congratulated the President for what they called "a technically sound, stringent, and innovative deal that will provide the necessary assurance . . . that Iran is not developing nuclear weapons." The scientists praised the agreement for its creative approach to verification and for the rigorous safeguards that will prevent Iran from obtaining the fissile material for a bomb.

Today, I will lay out the facts that caused those scientists and many other experts to reach the favorable conclusions that they have. I will show why the agreed plan will make the United States, Israel, the Gulf States, and the world safer. I will explain how it gives us the access that we need to ensure that Iran's nuclear program remains wholly peaceful, while preserving every option to respond if Iran fails to meet its commitments. I will make clear that the key elements of the agreement will last not for 10 or 15 years, as some are trying to assert, or for 20 or 25, but they will last for the lifetime of Iran's nuclear program. And I will dispel some of the false information that has been circulating about the proposal on which Congress is soon going to vote.

Now, for this discussion, there is an inescapable starting point—a place where every argument made against the agreement must confront a stark reality—the reality of how advanced Iran's nuclear program had become and where it was headed when Presidents Obama and Rouhani launched the diplomatic process that concluded this past July.

Two years ago, in September of 2013, we were facing an Iran that had already mastered the nuclear fuel cycle; already stockpiled enough enriched uranium that, if further enriched, could arm 10 to 12 bombs; an Iran that was already enriching uranium to the level of 20 percent, which is just below weapons-grade; an Iran that had already installed 10,000-plus centrifuges; and an Iran that was moving rapidly to commission a heavy water reactor able to produce enough weapons-grade plutonium for an additional bomb or two a year. That, my friends, is where we already were when we began our negotiations.

At a well-remembered moment during the UN General Assembly the previous fall, Israeli Prime Minister Netanyahu had held up a cartoon of a bomb to show just how dangerous Iran's nuclear program had become. And in 2013, he returned to that podium to warn that Iran was positioning itself to "rush forward to build nuclear bombs before the international community can detect it and much less prevent it." The prime minister argued rightly that the so-called breakout time—the interval required for Iran to produce enough fissile material for one bomb—had dwindled to as little as two months. Even though it would take significantly longer to actually build the bomb itself using that fissile material, the prime minister's message was clear: Iran had successfully transformed itself into a nuclear threshold state.

In the Obama Administration, we were well aware of that troubling fact, and more important, we were already responding to it. The record is irrefutable that, over the course of two American administrations, it was the United States that led the world in assembling against Tehran one of the toughest international sanctions regimes ever developed.

But we also had to face an obvious fact: sanctions alone were not getting the job done, not even close. They were failing to slow, let alone halt, Iran's relentless march towards a nuclear weapons capability. So President Obama acted. He reaffirmed his vow that Iran would absolutely

Kerry, John. "Remarks on Nuclear Agreement With Iran," U.S. Department of State, September 2015.

not be permitted to have a nuclear weapon. He marshaled support for this principle from every corner of the international community. He made clear his determination to go beyond what sanctions could accomplish and find a way to not only stop, but to throw into reverse, Iran's rapid expansion of its nuclear program.

As we developed our strategy, we cast a very wide net to enlist the broadest expertise available. We sat down with the IAEA and with our own intelligence community to ensure that the verification standards that we sought on paper would be effective in reality. We consulted with Congress and our international allies and friends. We examined carefully every step that we might take to close off each of Iran's potential pathways to a bomb. And of course, we were well aware that every proposal, every provision, every detail would have to withstand the most painstaking scrutiny. We knew that. And so we made clear from the outset that we would not settle for anything less than an agreement that was comprehensive, verifiable, effective, and of lasting duration.

We began with an interim agreement reached in Geneva—the Joint Plan of Action. It accomplished diplomatically what sanctions alone could never have done or did. It halted the advance of Iran's nuclear activities. And it is critical to note—you don't hear much about it, but it's critical to note that for more than 19 months now, Iran has complied with every requirement of that plan. But this was just a first step.

From that moment, we pushed ahead, seeking a broad and enduring agreement, sticking to our core positions, maintaining unity among a diverse negotiating group of partners, and we arrived at the good and effective deal that we had sought.

And I ask you today and in the days ahead, as we have asked members of Congress over the course of these last months, consider the facts of what we achieved and judge for yourself the difference between where we were two years ago and where we are now, and where we can be in the future. Without this agreement, Iran's so-called breakout time was about two months; with this agreement it will increase by a factor of six, to at least a year, and it will remain at that level for a decade or more.

Without this agreement, Iran could double the number of its operating centrifuges almost overnight and continue expanding with ever more efficient designs. With this agreement, Iran's centrifuges will be reduced by two-thirds for 10 years.

Without this agreement, Iran could continue expanding its stockpile of enriched uranium, which is now more than 12,000 kilograms—enough, if further enriched, for multiple bombs. With this agreement, that stockpile will shrink and shrink some more—a reduction of some 98 percent, to no more than 300 kilograms for 15 years.

Without this agreement, Iran's heavy-water reactor at Arak would soon be able to produce enough weapons-grade plutonium each year to fuel one or two nuclear weapons. With this agreement, the core of that reactor will be removed and filled with concrete, and Iran will never be permitted to produce any weapons-grade plutonium.

Without this agreement, the IAEA would not have assured access to undeclared locations in Iran where suspicious activities might be taking place. The agency could seek access, but if Iran objected, there would be no sure method for resolving a dispute in a finite period, which is exactly what has led us to where we are today—that standoff. With this agreement, the IAEA can go wherever the evidence leads. No facility—declared or undeclared—will be off limits, and there is a time certain for assuring access. There is no other country to which such a requirement applies. This arrangement is both unprecedented and unique.

In addition, the IAEA will have more inspectors working in Iran, using modern technologies such as real-time enrichment monitoring, high-tech electronic seals, and cameras that are always watching—24/7, 365. Further, Iran has agreed never to pursue key technologies that would be necessary to develop a nuclear explosive device.

So the agreement deals not only with the production of fissile material, but also with the critical issue of weaponization. Because of all of these limitations and guarantees, we can sum up by saying that without this agreement, the Iranians would have several potential pathways to a bomb; with it, they won't have any.

Iran's plutonium pathway will be blocked because it won't have a reactor producing plutonium for a weapon, and it won't build any new heavy-water reactors or engage in reprocessing for at least 15 years, and after that we have the ability to watch and know precisely what they're doing.

The uranium pathway will be blocked because of the deep reductions in Iran's uranium enrichment capacity, and because for 15 years the country will not enrich uranium to a level higher than 3.67 percent. Let me be clear: No one can build a bomb from a stockpile of 300 kilograms of uranium enriched only 3.67 percent. It is just not possible.

Finally, Iran's covert pathway to a bomb will also be blocked. Under our plan, there will be 24/7 monitoring of Iran's key nuclear facilities. As soon as we start the implementation, inspectors will be able to track Iran's uranium as it is mined, then milled, then turned into yellow cake, then into gas, and eventually into waste. This means that for a quarter of a century at least, every activity throughout the nuclear fuel chain will receive added scrutiny. And for 20 years, the IAEA will be monitoring the production of key centrifuge components in Iran in order to assure that none are diverted to a covert program.

So if Iran did decide to cheat, its technicians would have to do more than bury a processing facility deep beneath the ground. They would have to come up with a complete—complete—and completely secret nuclear supply chain: a secret source of uranium, a secret milling facility, a secret conversion facility, a secret enrichment facility. And our intelligence community and our Energy Department, which manages our nuclear program and our nuclear weapons, both agree Iran could never get away with such a deception. And if we have even a shadow of doubt that illegal activities are going on, either the IAEA will be given the access required to uncover the truth or Iran will be in violation and the nuclear-related sanctions can snap back into place. We will also have other options to ensure compliance if necessary.

Given all of these requirements, it is no wonder that this plan has been endorsed by so many leading American scientists, experts on nuclear nonproliferation, and others. More than 60 former top national security officials, 100—more than 100 retired ambassadors—people who served under Democratic and Republican presidents alike, are backing the proposal—as are retired generals and admirals from all 5 of our uniformed services. Brent Scowcroft, one of the great names in American security endeavors of the last century and now, served as a national security advisor to two Republican presidents. He is also among the many respected figures who are supporting it. Internationally, the agreement is being backed, with one exception, by each of the more than 100 countries that have taken a formal position. The agreement was also endorsed by the United Nations Security Council on a vote of 15 to nothing. This not only says something very significant about the quality of the plan, particularly when you consider that 5 of those countries are permanent members—and they're all nuclear powers, but it should also invite reflection from those who believe the United States can walk away from this without causing grave harm to our international reputation, to relationships, and to interests.

You've probably heard the claim that because of our strength, because of the power of our banks, all we Americans have to do if Congress rejects this plan is return to the bargaining table, puff out our chests, and demand a better deal. I've heard one critic say he would use sanctions to give Iran a choice between having an economy or having a nuclear program. Well, folks, that's a very punchy soundbite, but it has no basis in any reality. As

Dick said, I was chair of the Senate Foreign Relations Committee when our nation came together across party lines to enact round after round of economic sanctions against Iran. But remember, even the toughest restrictions didn't stop Iran's nuclear program from speeding ahead from a couple of hundred centrifuges to 5,000 to 19,000. We've already been there. If this agreement is voted down, those who vote no will not be able to tell you how many centrifuges Iran will have next year or the year after. If it's approved, we will be able to tell you exactly what the limits on Iran's program will be.

The fact is that it wasn't either sanctions or threats that actually stopped and finally stopped the expansion of Iran's nuclear activities. The sanctions brought people to the table, but it was the start of the negotiating process and the negotiations themselves, recently concluded in Vienna, that actually stopped it. Only with those negotiations did Iran begin to get rid of its stockpile of 20 percent enriched uranium. Only with those negotiations did it stop installing more centrifuges and cease advancing the Arak reactor. Only then did it commit to be more forthcoming about IAEA access and negotiate a special arrangement to break the deadlock.

So just apply your common sense: What do you think will happen if we say to Iran now, "Hey, forget it. The deal is off. Let's go back to square one"? How do you think our negotiating partners, all of whom have embraced this deal, will react; all of whom are prepared to go forward with it—how will they react? What do you think will happen to that multilateral sanctions regime that brought Iran to the bargaining table in the first place? The answer is pretty simple. The answer is straightforward. Not only will we lose the momentum that we have built up in pressing Iran to limit its nuclear activities, we will almost surely start moving in the opposite direction.

We need to remember sanctions don't just sting in one direction, my friends. They also impose costs on those who forego the commercial opportunities in order to abide by them. It's a tribute to President Obama's diplomacy—and before that, to President George W. Bush—that we were able to convince countries to accept economic difficulties and sacrifices and put together the comprehensive sanctions regime that we did. Many nations that would like to do business with Iran agreed to hold back because of the sanctions and—and this is vital—and because they wanted to prevent Iran from acquiring a nuclear weapon. They have as much interest in it as we do. And that's why they hoped the negotiations would succeed, and that's why they will join us in insisting that Iran live up to its obligations. But they will not join us if we unilaterally walk away from the very deal that the sanctions were designed to bring about. And they will not join us if we're demanding even greater sacrifices and threatening their businesses and banks because of a choice we made and they opposed.

So while it may not happen all at once, it is clear that if we reject this plan, the multilateral sanctions regime will start to unravel. The pressure on Iran will lessen and our negotiating leverage will diminish, if not disappear. Now, obviously, that is not the path, as some critics would have us believe, to a so-called better deal. It is a path to a much weaker position for the United States of America and to a much more dangerous Middle East.

And this is by no means a partisan point of view that I just expressed. Henry Paulson was Secretary of Treasury under President George W. Bush. He helped design the early stages of the Iran sanctions regime. But just the other day, he said, "It would be totally unrealistic to believe that if we backed out of this deal, the multilateral sanctions would remain in place." And Paul Volcker, who chaired the Federal Reserve under President Reagan, he said, "This agreement is as good as you are going to get. To think that we can unilaterally maintain sanctions doesn't make any sense."

We should pause for a minute to contemplate what voting down this agreement might mean for Iran's cadre of hardliners, for those people in Iran who lead the chants of "Death to America," "Death to Israel," and even "Death to Rouhani," and who prosecute journalists simply for doing their jobs. The evidence documents that among those who most fervently want this agreement to fall apart are the most extreme factions in Iran. And their opposition should tell you all you need to know. From the very beginning, these extremists have warned that negotiating with the United States would be a waste of time; why on Earth would we now take a step that proves them right?

Let me be clear. Rejecting this agreement would not be sending a signal of resolve to Iran; it would be broadcasting a message so puzzling most people across the globe would find it impossible to comprehend. After all, they've listened as we warned over and over again about the dangers of Iran's nuclear program. They've watched as we spent two years forging a broadly accepted agreement to rein that program in. They've nodded their heads in support as we have explained how the plan that we have developed will make the world safer.

Who could fairly blame them for not understanding if we suddenly switch course and reject the very outcome we had worked so hard to obtain? And not by offering some new and viable alternative, but by offering no alternative at all. It is hard to conceive of a quicker or more self-destructive blow to our nation's credibility and leadership—not only with respect to this one issue, but I'm telling you across the board—economically, politically, militarily, and even morally. We would pay an immeasurable price for this unilateral reversal.

Friends, as Dick mentioned in his introduction, I have been in public service for many years and I've been called on to make some difficult choices in that course of time. There are those who believe deciding whether or not to support the Iran agreement is just such a choice. And I respect that and I respect them. But I also believe that because of the stringent limitations on Iran's program that are included in this agreement that I just described, because of where that program was headed before our negotiations began and will head again if we walk away, because of the utter absence of a viable alternative to this plan that we have devised, the benefits of this agreement far outweigh any potential drawbacks. Certainly, the goal of preventing Iran from having a nuclear weapon is supported across our political spectrum and it has the backing of countries on every continent. So what then explains the controversy that has persisted in this debate?

A big part of the answer, I think, is that even before the ink on the agreement was dry, we started being bombarded by myths about what the agreement will and won't do, and that bombardment continues today.

The first of these myths is that the deal is somehow based on trust or a naive expectation that Iran is going to reverse course on many of the policies it's been pursuing internationally. Critics tell us over and over again, "You can't trust Iran." Well, guess what? There is a not a single sentence, not a single paragraph in this whole agreement that depends on promises or trust, not one. The arrangement that we worked out with Tehran is based exclusively on verification and proof. That's why the agreement is structured the way it is; that's why sanctions relief is tied strictly to performance; and it is why we have formulated the most far-reaching monitoring and transparency regime ever negotiated.

Those same critics point to the fact that two decades ago, the United States reached a nuclear framework with North Korea that didn't accomplish what it set out to do. And we're told we should have learned a lesson from that. Well, the truth is we did learn a lesson.

The agreement with North Korea was four pages and only dealt with plutonium. Our agreement with Iran runs 159 detailed pages, applies to all of Tehran's potential pathways to a bomb, and is specifically grounded in the transparency rules of the IAEA's Additional Protocol, which didn't even exist two decades ago when the North Korea deal was made because it was developed specifically with the North Korea experience in mind. Lesson learned.

The reality is that if we trusted Iran or thought that it was about to become more moderate, this agreement would be less necessary than it is. But we don't. We would like nothing more than to see Iran act differently, but not for a minute are we counting on it. Iran's support for terrorist groups and its contributions to sectarian violence are not recent policies. They reflect the perceptions of its leaders about Iran's long-term national interests and there are no grounds for expecting those calculations to change in the near future. That is why we believe so strongly that every problem in the Middle East—every threat to Israel and to our friends in the region—would be more dangerous if Iran were permitted to have a nuclear weapon. That is the inescapable bottom line.

That's also why we are working so hard and so proactively to protect our interests and those of our allies.

In part because of the challenge posed by Iran, we have engaged in an unprecedented level of military, intelligence, and security cooperation with our friend and ally Israel. We are determined to help our ally address new and complex security threats and to ensure its qualitative military edge.

We work with Israel every day to enforce sanctions and prevent terrorist organizations such as Hamas and Hizballah from obtaining the financing and the weapons that they seek—whether from Iran or from any other source. And we will stand with Israel to stop its adversaries from once again launching deadly and unprovoked attacks against the Israeli people.

Since 2009, we have provided $20 billion in foreign military financing to Israel, more than half of what we have given to nations worldwide.

Over and above that, we have invested some 3 billion in the production and deployment of Iron Dome batteries and other missile defense programs and systems. And we saw how in the last Gaza War lives were saved in Israel because of it. We have given privileged access to advanced military equipment such as the F-35 Joint Strike Fighter; Israel is the only nation in the Middle East to which the United States has sold this fifth-generation aircraft. The President recently authorized a massive arms resupply package, featuring penetrating munitions and air-to-air missiles. And we hope soon to conclude a new memorandum of understanding—a military assistance plan that will guide our intensive security cooperation through the next decade.

And diplomatically, our support for Israel also remains rock solid as we continue to oppose every effort to delegitimize the Jewish state, or to pass biased resolutions against it in international bodies.

Now, I understand—I understand personally there is no way to overstate the concern in Israel about Iran and about the potential consequences that this agreement—or

rejecting this agreement—might have on Israel's security. The fragility of Israel's position has been brought home to me on every one of the many trips I have made to that country.

In fact, as Secretary of State, I have already traveled to Israel more than a dozen times, spending the equivalent of a full month there—even ordering my plane to land at Ben Gurion Airport when commercial air traffic had been halted during the last Gaza War; doing so specifically as a sign of support.

Over the years, I have walked through Yad Vashem, a living memorial to the 6 million lost, and I have felt in my bones the unfathomable evil of the Holocaust and the undying reminder never to forget.

I have climbed inside a shelter at Kiryat Shmona where children were forced to leave their homes and classrooms to seek refuge from Katyusha rockets.

I visited Sderot and witnessed the shredded remains of homemade missiles from Gaza—missiles fired with no other purpose than to sow fear in the hearts of Israeli families.

I have piloted an Israeli jet out of Ovda Airbase and observed first-hand the tininess of Israel airspace from which it is possible to see all of the country's neighbors at the same time.

And I have bowed my head at the Western Wall and offered my prayer for peace—peace for Israel, for the region, and for the world.

I take a back seat to no one in my commitment to the security of Israel, a commitment I demonstrated through my 28-plus years in the Senate. And as Secretary of State, I am fully conscious of the existential nature of the choice Israel must make. I understand the conviction that Israel, even more than any other country, simply cannot afford a mistake in defending its security. And while I respectfully disagree with Prime Minister Netanyahu about the benefits of the Iran agreement, I do not question for an instant the basis of his concern or that of any Israeli.

But I am also convinced, as is President Obama, our senior defense and military leaders, and even many former Israeli military and intelligence officials, that this agreement puts us on the right path to prevent Iran from ever getting a nuclear weapon. The people of Israel will be safer with this deal, and the same is true for the people throughout the region.

And to fully ensure that, we are also taking specific and far-reaching steps to coordinate with our friends from the Gulf states. President Obama hosted their leaders at Camp David earlier this year. I visited with them in Doha last month. And later this week, we will welcome King Salman of Saudi Arabia to Washington. Gulf leaders share our profound concerns about Iran's policies in the Middle East, but they're also alarmed by Iran's nuclear program. We must and we will respond on both fronts. We will make certain that Iran lives up to its commitments under the nuclear agreement, and we will continue strengthening our security partnerships.

We're determined that our Gulf friends will have the political and the military support that they need, and to that end, we are working with them to develop a ballistic missile defense for the Arabian Peninsula, provide special operations training, authorize urgently required arms transfers, strengthen cyber security, engage in large-scale military exercises, and enhance maritime interdiction of illegal Iranian arms shipments. We are also deepening our cooperation and support in the fight against the threat posed to them, to us, and to all civilization by the forces of international terror, including their surrogates and their proxies.

Through these steps and others, we will maintain international pressure on Iran. United States sanctions imposed because of Tehran's support for terrorism and its human rights record—those will remain in place, as will our sanctions aimed at preventing the proliferation of ballistic missiles and transfer of conventional arms. The UN Security Council prohibitions on shipping weapons to Hizballah, the Shiite militias in Iraq, the Houthi rebels in Yemen—all of those will remain as well.

We will also continue to urge Tehran to provide information regarding an American who disappeared in Iran several years ago, and to release the U.S. citizens its government has unjustly imprisoned. We will do everything we can to see that our citizens are able to safely return to where they belong—at home and with their families.

Have no doubt. The United States will oppose Iran's destabilizing policies with every national security tool available. And disregard the myth. The Iran agreement is based on proof, not trust. And in a letter that I am sending to all the members of Congress today, I make clear the Administration's willingness to work with them on legislation to address shared concerns about regional security consistent with the agreement that we have worked out with our international partners.

This brings us to the second piece of fiction: that this deal would somehow legitimize Iran's pursuit of a nuclear weapon. I keep hearing this. Well, yes, for years Iran has had a civilian nuclear program. Under the Nonproliferation Treaty, you can do that. It was never a realistic option to change that. But recognizing this reality is not the same as legitimizing the pursuit of a nuclear weapon. In fact, this agreement does the exact opposite. Under IAEA safeguards, Iran is prohibited from ever pursuing a nuclear weapon.

This is an important point, so I want to be sure that everyone understands: The international community is not telling Iran that it can't have a nuclear weapon for 15 years. We are telling Iran that it can't have a nuclear weapon, period. There is no magic moment 15, 20, or 25 years from now when Iran will suddenly get a pass from the mandates of the Nuclear Nonproliferation Treaty—doesn't happen. In fact, Iran is required by this agreement to sign up to and abide by the IAEA Additional Protocol that I mentioned earlier that came out of the North Korea experience. And that requires inspections of all nuclear facilities.

What does this mean? It means that Iran's nuclear program will remain subject to regular inspections forever. Iran will have to provide access to all of its nuclear facilities forever. Iran will have to respond promptly to requests for access to any suspicious site forever. And if Iran at any time—at any time—embarks on nuclear activities that are incompatible with a wholly peaceful program, it will be in violation of the agreement forever. We will know of that violation right away and we will retain every option we now have to respond, whether diplomatically or through a return to sanctions or by other means. In short, this agreement gives us unprecedented tools and all the time we need to hold Iran accountable for its choices and actions.

Now, it's true some of the special additional restrictions that we successfully negotiated, those begin to ease after a period—in some cases 10 or 15, in others 20 or 25. But it would defy logic to vote to kill the whole agreement—with all of the permanent NPT restrictions by which Iran has to live—for that reason. After all, if your house is on fire, if it's going up in flames, would you refuse to extinguish it because of the chance that it might be another fire in 15 years? Obviously, not. You'd put out the fire and you'd take advantage of the extra time to prepare for the future.

My friends, it just doesn't make sense to conclude that we should vote "no" now because of what might happen in 15 years—thereby guaranteeing that what might happen in 15 years will actually begin to happen now. Because if this agreement is rejected, every possible reason for worry in the future would have to be confronted now, immediately, in the months ahead. Once again and soon, Iran would begin advancing its nuclear program. We would lose the benefit of the agreement that contains all these restrictions, and it would give a green light to everything that we're trying to prevent. Needless to say, that is not the outcome that we want, it is not an outcome that would be good for our country, nor for our allies or for the world

There is a third myth—a quick one, a more technical one—that Iran could, in fact, get away with building a covert nuclear facility because the deal allows a maximum of 24 days to obtain access to a suspicious site. Well, in truth, there is no way in 24 days, or 24 months, 24 years for that matter, to destroy all the evidence of illegal activity that has been taking place regarding fissile material. Because of the nature of fissile materials and their relevant precursors, you can't eliminate the evidence by shoving it under a mattress, flushing it down a toilet, carting it off in the middle of the night. The materials may go, but the telltale traces remain year after year after year. And the 24 days is the outside period of time during which they must allow access.

Under the agreement, if there is a dispute over access to any location, the United States and our European allies have the votes to decide the issue. And once we have identified a site that raises questions, we will be watching it continuously until the inspectors are allowed in.

Let me underscore that. The United States and the international community will be monitoring Iran non-stop. And you can bet that if we see something, we will do something. The agreement gives us a wide range of enforcement tools, and we will use them. And the standard we will apply can be summed up in two words: zero tolerance. There is no way to guarantee that Iran will keep its word. That's why this isn't based on a promise or trust. But we can guarantee that if Iran decides to break the agreement, it will regret breaking any promise that it has made.

Now, there are many other myths circulating about the agreement, but the last one that I'm going to highlight is just economic. And it's important. The myth that sanctions relief that Iran will receive is somehow both too generous and too dangerous.

Now, obviously, the discussions that concluded in Vienna, like any serious negotiation, involved a quid pro quo. Iran wanted sanctions relief; the world wanted to ensure a wholly peaceful nature of Iran's program. So without the tradeoff, there could have been no deal and no agreement by Iran to the constraints that it has accepted—very important constraints.

But there are some who point to sanctions relief as grounds to oppose the agreement. And the logic is faulty for several reasons. First, the most important is that absent new violations by Iran the sanctions are going to erode regardless of what we do. It's an illusion for members of Congress to think that they can vote this plan down and then turn around and still persuade countries like China, Japan, South Korea, Turkey, India—Iran's major oil customers—they ought to continue supporting the sanctions that are costing them billions of dollars every year. That's not going to happen. And don't forget that the money that has been locked up as the result of sanctions is not sitting in some American bank under U.S. control. The

money is frozen and being held in escrow by countries with which Iran has had commercial dealings. We don't have that money. We can't control it. It's going to begin to be released anyway if we walk away from this agreement.

Remember, as well, that the bulk of the funds Iran will receive under the sanctions relief are already spoken for and they are dwarfed by the country's unmet economic needs. Iran has a crippled infrastructure, energy infrastructure. It's got to rebuild it to be able to pump oil. It has an agriculture sector that's been starved for investment, massive pension obligations, significant foreign reserves that are already allocated to foreign-led projects, and a civilian population that is sitting there expecting that the lifting of sanctions is going to result in a tangible improvement in the quality of their lives. The sanctions relief is not going to make a significant difference in what Iran can do internationally—never been based on money. Make no mistake, the important thing about this agreement is not what it will enable Iran to do, but what it will stop Iran from doing—and that is the building of a nuclear weapon.

Before closing, I want to comment on the nature of the debate which we are currently engaged in. Some have accused advocates of the Iran agreement—including me—of conjuring up frightening scenarios to scare listeners into supporting it. Curiously, this allegation comes most often from the very folks who have been raising alarms about one thing or another for years.

The truth is that if this plan is voted down, we cannot predict with certainty what Iran will do. But we do know what Iran says it will do and that is begin again to expand its nuclear activities. And we know that the strict limitations that Iran has accepted will no longer apply because there will no longer be any agreement. Iran will then be free to begin operating thousands of other advanced and other centrifuges that would otherwise have been mothballed; they'll be free to expand their stockpile of low-enriched uranium, rebuild their stockpile of 20 percent enriched uranium, free to move ahead with the production of weapons-grade plutonium, free to go forward with weaponization research.

And just who do you think is going to be held responsible for all of this? Not Iran—because Iran was preparing to implement the agreement and will have no reason whatsoever to return to the bargaining table. No, the world will hold accountable the people who broke with the consensus, turned their backs on our negotiating partners, and ignored the counsel of top scientists and military leaders. The world will blame the United States. And so when those same voices that accuse us of scaremongering now begin suddenly to warn, oh, wow, Iran's nuclear activities are once again out of control and must at all costs be stopped—what do you think is going to happen?

The pressure will build, my friends. The pressure will build for military action. The pressure will build for the United States to use its unique military capabilities to disrupt Iran's nuclear program, because negotiating isn't going to work because we've just tried it. President Obama has been crystal clear that we will do whatever is necessary to prevent Iran from getting a nuclear weapon. But the big difference is, at that point, we won't have the world behind us the way we do today. Because we rejected the fruits of diplomacy, we will be held accountable for a crisis that could have been avoided but instead we will be deemed to have created.

So my question is: Why in the world would we want to put ourselves in that position of having to make that choice—especially when there is a better choice, a much more broadly supported choice? A choice that sets us on the road to greater stability and security but that doesn't require us to give up any option at all today.

So here is the decision that we are called on to make. To vote down this agreement is to solve nothing because none of the problems that we are concerned about will be made easier if it is rejected; none of them—not Iran's nuclear program, not Iran's support for terrorism or sectarian activities, not its human rights record, and not its opposition to Israel. To oppose this agreement is—whether intended or not—to recommend in its policy a policy of national paralysis. It is to take us back directly to the very dangerous spot that we were in two years ago, only to go back there devoid of any realistic plan or option.

By contrast, the adoption and implementation of this agreement will cement the support of the international community behind a plan to ensure that Iran does not ever acquire or possess a nuclear weapon. In doing so it will remove a looming threat from a uniquely fragile region, discourage others from trying to develop nuclear arms, make our citizens and our allies safer, and reassure the world that the hardest problems can be addressed successfully by diplomatic means.

At its best, American foreign policy, the policy of the United States combines immense power with clarity of purpose, relying on reason and persuasion whenever possible. As has been demonstrated many times, our country does not shy from the necessary use of force, but our hopes and our values push us to explore every avenue for peace. The Iran deal reflects our determination to protect the interests of our citizens and to shield the world from greater harm. But it reflects as well our knowledge that the firmest foundation for security is built on mobilizing countries across the globe to defend—actively and bravely—the rule of law.

In September 228 years ago, Benjamin Franklin rose in the great city of Philadelphia, right down there, to close debate on the proposed draft of the Constitution of the United States. He told a rapt audience that when people of opposing views and passions are brought together, compromise is essential and perfection from the perspective of any single participant is not possible. He said that after weighing carefully the pros and cons of that most historic debate, he said the following: "I consent, sir, to this Constitution because I expect no better, and because I am not sure that it is not the best."

My fellow citizens, I have had the privilege of serving our country in times of peace and in times of war, and peace is better. I've seen our leaders act with incredible foresight and also seen them commit tragic errors by plunging into conflicts without sufficient thought about the consequences.

Like old Ben Franklin, I can claim and do claim no monopoly on wisdom, and certainly nothing can compare to the gravity of the debate of our founding fathers over our nation's founding documents. But I believe, based on a lifetime's experience, that the Iran nuclear agreement is a hugely positive step at a time when problem solving and danger reduction have rarely been so urgent, especially in the Middle East.

The Iran agreement is not a panacea for the sectarian and extremist violence that has been ripping that region apart. But history may judge it a turning point, a moment when the builders of stability seized the initiative from the destroyers of hope, and when we were able to show, as have generations before us, that when we demand the best from ourselves and insist that others adhere to a similar high standard—when we do that, we have immense power to shape a safer and a more humane world. That's what this is about and that's what I hope we will do in the days ahead.

John Kerry is US Secretary of State. Previously he served as Senator from Massachusetts.

David E. Sanger and
Michael R. Gordon

NO

Future Risks of an Iran Nuclear Deal

WASHINGTON—As President Obama begins his three-week push to win approval of the Iran nuclear deal, he is confronting this political reality: His strongest argument in favor of passage has also become his greatest vulnerability.

Mr. Obama has been pressing the case that the sharp limits on how much nuclear fuel Iran can hold, how many centrifuges it can spin and what kind of technology it can acquire would make it extraordinarily difficult for Iran to race for the bomb over the next 15 years.

His problem is that most of the significant constraints on Tehran's program lapse after 15 years—and, after that, Iran is free to produce uranium on an industrial scale.

"The chief reservation I have about the agreement is the fact that in 15 years they have a highly modern and internationally legitimized enrichment capability," said Representative Adam B. Schiff, a California Democrat who supports the accord. "And that is a bitter pill to swallow."

Even some of the most enthusiastic backers of the agreement, reached by six world powers with Iran, say they fear Mr. Obama has oversold some of the accord's virtues as he asserts that it would "block" all pathways to a nuclear weapon.

A more accurate description is that the agreement is likely to delay Iran's program for a decade and a half—just as sanctions and sabotage have slowed Iran in recent years. The administration's case essentially is that the benefits over the next 15 years overwhelmingly justify the longer-term risks of what comes after.

"Of course there are risks, and they have to be acknowledged," said R. Nicholas Burns, who was undersecretary of state in the George W. Bush administration and has testified before Congress in favor of the deal. Mr. Obama's "most convincing argument," he added, "is that there is no better alternative out there."

In making the administration's case, Mr. Obama can underscore that economic sanctions on Iran begin to lift only as it reduces its current stockpile of low enriched uranium, to 300 kilograms, or 660 pounds. That is not enough to make a single nuclear weapon, and is a 98 percent reduction in its current stockpile of nearly 12 tons.

The accord also calls for regular inspections at Iran's nuclear installations and includes arrangements to reimpose international sanctions if the Iranians are caught cheating.

But the flip side is that after 15 years, Iran would be allowed to produce reactor-grade fuel on an industrial scale using far more advanced centrifuges. That may mean that the warning time if Iran decided to race for a bomb would shrink to weeks, according to a recent Brookings Institution analysis by Robert J. Einhorn, a former member of the American negotiating team.

Critics say that by that time, Iran's economy would be stronger, as would its ability to withstand economic sanctions, and its nuclear installations probably would be better protected by air defense systems, which Iran is expected to buy from Russia.

Some members of Congress and other experts are urging the administration to take fresh steps to deter Iran from edging dangerously close to a nuclear weapons capability after the main limits in the agreement expire.

"I believe it buys 15 years for real," said Dennis B. Ross, who served as a White House adviser on Iran during Mr. Obama's first term and has yet to decide if he will back the accord. "But I do see vulnerabilities that I feel must be addressed. The gap between threshold and weapons status after year 15 is small."

A Loss of Leverage

The duration of the agreement is the most important and complex issue. Under restrictions imposed by the accord, Iran would need a full year to produce enough nuclear material for a bomb; currently that timeline is two or three months, according to American intelligence agencies. But starting at year 10, that "breakout time" would begin to shrink again, as Iran gets more centrifuges into operation.

Administration officials argue that it would be obvious if Iran made weapons-grade fuel, and negotiators secured a permanent ban on the metallurgy needed to turn the fuel into a bomb.

Supporters of the agreement are betting that improved intelligence would deter Iran from racing for a bomb. Under the agreement, inspectors will be able to monitor the production of rotors and other centrifuge components for up to 20 years and can monitor Iran's stocks of uranium ore concentrate for 25 years.

Skeptics counter that, after 15 years, the United States would lose much of its leverage to stop a program. So Mr. Obama is trying to assure Congress that he and his successors will create that leverage.

In a letter last week to Representative Jerrold Nadler, a Democrat from New York, Mr. Obama detailed the expanded military support he has offered Israel and reaffirmed that the United States retains the option to use economic sanctions and even military force should Iran break out of its agreement.

But Mr. Obama's letter was mostly a repackaging of previous assurances made to lawmakers, to Israel and to diplomats from Arab nations by the Persian Gulf.

Some backers of the agreement are urging the White House and Congress to do more. Mr. Schiff and Mr. Ross suggested in interviews that the United States should put Iran on notice that its production of highly enriched uranium after the main provisions of the accord expire would be taken by American officials as an indication that Iran has decided to pursue nuclear weapons—and could trigger an American military strike.

And both said the United States should also be prepared to provide bunker-busting bombs to Israel to deter Iran from trying to shield illicit nuclear work underground. Others have called for a long-term congressional "authorization to use military force" if Iran violated the accord.

Mr. Ross has also urged the White House to specify the penalties for smaller violations of the accord, an idea Mr. Obama rejected in his letter, saying he wanted to maintain "flexibility" to decide what responses might be needed.

Energy Secretary Ernest J. Moniz told a House committee last month that any attempt by Iran to produce highly enriched uranium "at any time must earn a sharp response by all necessary means."

But some experts like Mr. Einhorn say that this warning should be conveyed directly, if privately, to Iran and that the United States should also increase intelligence sharing with the world's nuclear inspector, the International Atomic Energy Agency, about possible Iranian cheating.

"The way to address challenges not covered by the agreement is to supplement it, not renegotiate it," Mr. Einhorn said.

Accounting for the Past

One of the trickier issues for Mr. Obama, and for Congress, is how to assess whether Iran has truly come clean about its past nuclear activities, an enormously sensitive issue for the Iranians. And in the end, it is one that Secretary of State John Kerry decided not to press too hard during negotiations, for fear it would undermine the chances of getting stronger inspections for current and future activity.

The job of assessing past activities is up to the I.A.E.A. It must certify on Oct. 15 that Iran is complying with a "road map" for cooperation and report in December on the agency's conclusions—especially about Iran's alleged work developing nuclear triggers and designing warheads.

Critics of the accord note that Mr. Kerry and his chief negotiator, Wendy R. Sherman, said repeatedly that Iran must provide access to "people, places and documents" that would resolve those questions, something Iran has refused to do for years. But the I.A.E.A. has never publicly specified what it is asking, or whom it must meet.

Mr. Einhorn, in his analysis, concluded that "a full and honest disclosure by Iran of its past weaponization activities—which would contradict Tehran's narrative of an exclusively peaceful program as well the supreme leader's fatwa that Islam forbids nuclear weapons—was never in the cards."

That said, he concludes, that may not be a "serious obstacle" to concluding Iran's work has halted.

Accessing Nuclear Sites

While the accord calls for regular inspections at Iran's nuclear sites, the enforcement is of limited duration. For example, while the I.A.E.A. can request access to all declared nuclear sites under the agreements it has with all member states, the far more intrusive monitoring at Iran's main nuclear enrichment site at Natanz is not mandated after 15 years. At that point, Iran also would be free to carry out nuclear enrichment at other locations.

But the issue that has garnered the most attention is a "24-day" rule for resolving disputes if Iran refuses to give inspectors access to a suspicious site—another measure that expires after 15 years. (After that, inspectors can still demand to enter sites, but under the existing rules, which do not set a deadline for compliance.) Critics say that is far different from "anywhere, anytime" access—

a phrase Mr. Moniz and others in the administration used a few months ago, and have come to regret.

If Iran balks at an inspection, then a commission—which includes Iran—can decide on punitive steps, including a reimposition of economic sanctions. A majority vote of the commission suffices, so even if Iran, China and Russia objected, the sanctions could go into effect.

That is the theory. In practice, reimposing sanctions could be politically challenging. Iran has warned that if sanctions are reimposed it will no longer be bound by the accord. The I.A.E.A., perhaps fearing its inspectors would be kicked out, might hesitate to start the 24-day clock.

Mr. Moniz argues that the 24-day time frame is sufficient because Iran will not be able to cover up evidence of nuclear work during that period, since traces of nuclear materials could be expected well after three weeks. But some experts say that Iran could cover up smaller-scale illicit activities, including work on the specialized high-explosives that might serve as a trigger in a nuclear bomb.

DAVID E. SANGER is the chief Washington correspondent for *The New York Times*. He has reported from New York, Tokyo, and Washington.

MICHAEL R. GORDON is the chief military correspondent for *The New York Times*. He is the co-author of two books, *The Generals' War: The Inside Story of the Conflict in the Gulf* (about the Gulf War) and *Cobra II* (about the Iraq war).

EXPLORING THE ISSUE

Is the Iran Nuclear Program Agreement Good for America and for the World?

Critical Thinking and Reflection

1. Does Iran's history of developing a nuclear program lead one to be suspicious of Iran's stated desire to forego nuclear weapons?
2. Was the Obama Administration too eager to secure a deal that it made too many concessions?
3. Do the components of the agreement seem to provide a reasonable deterrent to Iran developing nuclear weapons?
4. Since the agreement was approved in July 2015, do you have any reason or evidence to feel less safe against Iran's possible aggressive moves in the Middle East and beyond?
5. Do you believe that the GOP opposition to the Iran deal is either purely political, or rooted in real substantive concerns, or a combination of both?

Is There Common Ground?

Much of the disagreement over the Iran nuclear program deal has focused on the non-technical aspects of the agreement, such as whether Iran could hide sites from the inspectors or whether Israel, or for that matter the rest of the Middle East and the United States, was at greater risk today and in 15 years as a consequence of Iran's having either nuclear weapons or much greater financial capability because of the lifting of economic and financial sanctions. There are two major areas of agreement, however. At the heart of a consensus are the science-related aspects of the nuclear technology. Both sides accepted the science behind the nuclear part of the agreement. Thus, there does appear to be common ground with respect to the fact that if Iran does carry out all its obligations under the agreement, it will not be able to produce a nuclear weapon in the next 15 years. Furthermore, if it complies with those aspects of the agreement that are in force after 15 years, the same result will occur in the post–1-year period.

Additional Resources

Borger, Julian, "Iran Nuclear Deal: The Key Points," *The Guardian* (July 14, 2015)

This article lists and briefly describes nine major components of the Iran nuclear program deal.

Bruno, Greg, "Iran's Nuclear Program," Council on Foreign Relations (March 10, 2015)

This comprehensive look at the Iran deal traces the history of Iran's nuclear program, provides some second thoughts on the deal, and lists some unanswered questions.

Castelvecchi, Davide, "Iranian Researchers Welcome Nuclear Deal," *Nature* (vol. 523, no. 7561, 2015)

This article reports that Iranian scientists have spoken in favor of the Iran agreement.

Einhorn, Robert, "Debating the Iran Nuclear Deal," *Brookings Institution* (August 2015)

This report provides a comprehensive discussion of the major areas of disagreement about the Iran nuclear program deal.

Katzman and Kerr, Paul K., "Iran Nuclear Agreement," *Congressional Record Service* (March 7, 2016)

This comprehensive report describes the background to the agreement, the plan itself, reaction to the agreement, and implications for the future of US-Iran relations.

Kerry, John, "Press Availability on Nuclear Deal with Iran,' US Department of State (Vienna, July 14, 2015)

This speech by Secretary of State John Kerry represents his initial comments immediately following the conclusion of negotiations on the Iran nuclear program deal.

McCarthy, Kevin, "21 Reasons the Iran Deal is a Bad Deal," Floor Protocols, US House of Representatives (July 22, 2015)

This statement by the House of Representatives GOP Majority Leader outlines his opposition to the Iran nuclear program deal.

Netanyahu, Benjamin, "Why the Deal is So Bad," *Vital Speeches International* (April 2015)

This speech by Israeli Prime Minister Netanyahu to the American Congress decisively attacks the Iranian deal.

Ploughshares Fund, "Why the Right Wing Is Angry That We Blocked War With Iran," *Huffpost Politics* (May 21, 2016)

This post describes efforts by the right wing to discredit the Ploughshares Fund's support of the Iran nuclear program agreement.

Samore, Gary, "The Iran Nuclear Deal: A Definitive Guide," Belfer Center for Science and International Affairs, Harvard Kennedy School (August 2015)

This 67-page report provides a comprehensive analysis of the Iran nuclear program deal.

Satloff, Robert, "What's Really Wrong with the Iran Nuclear Deal," The Washington Institute for Near East Policy, Reprinted in *The New York Daily News* (July 14, 2015)

The author argues that the agreement will result in Iran becoming a regional power with America's blessing.

The Economist, "The Iran Deal: Not Déjà vu All Over Again" (July 25, 2015)

The article compares the Iran deal to one made with North Korea in 1994, suggesting major differences.

Internet References . . .

Belfer Center for Science and International Affairs, Harvard Kennedy School

http://belfercenter.ksg.harvard.edu/publication/25599/iran_nuclear_deal.html

Cato Institute

www.cato.org/

The Iran Project

www.iranprojectfcsny.org/

Ploughshares Fund

www.ploughshares.org/

Site that lists many websites addressing the Iran nuclear program agreement

www.search.com/Iranian+Nuclear+Agreement

The US Department of State official website on the Iran agreement

www.state.gov

The White House office website for the Iran deal

https://www.whitehouse.gov/issues/foreign-policy/iran-deal

Selected, Edited, and with Issue Framing Material by:
James E. Harf, *Maryville University*
and
Mark Owen Lombardi, *Maryville University*

ISSUE

Is the European Union in Danger of Disintegrating?

YES: **John Feffer**, from "The European Union May be on the Verge of Collapse," *The Nation* (2015)

NO: **Kalin Anev Janse**, from "How the Financial Crisis Made Europe Stronger," *World Economic Forum* (2016)

Learning Outcomes
After reading this issue, you will be able to: • Understand the evolution of the EU as a political and economic entity.   • Identify the issues today that are dividing the EU.   • Speculate as to the impact of the Brexit vote on the future of the EU.   • Evaluate whether the EU is a net positive or negative for the people of Europe and indeed the rest of the world.

ISSUE SUMMARY

YES: Feffer points out that the latest crises facing Europe have caused some to question its viability. He sees an EU that may be one more challenge away from splintering.

NO: Janse argues that the economic crisis has brought the EU together, recognizing the shared benefits of unity.

One of the most enduring and profound outgrowths of the end of World War II and the fall of the Soviet Union has been the gradual development and formation of the European Union. Originally conceived in the halls of power in Britain and France, and even conceived of by Winston Churchill at the end of the war, the European Union began as an economic partnership among the states of Western Europe as a bulwark against the Soviet Union and its hegemony over Eastern Europe. It grew out of a recognition that after hundreds of years of interstate conflict and two world wars that were absolutely devastating to the European continent, Europe could no longer engage in such rivalry and survive. It also grew out of a recognition that the people of Western Europe had more in common with one another than they had differences.

Throughout the decades to come a series of agreements built a deeper and stronger European Union. From the treaty of Paris in 1951 to the Treaty of Rome in 1958 that codified the European Economic Community to the

Maastricht Treaty that created the EU in 1992, European states had engaged in largely economic negotiations designed to integrate economies, break down travel and other barriers to commerce and the movement of people, and more directly created governmental structures that oversee EU policy actions internally and externally.

Through amendments and iterations, the EU has withstood economic and political challenges and the shifting tides of popular opinion on the support for the EU, which in some states is as high as 80 percent while in others as low as 30 percent.

The economic collapse of 2008 and the subsequent crisis have shaken the EU to its very core. The fragility and near default of member states like Ireland, Portugal, and of course Greece, and the issues outstanding with Italy and Spain are just a few of the problem areas. Germany's dominant position as the guarantor of the EU has pitted "have against have not" states and internal cultures regarding spending, austerity, social programs and capitalism, economic development, and protectionism.

Layered on top of this economic instability and how best to respond has been the immigration issue where millions of refugees and foreign workers coming largely from Africa, the Middle East, and Asia have fractured internal states along cultural, linguistic, and religious lines. The terrorist attacks are the most pronounced manifestation of these divisions and less dramatic but equally vexing issues remain about who is a European and what rights should foreign workers and immigrants have. This has led to the rise of rightist parties in selected European countries and a backlash by leftist parties.

The default and exit talk within Greece and other such moves are directly related to the new found fragility of the EU. The recent Brexit vote in Britain that begins the process of their exit from the EU means that the future of the EU is even more precarious in the months ahead as other states explore what this means and their own options.

Feffer in his article sees the latest crisis as a series of threats to the EU that cumulatively over time have weakened it to such an extent that it may simply implode with one more threat, such as the exit of Great Britain.

Janse argues the opposite. Economic crisis and external threats have brought Europeans together in such a shared sense of commitment and recognition that they face and solve these issues as a single entity. Only the fractiousness within some states is the real cause for concern.

YES ⬅

John Feffer

The European Union May Be on the Verge of Collapse

The Complex Federal Project of the EU Has Proven Fragile in the Absence of a Strong External Threat

Europe won the Cold War.

Not long after the Berlin Wall fell a quarter of a century ago, the Soviet Union collapsed, the United States squandered its peace dividend in an attempt to maintain global dominance and Europe quietly became more prosperous, more integrated and more of a player in international affairs. Between 1989 and 2014, the European Union (EU) practically doubled its membership and catapulted into third place in population behind China and India. It currently boasts the world's largest economy and also heads the list of global trading powers. In 2012, the EU won the Nobel Peace Prize for transforming Europe "from a continent of war to a continent of peace."

In the competition for "world's true superpower," China loses points for still having so many impoverished peasants in its rural hinterlands and a corrupt, illiberal bureaucracy in its cities; the United States, for its crumbling infrastructure and a hypertrophied military-industrial complex that threatens to bankrupt the economy. As the only equitably prosperous, politically sound and rule-of-law-respecting superpower, Europe comes out on top, even if—or perhaps because—it doesn't have the military muscle to play global policeman.

And yet, for all this success, the European project is currently teetering on the edge of failure. Growth is anemic at best and socio-economic inequality is on the rise. The countries of Eastern and Central Europe, even relatively successful Poland, have failed to bridge the income gap with the richer half of the continent. And the highly indebted periphery is in revolt.

Politically, the center may not hold and things seem to be falling apart. From the left, parties like Syriza in Greece are challenging the EU's prescriptions of austerity. From the right, Euroskeptic parties are taking aim at the entire quasi-federal model. Racism and xenophobia are gaining ever more adherents, even in previously placid regions like Scandinavia.

Perhaps the primary social challenge facing Europe at the moment, however, is the surging popularity of Islamophobia, the latest "socialism of fools." From the killings at the Munich Olympics in 1972 to the recent attacks at *Charlie Hebdo* and a kosher supermarket in Paris, wars in the Middle East have long inspired proxy battles in Europe. Today, however, the continent finds itself ever more divided between a handful of would-be combatants who claim the mantle of true Islam and an ever-growing contingent who believe Islam—all of Islam—has no place in Europe.

The fracturing European Union of 2015 is not the Europe that political scientist Frances Fukuyama imagined when, I 1989, he so famously predicted "the end of history," as well as the ultimate triumph of liberal democracy and the bureaucracy in Brussels, the EU's headquarters, that now oversees continental affairs. Nor is it the Europe that British Prime Minister Margaret Thatcher imagined when, in the 1980s, she spoke of the global triumph of TINA ("there is no alternative") and of her brand of market liberalism. Instead, today's Europe increasingly harkens back to the period between the two world wars when politicians of the far right and left polarized public debate, economies went into a financial tailspin, anti-Semitism surged out of the sewer and storm clouds gathered on the horizon.

Another continent-wide war may not be in the offing, but Europe does face the potential for regime collapse: that is, the end of the Eurozone and the unraveling of regional integration. Its possible dystopian future can be glimpsed in what has happened in its eastern borderlands. There, federal structures binding together culturally diverse people have had a lousy track record over the last quarter-century. After all, the Soviet Union imploded in 1991; Czechoslovakia divorced in 1993; and Yugoslavia was torn asunder in a series of wars later in the 1990s.

If its economic, political and social structures succumb to fractiousness, the European Union could well follow the Soviet Union and Yugoslavia into the waste bin of failed federalisms. Europe as a continent will remain, its nation-states will continue to enjoy varying degrees of prosperity, but Europe as an idea will be over. Worse yet, if, in the end, the EU snatches defeat from the jaws of its Cold War victory, it will have no one to blame but itself.

The Rise and Fall of TINA

The Cold War was an era of alternatives. The United States offered its version of freewheeling capitalism, while the Soviet Union peddled its brand of centralized planning. In the middle, continental Europe offered the compromise of a social market: capitalism with a touch of planning and a deepening concern for the welfare of all members of society.

Cooperation, not competition, was the byword of the European alternative. Americans could have their dog-eat-dog, frontier capitalism. Europeans would instead stress greater coordination between labor and management and the European Community (the precursor to the EU) would put genuine effort into bringing its new members up to the economic and political level of its core countries.

Then, at a point in the 1980s when the Soviet model had ceased to exert any influence at all globally, along came TINA.

At the time, British Prime Minister Margaret Thatcher and American President Ronald Reagan were ramping up their campaigns to shrink government, while what later became known as globalization—knocking down trade walls and opening up new opportunities for the financial sector—began to be felt everywhere. Thatcher summed up this brave new world with her TINA acronym: the planet no longer had any alternative to globalized market democracy.

Not surprisingly, then, in the post-Cold War era, European integration shifted its focus toward removing barriers to the flow of capital. As a result, the expansion of Europe no longer came with an implied guarantee of eventual equality. The deals that Ireland (1973) and Portugal (1986) had received on accession were now, like the post-World War II Marshall Plan, artifacts of another era. The sheer number of potential new members knocking on Europe's door put a strain on the EU's coffers, particularly since the economic performance of countries like Romania and Bulgaria was so far below the European average. But even if the EU had been overflowing with funds, it might not have mattered, since the new "neoliberal" spirit of capitalism now animated its headquarters in Brussels where the order of the day had become: cut government, unleash the market.

At the heart of Europe, as well as of this new orthodoxy, lies Germany, the exemplar of continental fiscal rectitude. Yet in the 1990s, that newly reunified nation engaged in enormous deficit spending, even if packaged under a different name, to bring the former East Germany up to the level of the rest of the country. It did not, however, care to apply this "reunification exception" to other former members of the Soviet bloc. Acting as the effective central bank for the European Union, Germany instead demanded balanced budgets and austerity from all newcomers (an some old timers as well) as the only effective answer to debt and fears of a future depression.

The rest of the old Warsaw Pact has had access to some EU funds for infrastructure development, but nothing on the order of the East German deal. As such, they remain in a kind of economic halfway house. The standard of living in Hungary, 25 years after the fall of Communism, remains approximately half that of neighboring Austria. Similarly, it took Romania fourteen years just to regain the gross national product (GDP) it had in 1989 and it remains stuck at the bottom of the European Union. People who visit only the capital cities of Eastern and Central Europe come away with a distorted view of the economic situation there, since Warsaw and Bratislava are wealthier than Vienna, and Budapest nearly on a par with it, even though Poland, Slovakia and Hungary all remain economically far behind Austria.

What those countries experienced after 1989—one course of "shock therapy" after another—became the medicine of choice for all EU members at risk of default following the financial crisis of 2007 and then the sovereign debt crisis of 2009. Forget deficit spending to enable countries to grow their way out of economic crisis. Forget debt renegotiation. The unemployment rate in Greece and Spain now hovers around 25 percent, with youth unemployment over 50 percent and all the EU members subjected to heavy doses of austerity have witnessed a steep rise in the number of people living below the poverty line. The recent European Central Bank announcement of "quantitative easing"—a monetary sleight-of-hand to pump money into the Eurozone—is too little, too late.

The major principle of European integration has been reversed. Instead of Eastern and Central Europe catching up to the rest of the EU, pockets of the "west" have begun to fall behind the "east" The GDP per capita of Greece, for example, has slipped below that of Slovenia and, when measured in terms of purchasing power, even Slovakia, both former Communist countries.

The Axis of Illiberalism

Europeans are beginning to realize that Margaret Thatcher was wrong and there are alternatives—to liberalism *and* European integration. The most notorious example of this new illiberalism is Hungary.

On July 26, 2014, in a speech to his party faithful, Prime Minister Viktor Orban confided that he intended a thorough reorganization of the country. The reform model Orban had in mind, however, had nothing to do with the United States, Britain, or France. Rather, he aspired to create what he bluntly called an "illiberal state" in the very heart of Europe, one strong on Christian values and light on the libertine ways of the West. More precisely, what he wanted was to turn Hungary into a mini-Russia or mini-China.

"Societies founded upon the principle of the liberal way," Orban intoned,"will not be able to sustain their world competitiveness in the following years, and more likely they will suffer a setback, unless they will be able to substantially reform themselves." He was also eager to reorient to the east, relying ever less on Brussels and ever more on potentially lucrative markets in and investments from Russia, China and the Middle East.

That July speech represented a truly Oedipal moment, for Orban was eager to drive a stake right through the heart of the ideology that had fathered him. As a young man more than 25 years earlier, he had led the Alliance of Young Democrats—Fidesz—one of the region's most promising liberal parties. In the intervening years, sensing political opportunity elsewhere on the political spectrum, he had guided Fidesz out of the Liberal International and into the European People's Party, alongside German Chancellor Angela Merkel's Christian Democrats.

Now, however, he was on the move again and his new role model wasn't Merkel, but Russian President Vladimir Putin and his iron-fisted style of politics. Given the disappointing performance of liberal economic reforms and the stinginess of the EU, it was hardly surprising that Orban had decided to hedge his bets by looking east.

The European Union has responded by harshly criticizing Orban's government for pushing through a raft of constitutional changes that restrict the media and

compromise the independence of the judiciary. Racism and xenophobia are on the uptick in Hungary, particularly anti-Roma sentiment and anti-Semitism. And the state has taken steps to reassert control over the economy and impose controls on foreign investment.

For some, the relationship between Hungary and the rest of Europe is reminiscent of the moment in the 1960s when Albania fled the Soviet bloc and, in an act of transcontinental audacity, aligned itself with Communist China. But Albania was then a marginal player and China still a poor peasant country. Hungary is an important EU member and China's illiberal development model, which has vaulted it to the top of the global economy, now has increasing international influence. This, in other words, is no Albanian mouse that roared. A new illiberal axis connecting Budapest to Beijing and Moscow would have far-reaching implications.

The Hungarian prime minister, after all, has many European allies in his Euroskeptical project. Far right parties are climbing in the polls across the continent. With 25 percent of the votes, Marine Le Pen's National Front, for instance, topped the French elections for the European parliament last May. In local elections in 2014, it also seized twelve mayoralties, and polls show that Le Pen would win the 2017 presidential race if it were held today. In the wake of the *Charlie Hebdo* shootings, the National Front has been pushing a range of policies from reinstating the death penalty to closing borders that would deliberately challenge the whole European project.

In Denmark, the far-right People's Party also won the most votes in the European parliamentary elections. In November, it topped opinion polls for the first time. The People's party has called for Denmark to slam shut its open-door policy toward refugees and re-introduce border controls. Much as the Green Party did in Germany in the 1970s, groupings like Great Britain's Independence Party, the Finns Party and even Sweden's Democrats are shattering the comfortable conservative-social democratic duopoly that has rotated in power throughout Europe during the Cold War and in its aftermath.

The Islamophobia that has surged in the wake of the murders in France provides an even more potent arrow in the quiver of these parties as they take on the mainstream. The sentiment currently expressed against Islam—at rallies, in the media and in the occasional criminal act— recalls a Europe of long ago, when armed pilgrims set out on a multiple crusades against Muslim powers, when early nation-states mobilized against the Ottoman Empire, and when European unity was forged not out of economic interest or political agreement but as a "civilizational" response to the infidel.

The Europe of today is, of course, a far more multicultural place and regional integration depends on "unity in diversity," as the EU's motto puts it. As a result, rising anti-Islamic sentiment challenges the inclusive nature of the European project. If the EU cannot accommodate Islam, the complex balancing act among all its different ethnic, religious and cultural groups will be thrown into question.

Euroskepticism doesn't only come from the right side of the political spectrum. In Greece, the Syriza party has challenged liberalism from the left, as it leads protests against EU and International Monetary Fund austerity programs that have plunged the population into recession and revolt. As elsewhere in Europe, the far right might have taken advantage of this economic crisis, too, had the government not arrested the Golden Dawn leadership on murder and other charges. In parliamentary elections on Sunday, Syriza won an overwhelming victory, coming only a couple seats short of an absolute majority. In a sign of the ongoing realignment of European politics, that party then formed a new government not with the center-left, but with the right-wing Independent Greeks, which is similarly anti-austerity but also skeptical of the EU and in favor of a crackdown on illegal immigration.

European integration continues to be a bipartisan project for the parties that straddle the middle of the political spectrum, but the Euroskeptics are now winning votes with their anti-federalist rhetoric. Though they tend to moderate their more apocalyptic rhetoric about "despotic Brussels" as they get closer to power, by pulling on a loose thread here and another there, they could very well unravel the European tapestry.

When the Virtuous Turn Vicious

For decades, European integration created a virtuous circle—prosperity generating political support for further integration that, in turn, grew the European economy. It was a winning formula in a competitive world. However, as the European model has become associated with austerity, not prosperity, that virtuous circle has turned vicious. A challenge to the Eurozone in one country, a repeal of open borders in another, the reinstitution of the death penalty in a third—it, too, is a process that could feed on itself, potentially sending the EU into a death spiral, even if, at first, no member states take the fateful step of withdrawing.

In Eastern and Central Europe, the growing crew who distrust the EU complain that Brussels has simply taken the place of Moscow in the post-Soviet era. (The Euroskeptics in the former Yugoslavia prefer to cite Belgrade.) Brussels, they insist, establishes the parameters of economic policy that its member states ignore at their peril, while Eurozone members find themselves with ever less control over their finances. Even if the edicts coming from Brussels are construed as economically sensible and possessed of a modicum of democratic legitimacy, to the Euroskeptics they still represent a devastating loss of sovereignty.

In this way, the same resentments that ate away at the Soviet and Yugoslav federations have begun to erode popular support for the European Union. Aside from Poland and Germany, where enthusiasm remains strong, sentiment toward the EU remains lukewarm at best across much of the rest of the continent, despite a post-euro crisis rebound. Its popularity now hovers at around 50 percent in many member states and below that in places like Italy and Greece.

The European Union has without question been a remarkable achievement of modern statecraft. It turned a continent that seemed destined to wallow in "ancestral hatreds" into one of the most harmonious regions on the planet. But as with the portmanteau states of the Soviet Union, Yugoslavia, and Czechoslovakia, the complex federal project of the EU has proven fragile in the absence of a strong external threat like the one that the Cold War provided. Another economic shock or a coordinated political challenge could tip it over the edge.

Unity in diversity may be an appealing concept, but the EU needs more than pretty rhetoric and good intentions to stay glued together. If it doesn't come up with a better recipe for dealing with economic inequality, political extremism and social intolerance, its opponents will soon have the power to hit the rewind button on European integration. The ensuing regime collapse would not only be a tragedy for Europe, but for all those who hope to overcome the dangerous rivalries of the past and provide shelter from the murderous conflicts of the present.

JOHN FEFFER, co-director of Foreign Policy in Focus at the Institute for Policy Studies, is the author of *North Korea, South Korea: U.S. Policy at a Time of Crisis* (Seven Stories). His past essays, including the essay for Tomdispatch.com, can be read at his website.

Kalin Anev Janse

How the Financial Crisis Made Europe Stronger

Europe is back in the game. Having suffered the worst financial and economic crisis of the last 80 years, Europe took decisive action to improve its public finances, push through deep reforms, and establish new institutions to manage and prevent crises better. The changes are structural, long-lasting, and make Europe more competitive. Europe is stronger, better equipped and in the midst of ambitious new financial and economic initiatives.

Europe Is Stronger Than You Think

The global crisis hit Europe twice. The first strike came from abroad in 2007. In the United States, markets had ignored credit risk in subprime mortgage markets. A lack of financial supervision allowed opaque financial instruments to flourish, aggravating the problem. As a result, the U.S. banking system underwent a dramatic bail-out in September 2008. European banks suffered in the fallout. Two years later, a second crisis erupted in the euro area. Years of unsustainable government policies had caused deficits and debt burdens to mushroom and bloated pre-crisis wages and housing prices. As the situation worsened, Europe took courageous decisions to put the continent back on firm footing.

Five key responses combined to steer Europe out of the crisis (figure 1).

First, crisis-hit countries like Ireland and Spain pushed through badly needed reforms, improving public finances and increasing competitiveness.

Second, EU economic governance was strengthened. The European Commission received new powers to enforce the Stability and Growth Pact, issue country-specific recommendations (the 'European semester'), and underline obstacles that need to be removed to foster growth. Eurostat, Europe's statistical agency, became more powerful in cross-checking and challenging the data received from each country.

Third, euro area countries created the European Stability Mechanism (ESM), and its predecessor the European Financial Stability Facility. These institutions were a great success: no country was forced to leave the euro area, the cash-for-reform approach worked, and growth accelerated in countries that implemented reforms. Importantly, no European taxpayer money was spent on the rescue programmes. The ESM raises the funds needed in capital markets through highly-rated bonds, and passes on the low funding cost to programme countries at rates today of around 1–2%. This approach produces budget savings for programme countries, particularly Greece, which would pay much more if it were to tap capital markets independently.

Fourth, the European Central Bank's unconventional measures helped. The ECB expanded its balance sheet like the FED, Bank of Japan and Bank of England, provided unlimited liquidity for banks, started a bond purchasing programme to avoid low inflation, which also made it easier for banks to lend and boost investor sentiment. The euro weakened which helped to increase exports.

Finally, the Banking Union was established: new institutions were created to monitor macro-prudential risks and supervise banks, securities markets and insurance companies. The Single Supervisory Mechanism is the centrepiece of this initiative; it oversees the 130 largest euro area banks. During the crisis, EU banks also padded out their capital, increasing their capital base by €600 billion since 2008.

And the Results?

The results are impressive (figure 2). The crisis-hit countries implemented radical reforms. They now head multiple international rankings, earning the sobriquet 'reform champions'. Many of Europe's crisis countries—Greece, Ireland, Portugal, and Spain—ended up in the top five of 34 OECD members (a club for the most developed countries)

Figure 1

Europe's five key responses to crisis

1. Budget consolidation & structural reforms
   - Crisis countries are reform champions
   - Fiscal deficit reduction policies are paying of: Euro Area is doing better than US, Japan and UK
   - Internal devaluations are restoring competitiveness: nominal unit labor costs went down & make countries competitive

2. Closer economic coordination in currency union
   - Tightened surveillance of fiscal policies: Stricter Stability and Growth Pact & "Fiscal compact"
   - European semester: Country draft budgets submitted to the European Commission
   - New Procedure to avoid macroeconomic imbalances

3. Firewalls against crises: EFSF and ESM
   - Maximum lending capacity of €700 bn
   - Disbursed to five countries so far: €254.5 bn
   - Ireland, Portugal, Spain have exited programs
   - Cyprus will exit in March 2016
   - Greece is a special case: €86 billion new program in 2015

4. An active monetary policy
   - ECB expanded its balance sheet, provided long-term liquidity to banks, purchased government and other securities, introduced low interest rates and negative deposit rates
   - Banks incentive to lend more: boost investor appetite
   - Weaker euro improves external competitiveness

5. Strengthening the banking system
   - New Institutions: European Systemic Risk Board (ESRB) to monitor macro prudential risks, new EU supervisory bodies for banks, securities markets and insurance companies European Banking Authority (EBA), European Securities and Markets Authority (ESMA) and European Insurance and Occupational Pensions Authority (EIOPA).
   - Begin of Banking Union: Single Supervisory Mechanism
   - More bank capital: EU banks raised more than €600 billion capital since beginning of crisis in 2008

---

in recognition of their structural reforms. They did so by improving their public finances, reducing deficits, and cutting labour costs to make themselves more competitive. The euro area outperformed the US, Japan and the UK in fiscal terms: the aggregate euro area budget deficit was significantly smaller than these three peers. And finally, some of the crisis countries are becoming growth leaders. In 2015, Ireland hit a record high 6.9% GDP growth and Spain 3.2%. Ireland's growth matched China's, while Spain's is almost a third again as high as the US's (2.5%) over the same period. These are extraordinary achievements.

Three Big Initiatives to Shape Europe's Economic Future

During the crisis, Europe picked up the pace of policy reform and integration with three big initiatives that merit close attention. They will set the economic and financial agenda for the next five to 10 years and will be key to Europe's economic future (figure 3).

Completion of the Banking Union

Europe's Banking Union is moving forward. On 1 January 2016, the new European mechanism to resolve failing banks went live. The Single Resolution Mechanism's (SRM) goal is to resolve distressed banks at the lowest cost to taxpayers. It includes the participation of the private sector, such as shareholders, junior and senior creditors and unsecured and very large depositors, according to the bail-in rules under the Bank Recovery and Resolution Directive (BRRD). Additionally, SRM-covered banks will most likely need to provide €55 billion in funding over the next eight years to create the Single Resolution Fund (SRF). While this means higher costs for banks, it reduces the need for taxpayers to cover bank failures.

The final step to completing the Banking Union is a common deposit guarantee scheme similar to the US's Federal Deposit Insurance Corporation (FDIC). This year, a European Commission proposal to establish a European Deposit Insurance Scheme (EDIS) has been hotly debated. The aim is to guarantee individual deposits of up to

Figure 2

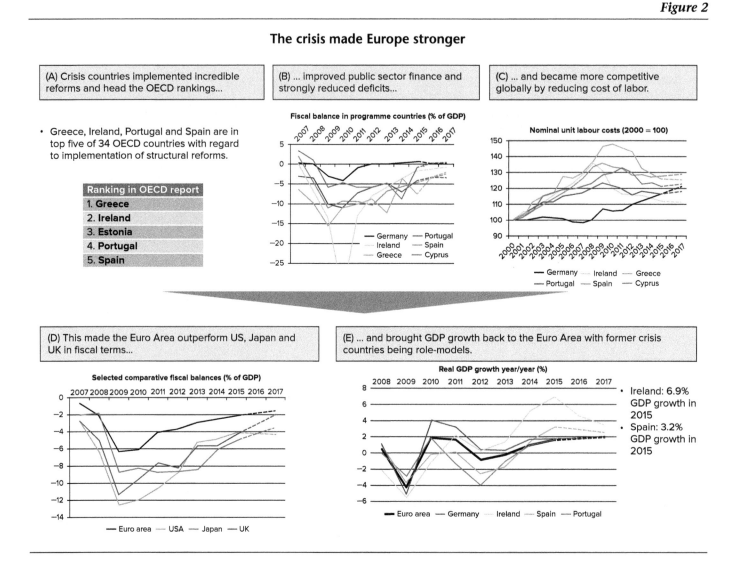

The crisis made Europe stronger

(A) Crisis countries implemented incredible reforms and head the OECD rankings...

(B) ... improved public sector finance and strongly reduced deficits...

(C) ... and became more competitive globally by reducing cost of labor.

- Greece, Ireland, Portugal and Spain are in top five of 34 OECD countries with regard to implementation of structural reforms.

Ranking in OECD report
1. Greece
2. Ireland
3. Estonia
4. Portugal
5. Spain

Fiscal balance in programme countries (% of GDP)

Germany — Portugal
Ireland — Spain
Greece — Cyprus

Nominal unit labour costs (2000 = 100)

Germany — Ireland — Greece
Portugal — Spain — Cyprus

(D) This made the Euro Area outperform US, Japan and UK in fiscal terms...

(E) ... and brought GDP growth back to the Euro Area with former crisis countries being role-models.

Selected comparative fiscal balances (% of GDP)

— Euro area — USA — Japan — UK

Real GDP growth year/year (%)

- Ireland: 6.9% GDP growth in 2015
- Spain: 3.2% GDP growth in 2015

— Euro area — Germany — Ireland — Spain — Portugal

€100,000 at all banks in the euro area—benefiting around 340 million citizens in 19 countries. It is far from a done deal, but if agreed and implemented, the US and Europe will share a similar approach to supervision, stress testing, resolution, and deposit guarantees.

A Deeper Capital Markets Union

With the Banking Union advancing at high speed, Europe launched another big project last year: the Capital Markets Union (CMU). The CMU's goal is to create deeper and more integrated capital markets. Traditionally, Europe has been dominated by bank financing: when an entrepreneur needed funding, banks were the go-to partner. During the

crisis, banks tried to reduce their risk exposure and, as a result, financing for entrepreneurs, small- and medium-sized enterprises (SMEs), and some corporations, dried up. The CMU aims to address this issue.

The differences on the two sides of the Atlantic are large. Banking Union bank assets total €30.2 trillion compared to just $13.4 trillion in the US, a gap explained by the source of financing (figure 4). Euro area banking credit reaches some 170% of GDP, against only 45% of GDP in the US. Stock market capitalisation and debt securities are respectively some 45% and 10% of GDP in the euro area compared to 110% and 45% in the US.

The CMU initiative could transform Europe's capital markets over the 2016–2019 period. The CMU

Figure 3

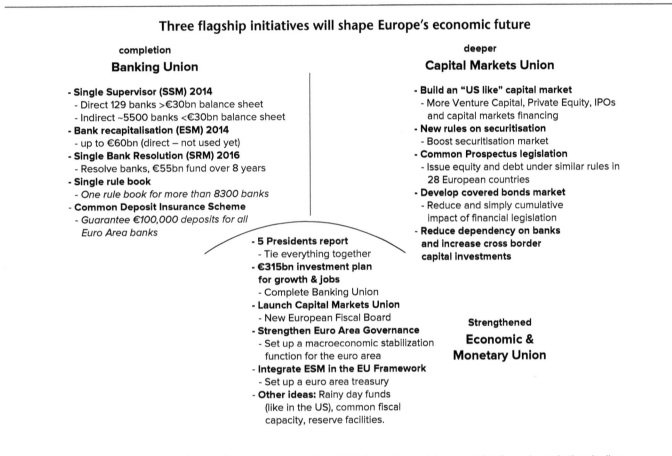

Three flagship initiatives will shape Europe's economic future

completion
Banking Union

- **Single Supervisor (SSM) 2014**
  - Direct 129 banks >€30bn balance sheet
  - Indirect ~5500 banks <€30bn balance sheet
- **Bank recapitalisation (ESM) 2014**
  - up to €60bn (direct – not used yet)
- **Single Bank Resolution (SRM) 2016**
  - Resolve banks, €55bn fund over 8 years
- **Single rule book**
  - *One rule book for more than 8300 banks*
- **Common Deposit Insurance Scheme**
  - *Guarantee €100,000 deposits for all Euro Area banks*

deeper
Capital Markets Union

- **Build an "US like" capital market**
  - More Venture Capital, Private Equity, IPOs and capital markets financing
- **New rules on securitisation**
  - Boost securitisation market
- **Common Prospectus legislation**
  - Issue equity and debt under similar rules in 28 European countries
- **Develop covered bonds market**
  - Reduce and simply cumulative impact of financial legislation
- **Reduce dependency on banks and increase cross border capital investments**

- **5 Presidents report**
  - Tie everything together
- **€315bn investment plan for growth & jobs**
  - Complete Banking Union
- **Launch Capital Markets Union**
  - New European Fiscal Board
- **Strengthen Euro Area Governance**
  - Set up a macroeconomic stabilization function for the euro area
- **Integrate ESM in the EU Framework**
  - Set up a euro area treasury
- **Other ideas:** Rainy day funds (like in the US), common fiscal capacity, reserve facilities.

Strengthened
Economic & Monetary Union

Dates indicate the dates the initiatives became operational, initiatives without dates are under discussion or in the pipeline

Source: 5 Presidents reports 2015, CMU Green Paper, CMU Action Plan European Commission, ESM Research

is designed to reduce dependency on banks. It should increase the flow of funding from Venture Capital to Private Equity, to IPOs and eventually to equity and debt capital market financing. It aims to standardise prospectus legislation across countries (currently there are large differences across European countries, not to mention the different language requirements), deepen the covered bond market, develop a stronger securitisation market, align bankruptcy laws, and stimulate more cross-border capital investments. Despite the challenges, Europe has strong foundations to build on. It is home to some of the largest pension funds in the world and countries with strong household savings habits such as Sweden, Germany, and the Netherlands.

Strengthen Economic and Monetary Union (EMU)

The Presidents of the European Central Bank, Eurogroup, European Commission, European Council and European Parliament presented a plan, the Five Presidents Report, to further deepen Economic and Monetary Union over the next 10 years, or through 2025. The goal is to pull all these initiatives together, complete the Banking Union, make progress on the Capital Markets Union, and introduce new initiatives for the euro area. These initiatives could include establishing a new fiscal council for the euro area, a euro area macroeconomic stabilisation function, or even a euro area treasury.

Combining these initiatives has one big advantage: it would allow for more risk reduction and better

Figure 4

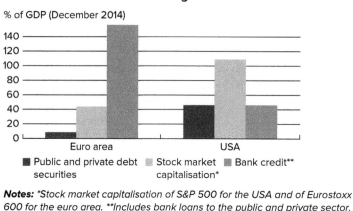

Sources of financing Euro Area vs US

% of GDP (December 2014)

- Public and private debt securities
- Stock market capitalisation*
- Bank credit**

Notes: *Stock market capitalisation of S&P 500 for the USA and of Eurostoxx 600 for the euro area. **Includes bank loans to the public and private sector.*

Source: "La Caixa" Research, based on data from Bloomberg, Sifma, the ECB and the Federal Reserve.

risk-sharing across the euro area and reduce the need for the public sector to use taxpayers' money for economic and financial shock absorption. In the US, economic and business cycles are smoothed out across the 50 states to a much greater degree than in Europe, with market mechanisms doing the lion's share of the work (figure 5). Fiscal transfers play only a limited role. About 75% of economic

and business cycle shocks in the US are smoothed, 62% by financial market transactions and only about 13% through fiscal transfers and taxpayers' contributions. This underscores how important it is for Europe to complete Banking Union, deepen CMU, and strengthen the EMU.

These three big initiatives, if implemented successfully over the next five to 10 years, will define Europe's economic future in the long-run.

Figure 5

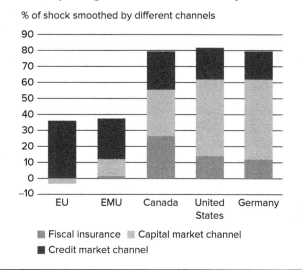

Deepening EMU and shock absorption

% of shock smoothed by different channels

- Fiscal insurance
- Capital market channel
- Credit market channel

Source: Bruegel and IMF

Can Europe Afford to Deviate from Its Path?

There are some headwinds, including the migrant crisis and the June 2016 UK referendum on EU membership. European policy makers, enterprises and citizens need to turn these events to their advantage.

Immigration can stimulate growth, with a small but positive short-term impact on EU GDP. It can also help compensate for weak long-term demographics. If executed well, the benefits could be significant. Take Germany for example. The country has around 81 million citizens today, but that figure is seen declining 13% to 67 million by 2060. This will hit economic growth. It will also undermine the sustainability of public finances, as the share of over-65-year-olds rises to 33% in 2060 from 20% and puts pressure on the pay-as-you-go pension scheme. According to the German Statistical agency, more immigration could bring the population to 73 million by 2060. At today's GDP per capita, that difference of six million people means

Figure 6

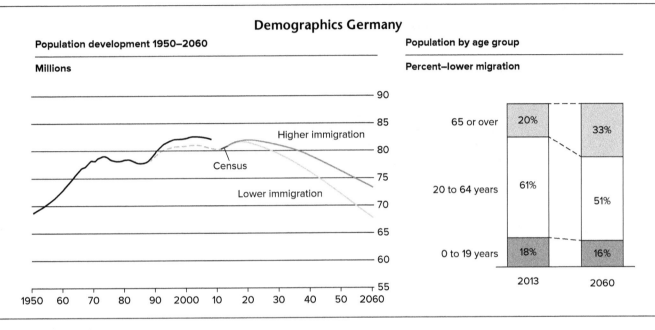

Demographics Germany

Population development 1950–2060

Millions

Population by age group

Percent–lower migration

Source: German Statistical Agency

€210 billion in GDP potential. However, the devil is in the detail. Europe must protect its outer borders well, bring in the right skill sets, and provide successful economic and social integration programmes. Managing the fear of immigration and growing populism present the biggest challenges. Countries that know how to manage most of these elements well, like Canada and Luxembourg, have reaped the rewards.

The UK referendum and similar discussions in other member states are an important part of a healthy democratic process, but it is essential to consider the implications. Can European countries really afford not to be financially and economically interconnected and integrated?

Europe's percentage of global GDP declined to 23% in 2010 from 32% in 1970 (figure 7) and, if no new policy initiatives are undertaken, could fall to 9% by 2050. This will have an enormous economic, financial, and geopolitical impact. Europe, and North America, will decline in size and influence. To counter this trend, further integration, not fragmentation, is needed.

Europe's best way forward is to stay together as a strong, well-integrated, and deeply interconnected continent. Over the past 70 years, Europe has achieved unprecedented peace, stability, and wealth creation. It is the world's largest single market. It is in its best interests to remain the global financial and economic leader.

A Bright Future

Europe not only endured the last crisis, it capitalised upon it. Europe's five key policy responses meant the continent emerged stronger. The results, such as in Ireland or Spain, are impressive. Completing the Banking Union, deepening the Capital Markets Union, and strengthening Economic and Monetary Union will make Europe's economy more competitive, diverse, and robust. It is crucial that Europe not deviate from this path and that it continues to turn headwinds, like immigration, into opportunities.

Europe has put together the building blocks for a bright future; its citizens can justly be proud of these achievements. Realising this ambitious agenda will make Europe stronger, more influential, and ensure it maintains its position as an economic powerhouse.

Figure 7

Can European countries really afford not to be financially and economically interconnected and integrated?

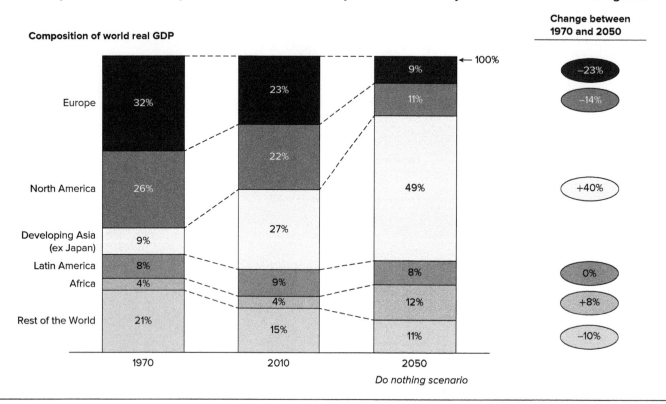

Source: IMF World Economic Outlook; Citi Investment Research and Analysis

KALIN ANEV JANSE is Secretary General of the European Stability Mechanism. He is also Secretary General of the European Financial Stability Facility, a position he has held since 2011. He is responsible for the Corporate Governance (Board of Directors and shareholder relations), Corporate Finance and Accounting, Operations, Human Resources, and Information Technology. He is also a member of the Board of Directors of the European Sovereign Bond Investment Facility.

EXPLORING THE ISSUE

Is the European Union in Danger of Disintegrating?

Critical Thinking and Reflection

1. What are the positive and negative implications of a weakened or eliminated European Union?
2. Does the end of the EU mean greater conflict and more economic and political instability?
3. Which states would see such a decline as an opportunity for greater power and influence in Europe?
4. Can outside actors influence this process and if so in what ways?

Is There Common Ground?

Given the fact that Europe has seen more destructive wars over the last 200 years than any other region on earth, the past decades of relative peace and security have been remarkable. Whether one is a proponent of the EU or one who sees it as positive but unnecessary or even those who wish it to go away completely, analysts can agree that the period of EU evolution and development has brought relative peace and economic security to millions of Europeans for the first time in centuries and that has been a positive development by any ideological or political measure.

Additional Resources

"2016 and the European Union-Getting Stronger Together?" *One Europe* (January 16, 2016)

A case is made for the EU.

Bentley, Guy, "Brexit the Movie Unmasks EU as a Dying Protectionist Bloc," *The daily Caller* (May 22, 2016)

The article makes the case for a dying EU.

Buchanan, Pat, "Is the European Union Dying?" *Chronicles* (March 13, 2015)

The author examines the trends in Europe.

Waterfield, Bruno, "The European Dream is Dying State by State," *Business Insider* (June 24, 2015)

The author discusses the decline of the EU.

Internet References . . .

Business Insider

www.businessinsider.com

Foreign Policy

http://foreignpolicy.com/

One Europe

http://one-europe.info/

Selected, Edited, and with Issue Framing Material by:
James E. Harf, *Maryville University*
and
Mark Owen Lombardi, *Maryville University*

ISSUE

Is China the Next Superpower?

YES: **Jonathan Watts,** from "China: Witnessing the Birth of a Superpower," *The Guardian* (2012)

NO: **Jonathan Adelman,** from "China's Long Road to Superpower Status," *U.S. News and World Report* (2014)

Learning Outcomes

After reading this issue, you will be able to:

- Understand China's rise to economic giant and how that has impacted the global system.
- Evaluate China's strengths and weaknesses as it relates to its potential superpower status.
- Understand what the traits of a superpower are.
- Evaluate whether China is a superpower or is on the road to such status eventually.

ISSUE SUMMARY

YES: Watts articulates the view that China is showing all the signs of emerging superpower status economically and militarily.

NO: Adelman details the areas where China lags far beyond the United States and other countries and shows no signs of solving.

Since the end of the cold war and fall of the Soviet Union in 1991, analysts, politicians, and observers have speculated whether another state will emerge as a challenger to American hegemony in the world. Much of that speculation has always centered on China. Since the rise of Deng Xiao Ping in the late 1970s in the aftermath of the death of Mao and the aborted coup of the Gang of Four, China has been engaged in a series of breathtaking economic and social reforms that have transformed the country. Once isolated, agrarian, communist, and averse to development, China has been the fastest growing economy on earth for nearly 25 years. Deng placed the Chinese economy on an export-oriented, manufacturing production footing that has seen double-digit economic growth from the 1980s through to 2015. No economy on earth has expanded more, produced more goods, or exported more products during that time. A few macro statistics tell part of the story. China's gross domestic product (GDP) per capita rose from less than $1000/year to over $11,000 per year. Real GDP growth has averaged 7–10 percent per year for over two decades.

With this economic expansion, China has expanded its military. It has upgraded its technology, expanded its naval capacity, and enhanced its military relationships with states across the globe. It is much better positioned than in the past to economically and strategically challenge states in its hemisphere and ultimately the United States. In fact, most analysts agree that if a state is capable of ultimately challenging the United States in this century it is or will be China.

Yet, China watchers also point to a host of internal issues facing the Chinese regime. While the standard of living has risen dramatically for some 350 million of its citizens another 1 billion still live in relative poverty. While its economic output is impressive, its level of inventiveness and technological sophistication still lags behind the United States and the West. Also, while its military continues to be modernized, it still lacks global power projection capability in the form of a global navy or air force capable of protecting and enforcing its interests. It is in many ways a regional military power only.

Jonathan Watts traveled extensively throughout China and argues that despite structural and political issues, China exhibits all of the signs of an emerging superpower. From its economic output to its commitment to development, from its ability to channel massive resources to its centrality in the global economy, China will emerge as the next superpower despite the challenges it must overcome.

Jonathan Adelman argues that China's rise has been impressive but its structural flaws will keep it from attaining superpower status. He contends that its centralized economic model, its political dictatorship of the party, and its inability to relinquish controls as witnessed by the Tiananmen uprising doom China to a second-tier status. Adelman also intimates that the move from manufacturing to high-technology global economy leaves China poorly positioned to take advantage of this shift and that will mean further erosion of its success.

YES ⤹

Jonathan Watts

China: Witnessing the Birth of a Superpower

When I moved to Beijing in August 2003, I believed I had the best job in the world: working for my favourite newspaper in the biggest nation at arguably the most dramatic phase of transformation in its history. In the past decade, it has given me a front-row seat to watch 200-odd years of industrial development playing at fast forward on a continent-wide screen with a cast of more than a billion.

That said, I am glad my daughters were young and easy to please back then or we might well have taken the first plane out of the country. As we drove from the airport to our apartment, I tried to maintain an upbeat chatter. "Look at all the kites," I said as we passed Chaoyang park, even though my heart sank at the tatty buildings, endless construction sites and stultifying haze. In my head, I asked myself: "Have I done the right thing for them?"

We had come from Japan—a democratic, comfortable, polite, hygiene-obsessed, orderly, first-world nation—to the grim-looking capital of a developing, nominally communist country that looked and sounded like a giant building site. For all my enthusiasm, my family must have felt we were taking a step backwards in lifestyle.

It required an adjustment of preconceptions. Like many newcomers, I delighted at discoveries of Chinese literature and Daoist philosophy, Beijing parks, the edgy eccentricity of Dashanzi and the glorious mix of classicism and obscenity in the Chinese language, though I never managed to master it. The mix of communist politics and capitalist economics appeared to have created a system designed to exploit people and the environment like never before. It was so unequal that Japan appeared far more socialist by comparison. And it was changing fast. As swaths of the capital were being demolished and rebuilt for the Olympics, there was an exhilarating (and sometimes disorientating) sense of mutability. Everything seemed possible.

Looking back over the stories that followed, it is hard to believe so much could be compressed into such a short span of time—the outbreaks of Sars and bird flu, the attempted assassination of the president of Taiwan, deadly unrest in Tibet, the devastating earthquake in Sichuan, murderous ethnic violence in Xinjiang, Tibetan self-immolations, as well as the huge regional stories: two tsunamis—in 2004 in the Indian Ocean and last year in the Pacific, a multiple nuclear meltdown in Fukushima, and the protracted rattling of nuclear sabres on the Korean peninsula.

. . . My focus has been on development and its impact on individuals and the environment. In 2003, China had the world's sixth-biggest GDP. It passed France in 2004, Britain in 2006, Germany in 2009, and Japan in 2011. On current course, it will replace the US as No. 1 within the next 15 years. It is already top in terms of internet population, energy consumption and the size of its car market.

A Decade of Dominance

The primary driver for change has been the movement of people. Over the past nine years 120 million Chinese people—almost twice the population of the UK—have moved from the countryside to the city. This mind-boggling shift has its problems, as I found in Chongqing, but for the most part, China appears to have avoided the worst of the poverty, crime and ghettoes seen in other rapidly urbanising countries.

Yet it also seems more brittle, perhaps because of the other big economic engine: infrastructure investment. In this period, China has been re-wired and re-plumbed. There has been an extraordinary expansion of power, transport and communication networks that have linked the nation like never before: west-east gas pipelines, south-north water diversion, hundreds of airports and a massive new electricity grid linking wind and solar power plants in the deserts to power-hungry consumers in cities and industrial plants.

This has been a decade of cement and steel, a time when economic development has pushed into the most remote corners of China with a series of prestige projects: the world's highest railway, the biggest dam, the longest bridge, putting a man into space, the most ambitious hydroengineering project in human history and, of course, hosting—and dominating—the Olympics for the first time.

It has been a privilege to watch this redistribution of wealth and power from the developed to the developing world. On a global level, such a shift will require nothing less than a grand accommodation—or a violent conflict. Beijing appears to be preparing for both. Other news during this period showed a hardening of China's military muscle: a breakthrough satellite-killing missile test, the launch of a first aircraft carrier, the development of a new stealth fighter and a steady increase in military spending.

Criticism has rarely been appreciated. All too often, there have been flare-ups of anti-foreign media hostility. Some of my colleagues in other media organisations have received death threats. I never expected China to be an easy place to work as a journalist. For political and cultural reasons, there is a huge difference in expectations of the media. For historical and geo-strategic reasons, there is a lingering distrust of foreign reporters.

Run-ins with the police, local authorities or thugs are depressingly common. I have been detained five times, turned back six times at roadblocks (including during several efforts to visit Tibetan areas) and physically manhandled on a couple of occasions. Members of state security have sometimes followed interviewees and invited my assistants "out for tea," to question them on who I was meeting and where I planned to visit. Censors have shut down a partner website that translated Guardian articles into Mandarin. Police have twice seized my journalist credentials, most recently on this year's World Press Freedom Day after I tried to interview the blind human-rights activist Chen Guangcheng in hospital. When that happened, I debated with another British newspaper reporter who was in the same position about whether to report on the confiscation. He argued that it was against his principles for journalists to become part of the story. I used to believe the same, but after nine years in China, I have seen how coverage is influenced by a lack of access, intimidation of sources and official harassment. I now believe reporters are doing a disservice to their readers if they fail to reveal these limitations on their ability to gather information.

Yes, there is often negative coverage and yes, many of the positive developments in China are underemphasised. But I don't think it does the country's international image any favours to clumsily choke access to what is happening on the ground. . . .

Foreign ministry officials often tell me China is becoming more open and, indeed, there have been steps in that direction. But restrictions create fertile ground for rumour-mongering. One of the biggest changes in this period has been the spread of ideas through mobile phones and social networks. The 513 million netizens in China (up from 68 million in 2003) have incomparably greater access to information than any previous generation and huge numbers now speak out in ways that might have got them threatened or detained in 2003. Microblogs are perhaps nowhere more influential than in China because there is so little trust of the communist-controlled official media. . . .

Heroism and Brutality

But there were also stories of success, heroism and inspiration as a nation embraced new wealth and battled for new ideas: the business empires built by enlightened philanthropists such as Yin Mingshan of Lifan auto, the internet fortunes accrued by entrepreneurs such as Jack Ma of Alibaba and Robin Li of Baidu.

Compared with nine years ago, people in China have more freedom to shop, to travel and to express their views on the internet. The Communist party tolerates a degree of criticism, but step over the invisible line of what is acceptable and the consequences are brutal. In my first years in China, I interviewed several outspoken opponents—Liu Xiaobo, Gao Zhisheng, Hu Jia and Teng Biao. I was impressed back then that they were at liberty to speak out. It seemed like the act of a confident government. But all of them have subsequently been locked up and, in at least two cases, tortured.

The blame for that surely lies with the authorities. But I have sometimes felt pangs of guilt. I first interviewed Ai Weiwei in the summer of 2007 for an Olympic preview. He was one of the creators of the "bird's nest" stadium and I was expecting him to tell me how proud he would be when it was unveiled at the opening ceremony. Instead, he told me he would not attend in protest at the "disgusting" political conditions in the one-party state and then launched into a withering assault on propaganda. It was the first time he had expressed such views to the foreign media—a great scoop, but also one fraught with risk. At the end of the interview, I cautioned the artist: "Are you sure you want to say this? It could get you into a great deal of trouble with the authorities."

"Absolutely," he replied. "I only wish I could say it more clearly."

Despite that confirmation and the similarly critical comments he subsequently made to other media organisations, I felt partly responsible when Ai was detained last year.

Whether the repression is getting better or worse has been a constant question with few clear answers. My feeling is that China has become a less tolerant country since 2008.

That was a coming of age of sorts, when China stopped seeming like a work in progress and started looking and behaving like a superpower. On the Beijing skyline, the scaffolding and cranes had been replaced by stunning architectural wonders. The ever-present sentiments of victimhood and nationalism found powerful outlets in the Tibetan uprising, torch relay protests and the Sichuan earthquake. Meanwhile, those who had supported moves towards a more open, liberal, internationalist China saw the value of their political stock plunge almost as fast as the Dow Jones index in the global financial crisis. With the western model apparently shattered, many in China understandably felt less inclined than ever to listen to outside advice.

In the four years since, China has become a more modern and connected nation, but—despite the official hubris—it also seems more anxious that the uprisings in the Middle East and North Africa may spread. The government now spends more on internal security than defence of its borders—a sign that it is more frightened of its own people than any external threat.

Little wonder. This has been an era of protest in China. The government stopped releasing figures a few years ago, but academics with access to internal documents say there are tens of thousands of demonstrations each year. The reasons are manifold—land grabs, ethnic unrest, factory layoffs, corruption cases and territorial disputes. But I have come to believe the fundamental cause is ecological stress: foul air, filthy water, growing pressure on the soil and an ever more desperate quest for resources that is pushing development into remote mountains, deserts and forests that were a last hold-out for bio and ethnic diversity.

This is not primarily China's fault. It is a historical, global trend. China is merely roaring along the same unsustainable path set by the developed world, but on a bigger scale, a faster speed and at a period in human history when there is much less ecological room for manoeuvre. The wealthy portion of the world has been exporting environmental stress for centuries. Outsourcing energy-intensive industries and resource extraction have put many problems out of sight and out of mind for western consumers. But they cannot be ignored in China.

The worst problems are found in the countryside: "cancer villages", toxic spills, pitched battles to block a

toxic chemical factory, health hazards from air pollution and water and the rapid depletion of aquifers under the north China plain—the country's bread-basket.

The implications are global. China has become the biggest greenhouse-gas emitter on the planet. This year, it will probably account for half the coal burned in the world. The number of cars on China's roads has increased fourfold since 2003, driving up demand for oil. Meanwhile, there is less and less space and respect for other species. For me, the most profound story of this period was the demise of the baiji—a Yangtze river dolphin that had been on earth for 20m years but was declared extinct in 2006 as a result of river traffic, pollution, reckless fishing and massive damming.

I switched my focus to environment reporting. It was not just the charismatic megafauna and the smog, though the concern about air quality never went away. It is really not funny to send your children off to school on days of high pollution with a cheery "Try not to breathe too much", knowing they will probably be kept in at breaktimes because the air outside is hazardous.

As I have noted at greater length elsewhere, I had come to fear that China may be where the 200-odd-year-old carbon-fuelled capital-driven model of economic development runs into an ecological wall. Britain, where it started, and China may be bookends on a period of global expansion that has never been seen before and may never be repeated again.

Developed nations have been outsourcing their environmental stress to other countries and future generations for more than two centuries. China is trying to do the same as it looks overseas for food, fuel and minerals to satisfy the rising demand of its cities and factories. This has been extremely good news for economies in Africa, Mongolia, Australia and South America I sympathise with China. It is doing what imperial, dominant powers have done for more than two centuries, but it is harder for China because the planet is running short of land and time.

With their engineering backgrounds, President Hu Jintao (a trained hydro-engineer) and premier Wen Jiabao (one of China's leading experts on rare earth minerals) are probably better aware than most global leaders about the challenge this poses. While there has been almost no political reform during their terms of office, there have been several ambitious steps forward in terms of environmental policy: anti-desertification campaigns; tree planting; an environmental transparency law; adoption of carbon targets; eco-services compensation; eco accounting; caps on water; lower economic growth targets; the 12th Five-Year Plan; debate and increased monitoring of

PM2.5 [fine particulate matter] and huge investments in eco-cities, "clean car" manufacturing, public transport, energy-saving devices and renewable technology. The far western deserts of China have been filled with wind farms and solar panels.

That is the most hopeful story of this grey era. If China could emerge from the smog with a low-carbon economy, it would be a boon for the world. But talk of the world's first green superpower remains as premature as the image of the "red menace" is outdated.

When my predecessor, John Gittings, left China after 25 years, he presciently foresaw how the old cold war stereotypes would be shattered by the country's speed of development. But that is just the start of realignment. In the future, I believe the most important political division will not be between left and right, but between conservers and consumers. The old battle of "equality versus competition" in the allocation of the resource pie will become secondary to maintaining the pie itself.

But the transition has some way to go. In the next 10 years, China is likely to build more dams than the US managed in its entire history and, despite the Fukushima disaster, it plans to construct more about 20 new nuclear power stations. But even with this huge expansion of non-fossil-fuel-based energy, if the economy continues to grow at its current pace China will require about 50% more coal than it currently burns. I expect there will be a slowdown before then as overseas markets contract and domestic investment suffers from the law of diminishing returns. Long-hidden environmental costs—over-depletion of key resources and under-regulation of waste—will force their way onto corporate balance books and national budgets in the form of turbulent commodity prices and higher clean-up expenses. China may well look back on the Hu and Wen era as a golden age of growth and perhaps a missed opportunity to put in place the reforms needed to adjust to leaner times.

Respect, Sympathy—and Pessimism

Meanwhile a new leadership—almost certainly to be headed by Xi Jinping and Li Keqiang—will take the helm at this autumn's party congress. They will have their work cut out. While the Hu-Wen era was one of construction, Xi and Li will have to put more effort into maintenance. This will require more than the creation of wealth and construction jobs; it will require a system with greater flexibility, efficiency and a new set of values. I expect that transition will be more turbulent than anything seen in the past 10 years. But success or failure, I believe it will remain the most important story in the world. . . .

JONATHAN WATTS is a reporter for *The Guardian* who spent the past decade living in and reporting on China.

Jonathan Adelman  **NO**

China's Long Road to Superpower Status

China Lacks the Political, Economic and Civil
Freedoms to become a World Leader

In the last decade, the notion of China becoming the world's next superpower has become almost an idee fixe for many. Compared to the other so-called BRICS—Brazil, Russia and India—China shines like the moon. Since Deng Xiaoping created the Four Modernizations in 1978, China has surged from being a marginal player on the global stage to a powerhouse that has attracted $2 trillion of foreign direct investment.

Its economy ranks first in the world in building modern infrastructure, global exports ($2.2 trillion), Internet usage (600 million people), college graduates (7 million per year), rate of economic growth (10 percent from 1980 to 2010), movement of peasants to the city (400 million from 1980 to 2013), high-speed rail under construction (40,000 miles) and major airports (43). By 2025, it will likely have the world's largest gross national product.

Given all this unprecedented growth, how can China miss becoming the world's next superpower in 10 or 20 years? Chinese officials have told me more than once that it will probably be 2050 before China becomes truly modern. How can this be?

A superpower also needs to develop political democracy, economic freedom, military power, legal system, quality of life and high tech creativity. In all these areas China lags far behind the United States.

Politically, the United States had the world's first democratic government in 1789 and expanded the franchise ever since. By contrast, after 65 years in power, the Chinese communist government has not even begun to make the transition towards a semi-democratic state. Rather, the government, whose think tanks in the 1990s used to talk of managing a democratic transition, has cracked down on movements in minority areas and in Hong Kong. There are no democratic elections at any level. Without this transition, the People's Republic of China faces the serious possibility of falling apart like the Soviet Union did in 1991.

Economically, while the United States has a strong, relatively open capitalist economy, Chinese economic freedom is so poor that the Heritage Foundation Index of Economic Freedom ranks China 137th in the world alongside Cameroon and Tajikistan. As a result, the Conference Board, citing the negative roles of state run capitalism and growth-fixated monetary policy, estimates Chinese economic growth to slide to only 4 percent by 2020.

Militarily, the United States has hundreds of bases around the world, 11 aircraft carrier battle fleets, tens of thousands of strategic and tactical nuclear weapons, well trained officers and numerous major allies around the world. China, whose military spending is less than 30 percent of American spending, is still working on its first aircraft carrier (bought from Ukraine), imports major weapon systems from Russia, and has a small strategic nuclear force. A large number of its officers are of peasant origins. It lacks any major allies.

The United States created a government of laws, an independent judiciary and the protection of civil liberties. In China, the government does not allow free speech, assembly, an independent judiciary or religion. Massive corruption has allowed high Chinese Communist government or party officials to reap fortunes of hundreds of millions, even billions, of dollars.

The United States has been a world leader in quality of life, with more than 60 percent of the population owning their own homes and over 90 percent owning cars. By contrast, with 50 percent of the population still living in dire conditions in the countryside and massive air, water and soil pollution killing 1.2 million people a year, Chinese quality of life is quite poor. A recent poll showed that more than 60 percent of the wealthiest Chinese want to leave the country.

Adelman, Jonathan. "China's Long Road to Superpower Status," *U.S. News & World Report*, November 2014. Copyright © 2014 by U.S. News & World Report. Reprinted with permission.

Finally, in a world increasingly dominated by advances in high technology, China lags far behind the United States. While the United States has the majority of leading high-tech companies—Google, Apple, Cisco, Hewlett-Packard, Microsoft, Oracle, and others—China has almost none. Neo-Confucianism and Communism have suppressed Chinese creativity. Since 1950, not a single Chinese scientist working in China has won a Nobel Prize in science. By contrast, the United States since 1945 has won a staggering 235 Nobel Prizes in science.

Overall, then China has come a long way but still has a long way to go to become a global superpower.

JONATHAN ADELMAN is a professor at the Josef Korbel School of International Studies at the University of Denver.

EXPLORING THE ISSUE

Is China the Next Superpower?

Critical Thinking and Reflection

1. Does China possess the necessary attributes of a superpower or are there areas where it has yet to achieve such status?
2. Does China have superpower ambitions or is it merely trying to maximize its influence in Asia alone?
3. What policies does China pursue that either prove its superpower status or call into question that position?
4. What are the implications of China as a superpower for the global community?

Is There Common Ground?

There seems to be a great debate about the classification of China as a full-fledged superpower given the traditional definitions of such. Analysts differ on the relative strengths and weaknesses of China in that role and appear that this will continue in the years ahead. What analysts can agree on is that China is a very influential economic actor in the global community both in terms of trade, productivity, expansion, and influence. That power does not seem to be a temporary or ephemeral thing but will grow and endure throughout the early half of the twenty-first century.

Additional Resources

Brooks, Stephan G. and Wohlforth, William C., "The Once and Future Superpower: Why China Won't Overtake the United States," *Foreign Affairs* (May/June 2006)

What will hold China back from overtaking the United States?

Graham-Harrison, Emma, Luhn Alec, Walker Shaun and Rice-Oxley, Mark, "China and Russia: The World's New Superpower Axis," *The Guardian* (July 7, 2015).

Kaplowitz, Seth M., "Why China Is Not a Superpower-Yet," *Fortune* (July 30, 2015)

The author discusses what's holding China back.

Masoud, Fahim, "Can China Become the Next Superpower?" *International Policy Digest* (September 13, 2014)

The author discusses what China needs to attain such status.

Internet References . . .

Foreign Affairs

www.foreignaffairs.com

International Policy Digest

www.intpolicydigest.org

Rand Corporation

www.rand.org

Selected, Edited, and with Issue Framing Material by:
James E. Harf, *Maryville University*
and
Mark Owen Lombardi, *Maryville University*

ISSUE

Should the International Community Pre-empt Against North Korea?

YES: **Patrick M. Cronin**, from "Time to Actively Deter North Korea," *The Diplomat* (2014)

NO: **Jeong Lee**, from "North Korea: Don't Pick a Fight We Can't Win," *Small Wars Journal* (2015)

Learning Outcomes

After reading this issue, you will be able to:

- Describe the state of the North Korean nuclear program.
- Understand its current capability and the implications of future advancement.
- List the options available to mitigate its nuclear power.
- Evaluate the threat that North Korea places on states within the region.

ISSUE SUMMARY

YES: Patrick M. Cronin believes that the West must be more aggressive in its approach to North Korea and raise the threat of preemption against their nuclear forces.

NO: Jeong Lee argues that the nature of the North Korean regime means diplomacy and calm pressure will work best at changing behavior.

North Korea is a pariah state. It is the last Stalinist regime on earth and it has been controlled by the Kim dynasty since 1948. Ruled successively by Kim Il-sung (1948–1994), Kim Jong-Il (1994–2011), and now Kim Jong-un (2011–?), North Korea is a closed society, centrally planned and dominated by a massive military and security apparatus designed to isolate the people from the outside world and maintain the control of the Kim dynasty. The North Korean state is largely inaccessible to the outside world, save its relations with China through which 90 percent of its trade flows. North Korea does engage in trade with a few selected other states usually on a product-for-product basis.

North Korea has been in a state of war with South Korea since the Korean War (1950–1953). The armistice signed in 1953 merely ceased hostilities but technically a state of war remains between the two nations. As a result, North Korea maintains a multimillion military force with an air force and navy while South Korea and the United States

maintain military readiness along the 38th parallel, the line of demarcation between the two Koreas. Since at least the 1980s North Korea has pursued nuclear power with an aim toward developing nuclear capability. In 2006, North Korea claimed that it had tested a nuclear bomb and this was confirmed by U.S. and other independent sources. Most analysts believe that North Korea possesses a small stockpile of atomic bombs and is now working on the rocket propulsion technology to be able to deliver those weapons over short, intermediate range and long-range distances, including the capability of reaching the United States.

While its testing of such technology has been fraught with as many successes as failures, most analysts now believe that North Korea has the rocket/missile technology to deliver nuclear weapons to most of East Asia and will soon have the capability to deliver such weapons to the western half of the United States. Analysts remain divided over how effective and accurate such technology will be under the stress of battle.

The central issue facing South Korea, Japan, the United States, China, Russia, and indeed the rest of the world is the closed, unpredictable, and volatile nature of the regime and its new leader Kim Jong-un. Little is known about this son of Kim Jong-Il. He emerged on the scene relatively late as his father's likely successor, and North Korean watchers are concerned about the increased number of purges and shake-ups within the senior leadership since his ascension. These acts may indicate a simple consolidation of power or they may reflect a growing paranoia among Kim Jong-un and his inner circle.

Recent reports during the spring of 2016 are that even the Chinese, the North's closest ally are disturbed by North Korea's drive for missile technology, its refusal to heed Chinese advice, and its own closed nature even to Chinese observers. The recent claim by North Korea in January, 2016 that it had tested a hydrogen bomb (not confirmed) caught Chinese officials off guard because they had received no prior warning from the North Korea regime.

This development leaves the rest of the world with a profound set of concerns. First, even though North Korea has already crossed the nuclear threshold, can the United States and its allies allow it to possess the capability to deliver missiles across the globe? Second, can the United States rely on the pragmatism of the regime and in particular Kim Jong-un to ensure sober dealing even during a crisis? Third, can the major powers including Russia, China, Japan, the United States, and South Korea work to mitigate and reform the regime while it possesses such capability and what if North Korea preempts against South Korea itself?

This has led to a growing debate around whether the West, in this case the United States, should preempt to remove the North Korean nuclear capability. In the YES selection, Cronin's article reflects the view that aggressive foreign policy and the threat of preemption, if not the actual act itself, can eliminate the North Korean capability and reduce tensions. In fact, it is the very nature of the North Korean regime and its ruling family that calls out for such action since pragmatism and sanity are not the likely responses to western action. In the NO selection, Lee argues that the nature of the North Korean regime and its leadership calls for measured, consistent, and not strident diplomacy to ensure stability and reduce the threat. This view is albeit the more long-term strategy designed to allow the regime fall of its own weight and internal issues.

YES ↵

Patrick M. Cronin

Time to Actively Deter North Korea

"It's time to make North Korea have to worry more about deterring us rather than the other way around."

It is only a matter of time before North Korea flaunts its ability to miniaturize a nuclear warhead, deploy intercontinental ballistic missiles and road-mobile missile launchers, and expand its plutonium nuclear arsenal with highly enriched uranium warheads. The cumulative failure of diplomacy to rein in Pyongyang's nuclear and missile programs begs the question as to whether it is time to turn the tables on North Korea. Rather than buy into a losing competitive strategy, hasn't the time come for the United States and the Republic of Korea (ROK), with the support of others, to pursue a strategy of active defense that alters the North's cost-benefit calculus?

The North Korean threat is inherently volatile and far more dangerous in the near-term than the sea skirmishes in the East and South China Seas. Because the North threatens to escalate, however, democracies are reluctant to accept risk. Former Secretary of Defense William Perry backed off possible strikes on Yongbyon reactor in 1994 because, as he put it, he was seeking to avert a general war rather than to cause it.

Without throwing caution to the wind, the U.S.-ROK alliance needs to introduce more risk into its approach. Our risk aversion grants North Korea wide latitude for mischief. Pyongyang uses an array of asymmetrical means—from unmanned aerial vehicles (UAVs) and cyber warfare to mini submarines and nuclear weapons–to poke holes through the superior conventional capabilities of the U.S.-ROK alliance. This is because the main purpose of the North's asymmetric threats is not to use these weapons but to coerce benefits by threatening to use them.

The United States and the Republic of Korea have responded to the growing uncertainty surrounding the Kim Family Regime under the young Kim Jong-un. They have doubled down on readiness, counter-provocation plans, exercises, deployments and missile defenses. These are important and welcome steps. But at some point the alliance needs to understand the math: North Korea is capable of posing more asymmetrical threats than the alliance can afford to counter with 100 percent effectiveness. Instead of trying to counter each specific threat, Seoul and Washington need to balance *deterrence by denial* with *deterrence by punishment*.

Such an active defense strategy has at least three essential defense components. The first is an upgraded intelligence, surveillance and reconnaissance (ISR) network, capable of early detection of ballistic missiles, as well as low-altitude cruise missiles and UAVs. The second is an upgraded missile defense system that includes deployment of better point defense systems (PAC-III) and wider-range defenses that include not only Standard Missiles on Aegis-equipped destroyers but also land-based Terminal High-Altitude Area Defense (THAAD) batteries. Third, there must be a stronger offensive capability, one that poses a "kill chain" threat capable of preempting missile launches before they happen. The aim is not to actually preempt but rather to pose the threat of preemption, thereby forcing North Korea to be more circumspect before threatening to turn South Korea into a sea of fire with every whim.

One way to threaten preemption even without missiles is to further develop a non-nuclear electromagnetic pulse (EMP) weapon that could neutralize missiles on the launcher. Because North Korea will soon develop road-mobile missiles capable of firing nuclear weapons, the further development of non-nuclear EMP systems capable of taking out, say, a 50-square-kilometer joint fire area, would also shift the cost-benefit calculus against North Korea.

Let's also consider the recent North Korean deployment of three UAVs into South Korea. It boggles the mind to imagine how much leverage North Korea appeared to gain in South Korea by deploying what amounted to toy drones that pose no direct threat to the country. After all, the United States and South Korea have two huge advantages over the North: they are both open democracies with advanced technology. North Korea, on the other hand, is a closed society fearful of information, which forces it to smuggle in whatever technology it can buy through clandestine channels and the black market.

In addition to direct defense investments, the UAV incident suggests that the alliance should be making far greater use of its information advantage. While North and South Korea agreed some years ago to forego psychological warfare against each other, the North is a flagrant purveyor of vitriol and falsehood. Surely the alliance can better saturate the North with uncomfortable facts—from pictures of Kim Jong-un luxury houses side by side with North Korean gulags, to video lectures by North Korean refugees who have managed to escape the world's most oppressive regime.

The bottom line is this: in addition to shoring up deterrence with more defense, the alliance can gain greater leverage against North Korean brinkmanship and coercion by adding an element of active defense and information warfare. It's time to make North Korea have to worry more about deterring us rather than the other way around.

Dr. Patrick M. Cronin is senior advisor and senior director of the Asia-Pacific Security Program at the Center for a New American Security in Washington, DC. He is the author of *If Deterrence Fails: Rethinking Conflict on the Korean Peninsula* (Center for a New American Security, 2014).

Jeong Lee    **NO**

North Korea: Don't Pick a Fight
We Can't Win

Summary

This paper examines potential covert operation scenarios against the DPRK (Democratic Republic of Korea, or North Korea) by dissecting retired Army Special Forces Colonel David S. Maxwell's latest essay entitled "Unification Options and Scenarios; Assisting A Resistance."[i] In his essay, Colonel Maxwell argues that the U.S.-ROK (Republic of Korea, or South Korea) alliance should embark upon a paramilitary uprising against the Kim regime to reunify the Korean peninsula and to preempt against potential insurgency threats that might emanate from the north.

After testing the inherent assumptions within the Maxwell article to draw potential scenarios, I argue here that we must not pursue covert operations to topple Kim Jong-un because doing so may trigger a global war. Lastly, I put forth two alternative courses of actions (COAs) to deal with the North Korean crisis.

The COAs I propose are:

a. Recognizing the DPRK as a nuclear state and engaging the DPRK through diplomacy and economic trade;

b. Maintaining troop presence in the Korean peninsula to deter the DPRK aggression, and at the same time, bolstering our missile defense at home to protect our citizens against potential North Korean missile attacks.

Background

To this day, the Korean Peninsula remains a veritable tinderbox where a full-scale war between two brethren states can erupt any moment. In the aftermath of Kim Jong-il's death and the succession of his son, Kim Jong-un, as the new leader in 2011, the DPRK remains unpredictable and dangerous as ever before. Indeed, another conflict on the Korean Peninsula will likely lead to catastrophic consequences with global repercussions.[ii]

Shortly after the 2010 sinking of the ROK corvette, the *Cheonan*, and the bombardment of the Yeonpyeong Island by the DPRK, the ROK Armed Forces have adopted swift and robust ROEs (rules of engagements) to counter any potential threats emanating from their northern brethren.[iii]

Also, for quite some time, the U.S.-ROK alliance has been planning covert operations deep inside the north. Retired Army Special Forces Colonel David S. Maxwell, who served as Chief of Staff of SOCKOR (Special Operations Command, Korea), and the chief author of OPLAN (Operational Plan) 5027-98 and CONPLAN (Concept of Operational Plan) 5029-99, has written extensively on covert operations against the DPRK.[iv] Explicit in Maxwell's writings on U.S.-ROK covert operations against the Kim Jong-un regime is the assumption that the U.S. armed forces and their ROK counterparts will encounter a grander version of the Iraqi and Afghan insurgencies, should Kim's regime suddenly collapse.

Methodology and Key Argument

The key method for this paper involves dissecting the underlying assumptions in Maxwell's article, laying out scenarios for what may transpire if we follow through with his recommendations, and coming up with alternative COAs if necessary.

The U.S.-ROK alliance must not pursue these risky COAs. This is because failed covert operations in the DPRK not only compromises requisite plausible deniability, but may also yield disastrous outcomes of a global magnitude.

Our Objectives in the DPRK Covert Operations

According to Maxwell, the rationales for joint U.S.-ROK covert operations against the DPRK are: to put an end to the DPRK's WMD (Weapons of Mass Destruction) programs; to effect a regime change within Pyongyang; to co-opt the potential insurgency within the DPRK in the event of the collapse of the Kim Jong-un regime; and finally, to reunify the Korean Peninsula under the ROKG (ROK Government) banner.[v]

In his paper, Maxwell discusses the so-called "pre-mortem analysis" whose purpose it is to expose flaws in supposedly failed OPLANs by "puncturing . . . the premises of a proposed course of action, its assumptions and tasks."[vi] Through this method, Maxwell writes that the Training and Doctrine Command (TRADOC) Red Team Leaders at Fort Leavenworth identified potential "internal resistance and insurgency" waged by the adherents of the Kim Jong-un regime and other North Koreans as the most significant barriers to the reunification of the Korean Peninsula.[vii] Basing his premises on the TRADOC pre-mortem analysis exercise results, Maxwell argues that we should embark upon covert operations that "focus on internal resistance."[viii]

Potential Scenarios

To ascertain the potential scenarios were we to follow through with fomenting resistance within the DPRK, we must dissect the inherent assumptions and rationales for covert operations within the north.

Kim Jong-un's Legitimacy Has Become Tenuous[ix]
Although many analysts and pundits point to Kim's health issues, relative youth, and lack of experience as evidence of instability within his regime, many of the so-called "indicators" are speculative.[x] We do not have accurate intelligence on Pyongyang, and Kim still appears to be in charge of his dynasty. Thus, were we to instigate an indigenous uprising, it will fail due to the totalitarian nature of the Kim regime. More importantly, the blowback from the failed paramilitary takeover will manifest itself immediately through compromised plausible deniability, and a full-scale war waged by Kim Jong-un in retaliation against the failed insurrection.

We Should Talk About the Paramilitary Option Regardless of Whether it Might Upset China[xi]
Even though China remains one of the key trading partners of the ROK, and is favorably disposed towards President Park Geun-hye's deft diplomacy, China will be anything but passive where their border security is concerned. In short, we will upset China if we embark upon the paramilitary operation to topple Kim Jong-un.

According to the Chinese People's Liberation Army (PLA) OPLANs that were leaked last year, in addition to setting up refugee camps along the Chinese border, the PLA units are ordered to detain DPRK political and military leaders in "special camps" where they are to be monitored closely by the PLA troops, lest they be co-opted by "foreign forces."[xii] Richard C. Bush III, the Director for the East Asian Policy Studies at the Brookings Institution, argues that what might explain the Chinese desire to preserve the *status quo* may be that China fears a reunified Korea that is friendly towards the United States, and thus, may eventually pose a threat to China's own authoritarian regime.[xiii]

We Must Identify and Co-opt "2nd Tier Leaders" Who are Opposed to the Kim Regime[xiv]
In his footnote, Maxwell defines 2nd tier leaders as "those who have regional political and military power and influence but who are not members of the core of the Kim family regime."[xv] Given the lack of human intelligence (HUMINT) on the ground, identifying such opposition leaders may be difficult and fraught with uncertainties. First, 2nd tier leaders will be difficult to vet. Will they remain loyal to the causes of the U.S.-ROK alliance? Are they competent leaders respected by their men? What are their operational capabilities? Second, how will we protect them in the event of a failed paramilitary uprising against the Kim regime? As Maxwell rightly acknowledges, one of our key weaknesses is our "inability or unwillingness to back up [our] words with deeds."[xvi] Third, and most important, how do we ensure that our North Korean paramilitary assets remain malleable to the wishes and demands of the U.S.-ROK alliance? Assuming that the paramilitary takeover even succeeds at all, our assets will likely demand rewards or some form of political recognition.

We Must Offer Protection for North Korean Scientists Involved in the Research and Development of WMDs [xvii]
Maxwell is correct to argue that the U.S.-ROK alliance should protect North Korean scientists involved in the development of WMDs if we embark upon covert operations against Kim Jong-un. There are two reasons for this: to prevent the proliferation of WMDs; and to cement their allegiance to the U.S.-ROK alliance. However, rather than allowing the North Korean scientists to work for the ROKG, we should offer them an asylum here in the United States. The reason is because if they work for the ROKG, the ROKG will employ them to develop their own nuclear

capabilities against Japan or China. Should the ROKG develop their own nuclear arsenals, they will indubitably destabilize Northeast Asia by triggering a nuclear arms race.

Our Mission is to Assist the ROK Forces with Stability Operations in the Aftermath of Kim's Demise[xviii]

As stated previously, the underlying premise behind this assumption is that the U.S.-ROK forces will face a grander version of the Iraqi and Afghan insurgencies, should the DPRK implode suddenly. To buttress his claims, Maxwell cites the "guerrilla ethos" with which the North Koreans are indoctrinated, and the fact that the DPRK fields the largest number of SOF (Special Operations Forces) operators in the world, as factors which may preclude a complete reunification under the ROKG banner.[xix] Army LTG John Johnson and COL Bradley T. Gericke also second Maxwell's claims when they argue that the DPRK troops will likely fight to the death should another Korean War break out.[xx] Thus, they believe that it is imperative that the U.S.-ROK forces must engage in stability operations.[xxi] But one must ask how a war in a different theater and with a different cultural setting will replicate another war in a distant land.

Where are the absolute indicators? This does not even take into account the fact that the ROK forces have not fought in counterinsurgency (COIN) campaigns since the Vietnam War. Although Maxwell suggests employing North Korean defectors to infiltrate back into their native land, one must ask why they would want to return after risking their lives to flee their homeland.[xxii] Furthermore, infiltrating North Korean defectors may lead to another Bay of Pigs-type fiasco, which, in turn, might damage our international image abroad. Last but not least, the American public will not likely tolerate yet another protracted conflict in a foreign country.

In short, at a glance, Maxwell's proposal adheres to the range of covert operations laid out in Mark M. Lowenthal's book, *Intelligence: From Secrecy to Policy*, to include political and economic activities, coups, and paramilitary actions.[xxiii] However, absent in his piece are the joint decision-making dynamics involved among the U.S. Executive, Congress, and the various SOF units and the IC (Intelligence Community) responsible for implementing covert actions as a policy instrument.[xxiv] Nor does the article offer an approximate time frame despite the advice that "it must be fully resourced and given the time to develop." Third, the article does not mention the likely

responses by China, were the USFK (U.S. Forces Korea) to embark upon secret operations against the north.[xxv] Last but not least, the pre-mortem analysis method may be flawed because we have no way of ascertaining how the TRADOC Red Team members arrived at their conclusion. How did they supposedly "puncture" the fallacies inherent in the extant OPLANs? More importantly, what exactly were the "flawed" assumptions within the OPLANs themselves prior to the exercises?

Decision: Don't Pick a Fight We Can't Win

After having laid out the potential scenarios based on the assumptions within Maxwell's writings, I have no other recourse but to recommend that we discard our plans for covert operations against the Kim Jong-un regime. Such endeavors are too risky, if not impossible to achieve. Regardless of the covert methods we adopt against the Kim regime, they will lead to a calamitous global war involving not only the DPRK, but also other East Asian actors such as China and Japan. Nor will the American public remain passive in the face of a protracted war in a foreign country.

Nonetheless, there are two viable COAs to deal with the DPRK. The first COA entails recognizing the DPRK as a nuclear state and attempting to engage the DPRK through diplomacy and economic trade. Even though we will not be able to completely rein in the reclusive regime, we still can make it malleable to our wishes and international norms. This option is pragmatic because the DPRK desperately needs trading partners to stave off famine and poverty in order to guarantee its own regime survival.

The second COA involves maintaining troop presence in the Korean peninsula to deter the DPRK aggression, and at the same time, bolstering our missile defense at home to protect our citizens against potential North Korean missile attacks. Although costly, maintaining our military presence in the Korean peninsula and bolstering our missile defense enable us to control the strategic calculus and the tempo in Northeast Asia.

Under no circumstances should we embark upon covert operations within the DPRK.

This position paper is based on an assignment for my Intelligence & National Security class at the Josef Korbel School of International Studies at the University of Denver. My hat is off to my instructor, Cmdr. Steve Recca (USN Ret.), for his invaluable insights on covert actions.

End Notes

[i] COL David S. Maxwell's (USA Ret'd) "Unification Options and Scenarios; Assisting A Resistance," *International Journal of Korean Unification Studies* [24: 2], 2015, p. 127–152

[ii] LTG John Johnson (USA) and COL Bradley T. Gericke "Spinning the Top; American Land Power and the Ground Campaigns of a Korean Crisis," *Joint Forces Quarterly* [78:3], 2015, pp. 99

[iii] Victor Cha and David Kang "Think Again: North Korea," *Foreign Policy*, March 25, 2013 (http://foreignpolicy.com/2013/03/25/think-again-north-korea/)

[iv] According to *Globalsecurity.org*, OPLAN 5027-98 is the U.S.-ROK Combined Forces war plan in the event of a full war against the DPRK. (See "OPLAN 5027 Major Theater War—West," Globalsecurity.org, May 7, 2011 http://www.globalsecurity.org/military/ops/oplan-5027.htm) CONPLAN 5029-99, on the other hand, is the U.S.-ROK Combined Forces war plan to prepare for the eventual collapse of the DPRK. (See "OPLAN 5029 - Collapse of North Korea," Globalsecurity.org, May 7, 2011 http://www.globalsecurity.org/military/ops/oplan-5029.htm)

[v] Maxwell, pp. 127–130.

[vi] "Red Team Leader Training (http://usacac.army.mil/organizations/ufmcs-red-teaming/classes)" cited in Maxwell, pp. 134–5

[vii] Maxwell, pp. 135

[viii] *Ibid.*, pp. 130–131

[ix] *Ibid.*, pp. 137

[x] One good example of such speculative commentaries is Victor Cha's "Why I'm Worried About Kim Jong-un" in *Politico*, October 10, 2014 (http://www.politico.com/magazine/story/2014/10/why-im-worried-about-kim-jong-un-111778).

[xi] Maxwell, pp. 130

[xii] Julian Ryall "China plans for North Korean regime collapse leaked," *The Telegraph*, May 5, 2014 (http://www.telegraph.co.uk/news/worldnews/asia/northkorea/10808719/China-plans-for-North-Korean-regime-collapse-leaked.html). By "foreign forces," the PLA means the U.S.-ROK forces.

[xiii] Richard C. Bush III "China's Response to Collapse in North Korea," The Brookings Institution, November 14, 2013 (http://www.brookings.edu/research/speeches/2013/11/14-china-north-korea- collapse-bush)

[xiv] Maxwell, pp. 138

[xv] Maxwell, pp. 138

[xvi] *Ibid.*, pp.148

[xvii] *Ibid.*, pp.138

[xviii] *Ibid.*

[xix] Maxwell, pp. 135–136. See also his "Irregular Warfare on the Korean Peninsula Thoughts on Irregular Threats for north Korea Post-Conflict and Post-Collapse: Understanding Them to Counter Them," *The Small Wars Journal*, November 30, 2010 (http://www.smallwarsjournal.com/blog/journal/docs-temp/609-maxwell.pdf).

[xx] LTG Johnson and COL Gericke "Spinning the Top; American Land Power and the Ground Campaigns of a Korean Crisis," J*oint Forces Quarterly* [78:3], 2015, pp. 103–104.

[xxi] *Ibid.*

[xxii] Maxwell, pp. 150

[xxiii] Mark M. Lowenthal *Intelligence: From Secrecy to Policy* CQ Press: Washington, D.C., 2000, pp. 111–112

[xxiv] *Ibid.*, pp. 109–111

[xxv] Maxwell, pp. 150

JEONG LEE is a freelance writer and an MA candidate in International Security Studies Program at the Josef Korbel School of International Studies at the University of Denver. His writings on U.S. defense policy issues and inter-Korean affairs have appeared on various online publications including the *Small Wars Journal*.

EXPLORING THE ISSUE

Should the International Community Pre-empt Against North Korea?

Critical Thinking and Reflection

1. Is preemption against the North Korean nuclear capability a viable military option?
2. What will the Chinese do in response?
3. What happens if the North Korean military launches an attack on South Korea in response?
4. Can the North Korean regime be changed through sanctions and patient negotiation?

Is There Common Ground?

The one thing that all actors in the region and indeed the world can agree on is that a nuclear North Korea with medium- and long-range missiles is a destabilizing force in East Asia. For the United States, it threatens its allies, it maintains the need for a large military presence, and it ensures higher defense expenditures for the foreseeable future. For China, it ensures issues with supporting the regime and with a strong American presence that is a counterweight to China's goals in the region. This mutual agreement albeit for different motives is the basis for building a coordinated policy to disarm North Korea of its nuclear capability.

Additional Resources

Council on Foreign Relations, "North Korea Crisis," Council on Foreign Relations (June 12, 2016).

Mizokami, Kyle, "Welcome to North Korean Nuclear Weapons 101," *The National Interest* (September 26, 2015).

Pry, Peter and Huessy, Peter, "North Korea's Missile Threat: Very bad News," Gatestone Institute (February 29, 2016).

Sang-Hun, Choe, "North Korea's Missile Tests Timed to Bolster Standing with China, Analysts Says," *The New York Times* (June 2, 2016).

Stratfor, "How the US Would Destroy North Korea's Nuclear Weapons," *Stratfor* (March 26, 2016).

Internet References . . .

Arms Control Association

www.armscontrol.org

Council on Foreign Relations

www.cfr.org

Gatestone Institute

www.gatestoneinstitute.org